CULTURE AND COSMOS
http://www.CultureAndCosmos.org

Culture and Cosmos is published twice a year, in northern spring/summer and autumn/winter, in association with the Sophia Centre for the Study of Cosmology in Culture, University of Wales Trinity Saint David.
Contributions and editorial correspondence should be addressed to: Editors@cultureandcosmos.org

Editor: Dr Nicholas Campion, the Editor of *Culture and Cosmos*, University of Wales Trinity Saint David, Lampeter, Ceredigion, Wales, SA48 7ED, UK.
E Mail **n.campion@uwtsd.ac.uk**
Deputy Editor: Dr Jennifer Zahrt
Editorial Board: Dr Silke Ackermann, Professor Anthony F. Aveni, Dr David Brown, Professor Charles Burnett, Dr Hilary M. Carey, Dr John Carlson, Dr Patrick Curry Professor Robert Ellwood, Dr Germana Ernst, Dr Ann Geneva, Professor Joscelyn Godwin, Dr Dorian Greenbaum, Dr Jacques Halbronn, Dr Robert Hand, Dr Jarita Holbrook, Professor Michael Hunter, Professor Ronald Hutton, Dr Peter Kingsley, Dr Edwin C. Krupp, Dr J. Lee Lehman, Dr Lester Ness, Professor P. M. Rattansi, Professor James Santucci, Robert Schmidt, Dr Fabio Silva, Dr Lorenzo Smerillo, Professor Richard Tarnas, Dr Graeme Tobyn, Dr David Ulansey, Robin Waterfield, Dr Charles Webster, Dr Graziella Federici Vescovini, Dr Angela Voss, Dr Paola Zambelli, Robert Zoller.
Special Issue Editors: Liz Henty, Bernadette Brady, Darrelyn Gunzburg, Frank Prendergast and Fabio Silva
Web: Dr Frances Clynes

Contributors Guidelines: Please see http://www.cultureandcosmos.org/submissions.html

Copying: Apart from fair dealing for the purposes of research or private study, or criticism or review, as permitted under the Copyright, Designs and Patents Act 1988, no part of this publication may be reproduced, stored or transmitted in any form or by means without the prior permission of the Publisher.

Front cover: Sunrise in Late Neolithic Stonehenge. Panorama by LBI ArchPro's media partner 7reasons based on LBI ArchPro's image-based model. Simulation with Stellarium 0.15.0.

Published by Culture and Cosmos, Dr Nicholas Campion, Faculty of Humanities and the Performing Arts, University of Wales Trinity Saint David, Lampeter, Ceredigion, Wales, SA48 7ED, UK.

© **Culture and Cosmos 2017**
Printed by Lightning Source.

The Sophia Centre
http://www.uwtsd.ac.uk/sophia

The Centre for the Study of Cosmology in Culture is an academic centre within the Faculty of Humanities and the Performing Arts at the University of Wales Trinity Saint David.

The Centre's academic goals are

- 'to pursue research, scholarship and teaching in the relationship between astrological, astronomical and cosmological beliefs and theories, and society, politics, religion and the arts, past and present' and
- 'to undertake the academic and critical examination of astrology and its practice'.

The Centre's wider goal is stated in its title – to 'study cosmology in culture'. In a traditional sense, a cosmology is a world view, an understanding of the cosmos which informs individual and social action and ideology. The Centre promotes research in the subject area, holds seminars and conferences, publishes scholarly material, is associated with Sophia Centre Press and supervises PhD students.

The Centre's teaching is focused on the MA Cultural Astronomy and Astrology. For further information see
http://www.uwtsd.ac.uk/ma-cultural-astronomy-astrology

THE MARRIAGE OF ASTRONOMY AND CULTURE: THEORY AND METHOD IN THE STUDY OF CULTURAL ASTRONOMY

A special issue of *Culture and Cosmos*
Vol. 21 no. 1
Spring/Summer 2017

Copyright © 2017 Culture and Cosmos
All rights reserved

Published by Culture and Cosmos
& Sophia Centre Press
England

www.cultureandcosmos.org

In association with the
Sophia Centre for the Study of Cosmology in Culture,

University of Wales Trinity Saint David,
Faculty of Humanities and the Performing Arts
Lampeter, Ceredigion, Wales, SA48 7ED, UK

British Library Cataloguing in Publication Data
A catalogue card for this book is available from the British Library

All rights reserved. No part of this book may be reproduced or utilized in any form or by any means, electronic or mechanical, including photocopying, recording or by any information storage and retrieval system, without permission in writing from the Publishers.

ISSN 1368-6534

Printed in Great Britain by Lightning Source

Culture and Cosmos

CONTENTS

Liz Henty, Bernadette Brady, Darrelyn Gunzburg, Frank
Prendergast and Fabio Silva, *Editorial*   1

ARCHAEOLOGY

Lionel Sims and David Fisher
Through the Gloomy Vale: Underworld Alignments at Stonehenge   11

Claude Maumené
*The Bush Barrow and Clandon Barrow Gold Lozenges and the Upton
Lovell Golden Button: A Possible Calendrical Interpretation*   31

J. Anna Estaroth
Clava Cairns, Midwinter Sunset and the Minor Lunar Limit   51

Marianna P. Ridderstad
*Orientations of Late Neolithic to Bronze Age and
Iron Age Long Cairns in Coastal Finland*   73

A. César González García, Felipe Criado-Boado and Benito Vilas
Megalithic Skyscapes in Galicia   87

Tore Lomsdalen
*Different Approaches to Cosmology in Archaeology and
Their Application to Maltese Prehistory*   105

Juan Antonio Belmonte and A. César González García
*Petra Revisited: An Astronomical Approach to the
Nabataean Cultic Calendar*   131

Andrew M. Munro, Tony Hull, J. McKim Malville,
F. Joan Mathien and Cherilynn Morrow
*Investigation of Solstice Horizon Interactions at
Chacoan Monumental Architecture*   151

J. McKim Malville and John L. Ninnemann
*Archaeoastronomy and Cultural Astronomy as
Scientific Disciplines: Falsifiability and Photo Documentation*   173

Steven R. Gullberg and J. McKim Malville
Caves, Liminality and the Sun in the Inca World   193

Stanisław Iwaniszewski
*Rethinking Nahualac, Iztaccíhuatl, Mexico: Between Animism to
Analogism in Mesoamerican Archaeoastronomy*   215

ETHNOGRAPHY

Nasser B. Ayash
*Evolution of Arabic Astronomy in Relation with the
Translation Movement in the Early Abbasid Era*   233

Roslyn M. Frank
*Metrology, Memory and Long-Term Landscape Inhabitation:
Evidence for the Septarian Package on the Atlantic Façade*   249

IMAGES

Georg Zotti, Florian Schaukowitsch and Michael Wimmer
The Skyscape Planetarium   269

Liana De Girolami Cheney
Edward Burne-Jones's The Planets: Luna, A Celestial Sphere   283

Dragana Van de moortel-Ilić
*An Examination of the Images of the Sun and the Moon in the
Visoki Dečani Monastery in Kosovo*   301

NOTES ON CONTRIBUTORS   321
BACK ISSUES   329

CULTURE AND COSMOS

www.CultureAndCosmos.org

Editor: Nicholas Campion
Guest Editors: Liz Henty, Bernadette Brady,
Darrelyn Gunzburg, Frank Prendergast and Fabio Silva
Vol. 21 No. 1 Spring/Summer 2017
ISSN 1368-6534

Published in Association with
The Sophia Centre for the Study of Culture in Cosmology,
Faculty of Humanities and the Performing Arts
University of Wales Trinity Saint David
http://www.uwtsd.ac.uk/sophia/

Editorial

This volume of *Culture and Cosmos* draws together a selection of papers delivered at the 24th annual conference of the European Society for Astronomy in Culture (SEAC). Titled 'The Marriage of Astronomy and Culture: Theory and Method in the Study of Cultural Astronomy', the meeting occurred between the 12th and the 16th September 2016 and was held at The Bath Literary and Scientific Institution (BRLSI), which has been hosting research endeavours since it foundation in 1824. BRLSI is now located in Queens Square, which was designed by John Wood in c.1728, the architect who also planned the famous Bath Circle. With its medieval abbey, restored in the seventeenth century, and Roman Baths close by, Bath is therefore full of archaeological interest. It also has archaeoastronomical significance because Wood believed that the seven hills of Bath were dedicated to the heavenly bodies.[1] He drew a planetary model overlaid onto plans of the monuments near Stonehenge so that for him Silbury was the hill of Mars and Stonehenge was an oracular lunar temple of the moon. With conference tours to Stonehenge, Avebury and Bath Abbey we could really believe we were 'standing on the shoulders of giants'.

---

[1] John Michell, [1977] *A Little History of Astro-Archaeology: Stages in the Transformation of a Heresy* (New York: Thames & Hudson Inc., 1989), p. 12.

SEAC 2016 combined history with the latest in twenty-first century developments and, for the very first time, was webcast to SEAC members who could not attend, in addition to students of the MA in Cultural Astronomy and Astrology. In total, twenty-five people took advantage of this facility and were able to engage in the live sessions and also ask questions of the speakers. The sessions were webcast and recorded using Cisco WebEx and held on the Sophia Centre's server for a few weeks to allow delegates to access the recordings after the conference. Delegates who registered for the online webcasts could choose fees varying from £25 for a half-day session to £120 for the full conference. Apart from allowing a wider audience to partake of the many excellent lectures given, the physical presence of around 90 delegates was boosted by these online attendees. The opening address and welcome speech for the five day conference was given by SEAC President Michael Rappenglück. This was followed with a presentation by Nicholas Campion of the Sophia Centre for the Study of Cosmology in Culture (University of Wales Trinity Saint David). He considered the state of cultural astronomy today and how it has moved on from Lockyer's early twentieth century speculations and the 1960s' controversial publications of Gerald Hawkins and Alexander Thom who brought the field to the public's attention. For many years the key role of the SEAC conferences has been the promotion of archaeoastronomy and cultural astronomy research, yet its research papers only tend to reach a limited audience. From this pioneering effort archaeoastronomy has developed in Britain and in the last few years has found a place in academia with the Sophia Centre's MA in Cultural Astronomy and Astrology which introduced a dedicated archaeoastronomy module in 2010. Skyscapes sessions at the Theoretical Archaeology Group conferences, the National Astronomy Meetings and the new *Journal of Skyscape Archaeology* have built upon this impetus.

Conference highlights
The conference offered an innovative programme, so in addition to the many research papers, a 'Skills' session, specifically dedicated to developing the skill sets of all attendants, was trialled. There were three presentations in this session focusing on different technical aspects. Frank Prendergast covered the basics of coordinate reference systems, map projections and coordinate conversion. Victor Reijs, whose presentation is freely available online (http://tinyurl.com/j8myvfb), spoke of virtual fieldwork tools, as well as common smartphone and tablet survey apps, especially highlighting their pitfalls. Finally, Georg Zotti, elaborated on

new Stellarium functionalities that include the import of Digital Terrain Models and/or 3D models of structures for accurate simulation of the natural and built environments. Altogether, we believe this was a very successful session, new to SEAC, with a didactic tone, but focussing on skills that are new and of use to even the most weathered cultural astronomer.

Following the lecture presentations, a Round Table Forum was held to discuss the key themes that could be used to define the discipline of cultural astronomy. A mind map, projected onto the presentation screen, was interactively built as the discussion developed. Debates surged around methodologies, as well as on some of the problems encountered within the scholarship of cultural astronomy and how the subject was perceived both by other scholars and the general public.

The ensuing mind map (Fig.1) shows the five commonly agreed points concerning cultural astronomy directly linked to the central topic. The delegates considered that cultural astronomy covers such disciplines as anthropology, astronomy, archaeology, history, art history, and religious studies and, consequently, that it requires an engagement with its primary sources using both qualitative and quantitative research techniques. Importantly it was recognised that scholars often needed to reach beyond an individual discipline and adopt a trans-disciplinary approach, collaborating with others, sharing resources, and integrating other methodologies into how primary sources were handled. Finally, the issue of recognition and acceptance of cultural astronomy as its own field by other academic fields, as well as by general society, was also discussed and regarded as being of primary importance to ensure the survival of the discipline.

Ancillary topics, also debated, are shown as orange boxes in Fig. 1. These focused on the practical approach to engaging in the study of cultural astronomy. Primarily it was acknowledged that there was a need to consider alternate ontologies, while, at the same time in terms of field work based studies, there was a strong need to engage physically in the landscape and skyscape. Finally, the problems of presenting such a trans-disciplinary subject to the wider social and academic communities, the usual clash between those trained in the humanities and the natural sciences, as well as the isolation of the cultural astronomy community were also debated. The resulting mind map is a record of one stage of this discussion among SEAC 2016 attendees. It is hoped that this map can form the basis for future discussions and will be revisited, updated, and developed over the coming years.

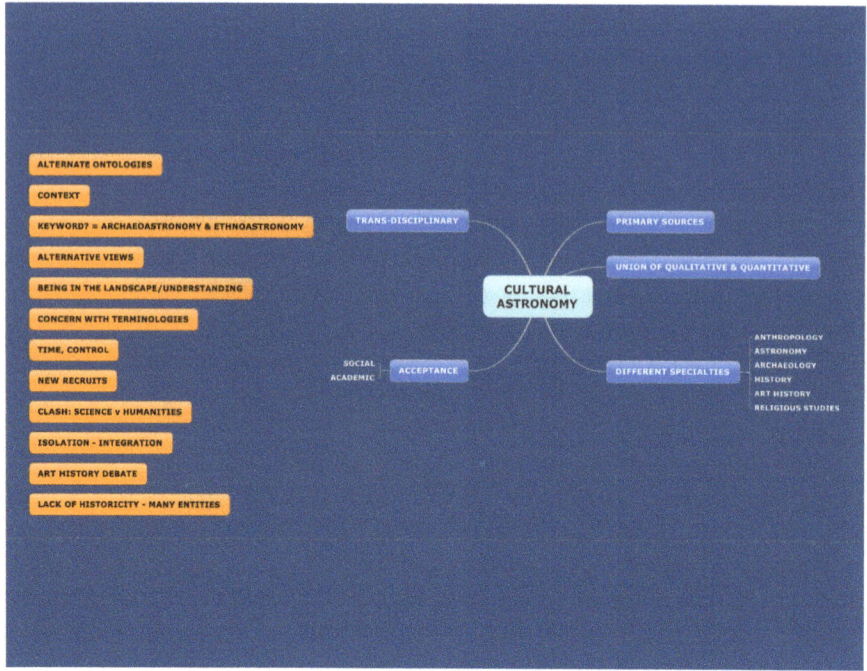

Fig. 1. The Mind Map composed by the delegates of SEAC2016, Bath, UK.

The well-attended public lecture, 'Exploring the Monuments and Cosmology of the Boyne Valley. What's the Bigger Picture?', was given by Frank Prendergast. He explored how, by about 3200 BCE, organised communities were farming in the Boyne Valley and constructing enormous burial chambers. These have yielded characteristic grave goods, with many of the structural stones being embellished with elaborate incised art. Their siting and axial orientations are believed to reflect societal concerns with hierarchy, religious beliefs and cosmology. Understanding these monuments may give insights in a regional and broader European context.

SEAC conferences, which are held in a different European city each year, make the most of the local attractions for socialising and tours. On Monday evening the University of Wales Trinity Saint David hosted a 'Wine and Nibbles' event at the Herschel Museum of Astronomy in Bath which is dedicated to the many achievements of William and Caroline Herschel. Apart from family memorabilia, the museum boasts a collection of telescopes, planetaria and globes. After a long day of thought-provoking presentations, delegates were also able to relax on Wednesday evening at a reception hosted by Sophia Centre Press where they were amply plied with drinks and finger food and had the opportunity to browse the bookstall. The occasion was to mark the launch of three new key texts in cultural astronomy. Firstly, there were the two Sophia Centre Press volumes: *Heavenly Discourses* edited by Nicholas Campion (Lampeter, 2016) and *The Materiality of the Sky*, Proceedings of the 22nd Annual SEAC Conference, 2014 (Lampeter, 2016), edited by Fabio Silva, Kim Malville, Tore Lomsdalen and Frank Ventura. Valerie Hall of Equinox Publishing Ltd was also on hand for the launch of *The Imagined Sky* (Sheffield, 2016), edited by Darrelyn Gunzburg, and to promote the *Journal of Skyscape Archaeology*, edited by Fabio Silva and Liz Henty.

On Thursday morning two optional excursions were offered to delegates. The first was a tour of Bath Abbey with architectural historian, writer, and Keeper of the Fabric at Bristol Cathedral, Jon Cannon The original building of the 1090s was one and a half times more extensive than its rebuild, begun in the 1480s and completed after the Reformation under the vision of Oliver King, bishop of Bath and Wells in 1500. The medieval abbey came alive once more with Jon's narrative. The second excursion was a trip to Stonehenge organised by Lionel Sims, for those who wished to witness sunrise at this iconic British monument. Because our access to the monument was approved by English Heritage for access at dawn, delegates had the unprecedented and no longer allowed privilege of entering the stone circle. Lionel has a deep understanding of the site's monumental landscape and history and accordingly, his onsite narratives and quizzes entertained and challenged his audience. It was hailed as an unforgettable and magical experience. Lionel also organised a tour to the Avebury Neolithic monuments about 20 miles north of Stonehenge, taking in Silbury Hill, the largest prehistoric mound in Europe, and the Avebury Stone Circle.

At the Annual General Meeting of SEAC, President Michael Rappenglück presented the Fifth Carlos Jaschek Award to Fabio Silva, the youngest recipient in SEAC's history, to fond and warm applause, as all present were well aware of the outstanding contribution he is making to the field of cultural astronomy.

Theory and method in the study of Cultural Astronomy
The conference organisers decided to highlight the importance of theory and method, as reflected in its title, and specifically asked contributors to acknowledge this both in their abstracts and presentations. Furthermore, the programme itself was devised so that papers were grouped by the type

of primary source used – archaeology, ethnography, images and music and texts and archives – and hence reflective of the range of sources available to cultural astronomy. It also provided for some interesting discussions that linked papers presented in different sessions. Given the breadth of the conference theme, a decision was made to publish two sets of conference proceedings. General papers and case studies are included in this *Culture and Cosmos* volume, while theory and method papers will appear in a specially themed issue of the *Journal of Skyscape Archaeology*, due to be published in December 2017. These proceedings can only give a flavour of the many varied and inspiring presentations, co-chaired by Nicholas Campion and Lionel Sims and other members of the organising committee and sadly cannot reproduce the warmth of meeting old and new friends, the often spectacular PowerPoint presentations, the posters, and the subsequent panel discussions. Nevertheless, we hope our selections below are enjoyable.

Archaeology
The archaeology papers cover a wide geographical span from sites in England and Scotland to those further afield in Europe, Jordan and the Americas. In their paper entitled 'Through the Gloomy Vale: Underworld Alignments at Stonehenge', Lionel Sims and David Fisher compare and test three interpretative explanatory design models for the Neolithic/Early Bronze Age palisade in the Stonehenge monument complex. Each model independently attests to there being dual design properties in some Neolithic and Bronze Age monuments as in the use of wood and stone, the mirror-imaging of landscape setting or, as in the case of this paper, a diacritical model involving materiality, landscape and lunar-solar alignments. Artefacts found within the Stonehenge monumental landscape are investigated by Claude Maumené in 'The Bush Barrow and Clandon Barrow Gold Lozenges and the Upton Lovell Golden Button: A Possible Calendrical Interpretation'. These decorative objects were most frequently thought to be ornamental breastplates designed to show the high-level political or religious status of the wearers, but the author proposes that they were crafted to facilitate counting, memorisation and transmission of the numbers of days of one or several synodic cycles of Venus, Mars and Jupiter, in agreement with a number of Lunar and Solar cycles. Additionally, the author notes that the symbolic lozenge shape associated with fertility and fecundity appeared in many of the cultural areas of ancient Europe. Further north in Inverness-shire in Scotland, J. Anna Estaroth conducts new research at the Early Bronze Age cairns at

Balnuaran of Clava in 'Clava Cairns, Midwinter Sunset and the Minor Lunar Limit'. New alignments are investigated in order to verify skyscape phenomena and integrate them with previous academic research but her investigation of the importance of seasonal alterations in light and darkness leads to a new interpretation of these monuments.

Continuing the focus on cairns, Marianna P. Ridderstad, in 'Orientations of Late Neolithic to Bronze Age and Iron Age Long Cairns in Coastal Finland', analyses the orientations, landscape character and typology of 138 long cairns in coastal Finland. Importantly, she argues for there being cultural continuity between the Neolithic and Middle Bronze Age societies. Additionally, Ridderstad offers the idea that outside cultural influences, possibly connected to central ideologies of the Nordic Bronze Age, were a possible reason to explain the Type 2 ship-formed cairns encountered in the Ostrobothnia region. By combining micro and mesoscale analyses, A. César González García, Felipe Criado-Boado and Benito Vilas Estévez summarise current work exploring the 'Megalithic Skyscapes in Galicia'. They look at two unique passage graves – Dombate and Forno dos Mouros – as well as two megalithic clusters – Barbanza and Leboreiro – in search of patterns that go beyond the orientation of their entrances. Illumination effects, inter-visibility and the relation between monument and landscape are explored, leading to the first tentative glimpses of the Neolithic cosmology of Galicia. In 'Different Approaches to Cosmology in Archaeology and Their Application to Maltese Prehistory', Tore Lomsdalen outlines the different ways archaeologists have engaged with notions of cosmology and worldview. He explains how specific approaches, such as Site Catchment Analysis, Fragmentation, Access Analysis and the Archaeology of Death for example, can be combined with archaeoastronomical research, to provide a more holistic picture of meaning, using Maltese sites as examples. In their chapter 'Petra revisited: an astronomical approach to the Nabataean cultic calendar', Juan Antonio Belmonte and A. César González García also use a multi-source approach to gain insights into the Nabataean calendar. Their work indicates the importance of the winter solstice and a link between this period of the solar cycle with festivals and associated pilgrimages dedicated to the main deities of that time.

Four of the archaeology papers draw on the variety of New World cultural expressions of astronomy, from the American Southwest to the Inca World. In 'Investigation of Solstice Horizon Interactions at Chacoan Monumental Architecture', Andrew M. Munro, Tony Hull, J. McKim Malville, F. Joan Mathien and Cherilynn Morrow explore the Chacoan

Great Houses of the American Southwest. Their approach was to look at them diachronically, seeking changes in the number of solstice foresights in the horizon across multiple time periods. They observe a significant rise during the Late Bonito phase, which adds weight to the argument for celestially-adept centralised leadership and/or a religious revival following the severe drought of the 1090s CE, illustrating how the skyscape could be appropriated for political and religious purposes. Photographs are usually a feature of any archaeological paper but J. McKim Malville and John L. Ninnemann demonstrate how they can be an important research tool in their paper 'Archaeoastronomy and Cultural Astronomy as Scientific Disciplines: Falsifiability and Photo Documentation'. They recognise that digital photography is one of several tools in the field that bridges the gap between observation, essential documentation, and a search for meaning and present case studies to back up their arguments. Steven R. Gullberg and J. McKim Malville look at the role of light hierophanies in caves in the Inca world. In 'Caves, Liminality, and the Sun in the Inca World' they report on fieldwork in a great number of caves, the results of which reveal a widespread interest in combining solar illuminations on key dates with artificially altered sections or features of the cave structures, such as steps or niches. The authors suggest that these caves were specifically chosen because sunlight already entered into them, and it is likely that the caves were perceived to be animated by the interaction of sunlight, water and darkness. In 'Rethinking Nahualac, Iztaccíhuatl, Mexico: Between Animism to Analogism in Mesoamerican Archaeoastronomy', Stanisław Iwaniszewski addresses the ritual and worldview of a culture that created a rectangular stone sanctuary on the western slopes of a prominent volcano in Central Mexico. Using the claimed astronomical alignments of the site, Iwaniszewski explains how Postclassic societies in Central Mexico conceptualised their relationship with the environment.

Ethnography
The use of ethnography, history and folklore adds another dimension to understanding cultural practices related to the sky, as the papers in this section demonstrate. Taking an historical and ethnographical approach Nasser B. Ayash presents his paper 'Evolution of Arabic Astronomy in Relation with the Translation Movement in the Early Abbasid Era'. In this work Ayash draws a thread from the Greek and Persian perceptions of the sky to its becoming embedded into the pre-Islamic astrological notions of Anwa and Lunar Mansions and, as a result, the Greek Uranography

became a part of the foundations of the astronomy of the Abbasid era of the mid-eight to mid-ninth century CE.

Roslyn M. Frank has been researching the Basque stone octagons for nearly forty years and in her paper 'Metrology, Memory and Long-Term Landscape Inhabitation: Evidence for the Septarian Package on the Atlantic Façade' she brings new insights to this field. She examines the possible association of the Septarian Package with agro-pastoral practices and shows how its metrological diffusion can shed light on the cultural conceptualisations and practices that might have been associated with megalithic structures found along the Atlantic façade.

Images

Astronomical images seem to mirror the emphasis of the sky in many cultures, examples of which include rock art, petroglyphs, cup-marks and so on. Yet the papers in our 'Images' section show that this field can be imaginative, far-reaching and diverse enough to include virtual landscapes and skyscapes, stained glass windows and frescos. Georg Zotti, Florian Schaukowitsch and Michael Wimmer report on the creation of 'The Skyscape Planetarium', a visual installation at the MAMUZ museum for prehistory in Austria. The exhibition draws on results of the Stonehenge Hidden Landscapes Project, and includes a huge curved screen upon which a Stellarium simulation is projected, virtually recreating the experience of standing within the stone circle and looking out into the surrounding landscape and skyscape. The authors describe the planning and implementation of this innovative installation and outreach experience. In her paper 'Edward Burne-Jones's The Planets: Luna, A Celestial Sphere', Liana De Girolami Cheney examines how this Pre-Raphaelite artist became fascinated with, and inspired by, astronomy and created themed stained-glass windows in a body of work known as the *Planets* cycle. Concentrating on *Luna*, Cheney first discusses the history of the artistic commission and then explains the its cultural sources. In 'An examination of the images of the Sun and the Moon in the Visoki Dečani monastery in Kosovo', Dragana Van de moortel-Ilić contextualises the the tear-shaped images of the Sun and the Moon framing The Crucifixion of Christ fresco in the Visoki Dečani monastery in Kosovo. By placing these images into a cultural and contemporary cosmological context, she concludes that these personifications of the Sun and the Moon may be explained by the synergy of Hellenistic and Christian thought.

Acknowledgements
The conference, and this volume, would not have come to fruition without the help of the Sophia Centre for the MA in Cultural Astronomy located at the University of Wales Trinity Saint David, the organising committee, the reviewing committee, and a dedicated band of volunteers, so our grateful thanks must go to the following:

Local Organising Committee: Nicholas Campion, Lionel Sims, Pamela Armstrong, Bernadette Brady, Frances Clynes, Darrelyn Gunzburg, Liz Henty, Frank Prendergast and Fabio Silva. Additional Reviewing Committee: John Steele, Emilia Pásztor, Kim Malville, Stanislaw Iwaniszewski, Juan Antonio Belmonte and Michael Rappenglück. Conference front desk and refreshments: Jennifer Fleming. Book Stall: Jennifer Zahrt. Conference volunteers: Ada Blair, Ilaria Cristofaro, Karine Dilanyan, Anna Estaroth, Morag Feeney-Beaton, Stavroula Konstantopoulou, Tore Lomsdalen, Chris Mitchell, Hanne Skagen, Melanie Sticker and Kathleen White.

Liz Henty, Bernadette Brady,
Darrelyn Gunzburg, Frank Prendergast
and Fabio Silva

Through the Gloomy Vale: Underworld Alignments at Stonehenge

Lionel Sims and David Fisher

Abstract: Three recent independently developed models suggest that some Neolithic and Bronze Age monuments exhibit dual design properties in monument complexes by pairing obverse structures. Parker Pearson's[1] materiality model proposes that monuments of wood are paired with monuments of stone, these material metaphors respectively signifying places of rituals for the living with rituals for the dead. Higginbottom's[2] landscape model suggests that many western Scottish megalithic structures are paired in mirror-image landscape locations in which the horizon distance, direction and height of one site is the topographical reverse of the paired site – all in the service of ritually experiencing the liminal boundaries to the world. Sims'[3] diacritical model suggests that materials, landscapes and lunar-solar alignments are *diacritically* combined to facilitate cyclical ritual processions between paired monuments through a simulated underworld. All three models combine in varying degrees archaeology and archaeoastronomy and our paper tests them through the case study of the late Neolithic/EBA Stonehenge Palisade in the Stonehenge monument complex.

The Stonehenge Palisade
The late Neolithic/Early Bronze Age (EBA) Stonehenge Palisade[4] was a two kilometre fence of timber posts close to Stonehenge Avenue. Fig. 1 shows how the northern end of this fence came close to the Great Cursus, and then swung south-west in an elbow turn within two metres of the Avenue elbow, and in a diverging angle continued uphill alongside the final section of the Avenue to end nearly 500 m past the Stonehenge monu-

---

[1] M. Parker Pearson, *Stonehenge,* (London: Simon & Schuster, 2012).
[2] G. Higginbottom, 'The World Begins, the World Ends Here', at https://www.academia.edu/22473630 [accessed 3 Jan. 2017].
[3] L. Sims, 'Entering, and returning from, the underworld: reconstituting Silbury Hill by combining a quantified landscape phenomenology with archaeoastronomy', *Journal of the Royal Anthropological Institute* 15, no. 2 (2009): pp. 386-408.
[4] R.M.J. Cleal, K.E. Walker, and R. Montague, *Stonehenge in its Landscape* (London: English Heritage, 1995).

Lionel Sims and David Fisher, 'Through the Gloomy Vale: Underworld Alignments at Stonehenge', *The Marriage of Astronomy and Culture,* a special issue of *Culture and Cosmos*, Vol. 21, nos. 1 and 2, 2017, pp. 11–22.
www.CultureAndCosmos.org

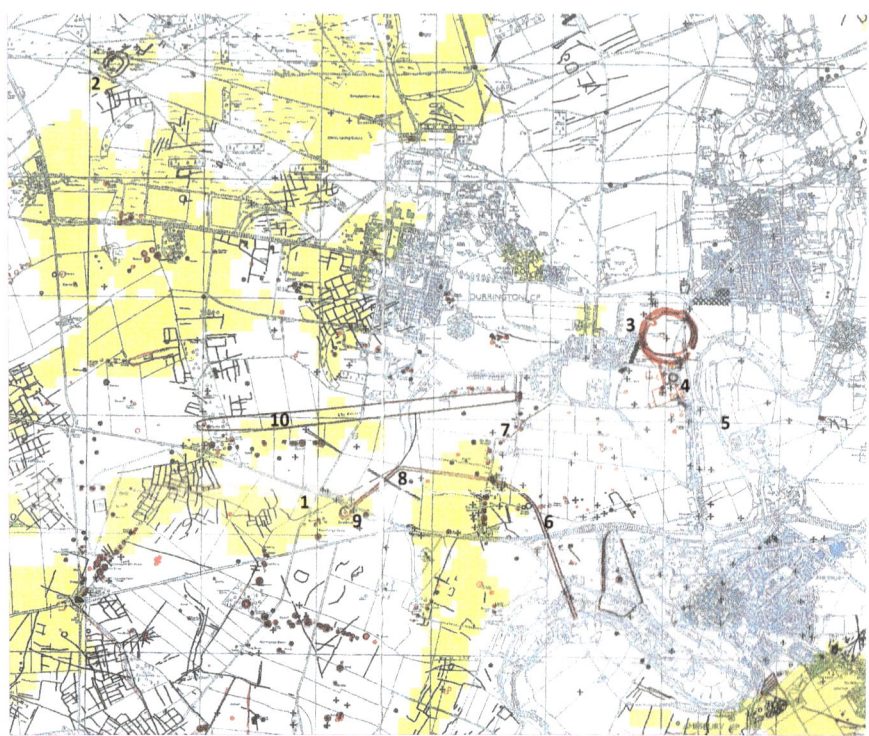

Fig. 1. The Stonehenge Palisade in its Landscape. Key: 1 Stonehenge Palisade; 2 Robin Hood's Ball; 3 Durrington Walls; 4 Woodhenge; 5 River Avon; 6 Stonehenge Avenue; 7 King Barrow Ridge; 8 Stonehenge Bottom; 9 Stonehenge; 10 Great Cursus; Yellow - Viewshed from Robin Hood's Ball; Grid lines are 1 km grid squares, the vertical grid lines indicate grid north which at the centre of the map is 0° 06' west of true north (Adapted from Cunliffe & Renfrew 1997: Plan 1).

ment. By utilising the details of the Vatcher's excavation of the old subway section of the Palisade ditch[5] various multipliers have been applied to the reported average 4½ foot posthole depth. Estimates vary from a 6.1 to a 3.4 m high Palisade.[6] Since this lower estimate poses a greater challenge to our landscape phenomenology and archaeoastronomy methods below, we will assume an average Stonehenge Palisade height of about 3.7 m. Air survey,

---

[5] Cleal, *Stonehenge*, p. 155.
[6] J. North, *Stonehenge: Neolithic Man and the Cosmos* (London: Harper Collins, 1996), pp. 353–56, 365–70.

site excavation and remote sensing[7] have all revealed that the Palisade was discontinuous with one two metre gap next to the Great Cursus and a large gap beginning about 150 m before the Heel Stone when progressing up the Avenue towards Stonehenge and then for 250 m extending to the old subway terminal excavated by the Vatcher's[8] before the Palisade recommences in the same line for a further 500 m.

Methodology
Three recent independently developed models suggest that some Neolithic/EBA monument complexes exhibit dual design properties by pairing obverse structures – the materiality,[9] landscape[10] and diacritical[11] models. In this paper we will use each model and the methodologies of site archaeology, landscape phenomenology and archaeoastronomy to interpret these presently known properties of the Stonehenge Palisade. To the extent that each model is inconsistent with the available evidence we will preserve those parts of it which survive such test. To the extent that each model is inconsistent with the tried, tested and confirmed parts of the other two models, we will preserve those parts of the initial model which survive this test. To the extent that it is then possible to integrate those parts of the three models which have survived this procedure will be a measure of not just triangulation by three independent models,[12] but of emergence to a higher level of meaning than possible with any one single model.

The Materiality Model
Parker Pearson's materiality model proposes a binary separation between monuments of wood and monuments of stone, each of which provides a focus for a domain of the living separated from a domain for the dead. The evidence of a timber palisade alongside Stonehenge is anomalous to this model. Parker Pearson suggests that the Stonehenge Palisade was never alongside *Neolithic/EBA* Stonehenge, but was a mid-Bronze age field boundary fence. However contrary to his claim that no Neolithic

---

[7] C. Gaffney and V. Gaffney, 'The Stonehenge Hidden Landscapes Project', *Archaeological Prospection* 19 (2012): pp. 147–55.
[8] Cleal, *Stonehenge*, pp. 155–61.
[9] Parker Pearson, *Stonehenge*.
[10] Higginbottom, 'The World'.
[11] Sims, 'Entering'.
[12] Wylie, *Thinking from Things* (London: University of California Press, 2002), pp. 176, 192, 207.

archaeology is associated with the Palisade,[13] previous excavation has found a Neolithic chalk plaque in the base of the ditch, and late Neolithic arrowheads, Grooved Ware pottery and cattle, sheep and red deer bones all suggesting a late Neolithic date.[14] Parker Pearson's suggestion that worked flint found in the Palisade trench is mid-Bronze Age is undermined by his own admission that it is indistinguishable from Neolithic worked flint and that they were found in the surface or upper to mid-levels of the trench.[15] The architecture of the Stonehenge Palisade strongly indicates a Neolithic/EBA feature, in which facades commonly shield certain viewpoints and prescribe others, such as also at the West Kennet Palisades in the Avebury monument complex.[16]

The findings of the Stonehenge Riverside Project have demonstrated that the complex timber circles at Durrington Walls are linked to Stonehenge by two avenues connected to the intervening River Avon. While linked they were not inter-visible yet their dimensions and layout mirror each other.[17] Furthermore, their materials were not exclusively timber or stone as suggested by the materiality model. A stone 'Heel' stone stood in the Durrington Walls short Avenue approaching uphill to the southern timber circle from the River Avon.[18] Similarly at Stonehenge the sarsen lintels were linked by wood-working joints when their inertial mass alone guaranteed stability and stone 16, hidden from its Heel Stone entrance behind the Grand Trilithon stone 56, had its surface rendered as oak bark.[19] Furthermore, the large gap between stones 10 and 12 above the space left by the half-height and half-width stone 11 was probably spanned by a timber lintel (Fig. 2). '…Stonehenge was built to look as if it was made of wood'.[20] In short, each timber and stone monument cryptically

---

[13] Parker Pearson, *Stonehenge*, p. 236.
[14] Cleal, *Stonehenge*, pp. 437, 482, 448.
[15] Parker Pearson, *Stonehenge*, p. 236.
[16] Cleal, *Stonehenge*, p. 159.
[17] J. Thomas, 'The Internal Features at Durrington Walls: Investigations in the Southern Circle and Western Enclosures 2005–6', in M. Larsson and M. Parker Pearson, eds., *From Stonehenge to the Baltic* (Oxford: BAR International Series 1692, 2007): pp.145–57.
[18] Parker Pearson, *Stonehenge*, p. 96.
[19] M. Pitts, *Hengeworld* (London: Arrow, 2001): pp. 268, 264.
[20] Parker Pearson, *Stonehenge*, p. 334.

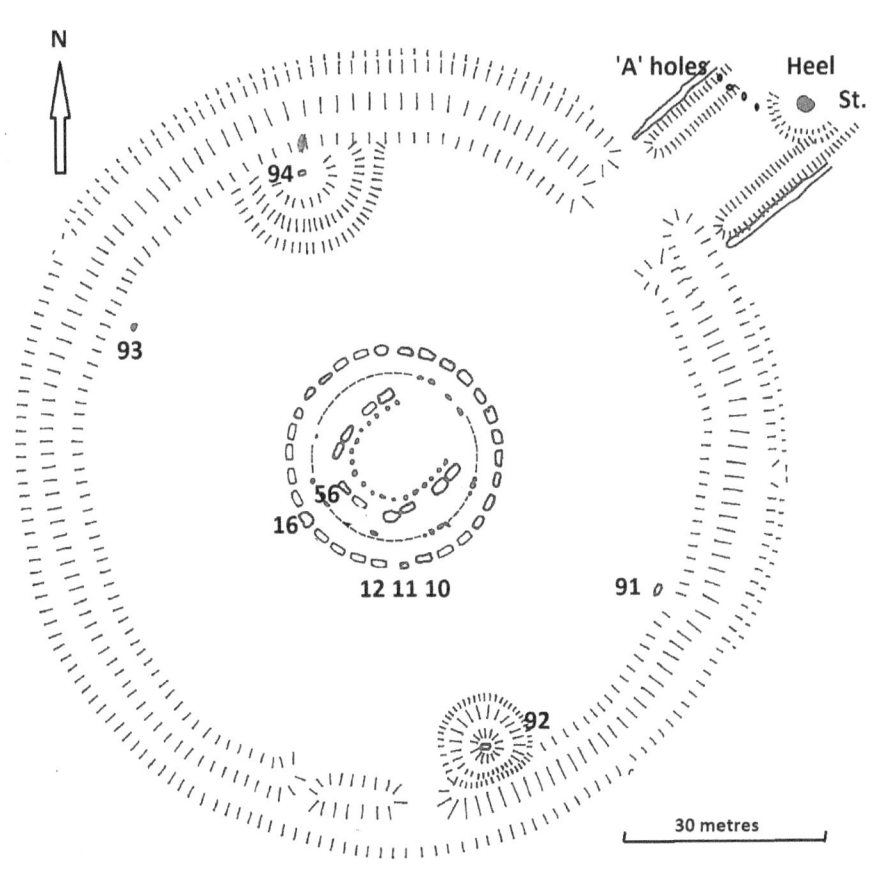

Fig. 2 Stonehenge (Phase 3v: Cleal 1995, 167 OR Phase 4: Parker Pearson 2012, 311) in plan view with features identified in text. NB: North East entrance; outer ditch and inner bank; Stone 11 is half height and half width of other stones remaining In outer sarsen circle; Stone 16 is rendered as oak bark and hidden behind Stone 56 when viewing from Heel Stone; Stone 56 is the one remaining upstanding stone of the Grand Trilithon.

included in a minor register a reference to its 'opposite' materiality suggested by Parker Pearson's model. Similarly the timber Palisade is a mirror image of the last two sections of the Stonehenge Avenue in the domain of stone, in which the Palisade 'elbow' closely abuts the Avenue 'elbow' and the line of timbers diverge from this point at roughly equal angles away from the angled lines of the Avenue (see Fig. 1). In sum, the monuments' designs, excavated archaeology and landscape positioning all suggest that timber and stone structures were contemporary and in dualistic mirroring dialogue with each other. Instead of a categorical distinction between these two building materials then whatever symbolic loadings were placed on each of them by the monument builders *in combination* they were probably utilised for some fused *diacritical* level of meaning. Parker Pearson suggests that Stonehenge has horizon alignments past the Heel Stone on the winter solstice sunsets and the summer solstice sunrises, the northern major moonrises amongst the north-east entrance 'A' holes, southern major full moon risings and northern major full moon settings from station stones 92 and 94, and summer solstice sunrise from 93 to 94 and 92 to 91.[21] For Durrington Walls he proposes a summer solstice sunset alignment along its entrance avenue, a winter solstice sunrise from the southern circle's entrance, a winter solstice sunset from the southern circle's inner horseshoe and a winter solstice sunrise from the northern circle's inner four-post square setting (Fig. 3).[22]

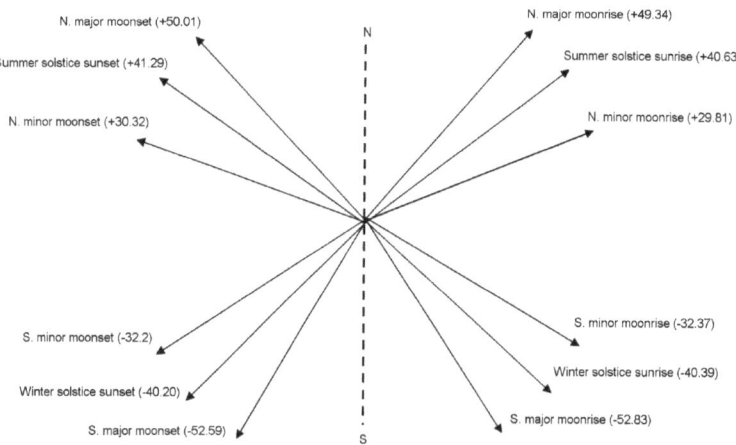

Fig. 3 Plan view of natural horizon alignments on sun's solstices and moon's standstills at Stonehenge 2500 BC (data from North 1996, appendix 3). Key: N = north, S = south. Bracketed numbers are degrees relative to the west-east axis. Horizon alignment on the upper limb of sun and moon. Natural horizon altitudes in degrees at Stonehenge: NE 0.6, SE 0.7 SW 0.6 NW 0.3

---

[21] Parker Pearson, *Stonehenge*, pp. 47–49.
[22] Parker Pearson, *Stonehenge*, pp. 79–82.

There are grounds for rejecting some of these claims.[23] For Stonehenge the 'A' holes were in use only in Phase 1, the summer solstice sunrise alignment over the Heel Stone lacks a backsight, while those from station stones 93 and 92 are too short a distance to provide a safe azimuth. For Durrington Walls winter solstice sunrise from the southern circle's entrance fails to take account of the horizon altitude whereas the summer solstice sunrise does.[24] Further, this portfolio of alignments does not fit the symbolic structure predicted by the materiality model. If Stonehenge is the focus for the domain of the dead we would not expect a summer solstice sunrise alignment for the start of the longest brightest day but a dark Moon during the week of the winter solstice sunsets at the start of the longest darkest night.[25] And if Durrington Walls is a structure that mirrors Stonehenge, we would also expect reversed but matching lunar alignments for Stonehenge.[26] Anthropology accepts that all rituals are characterised by a multi-media redundancy and amplification of symbols, and that structures that we find in one dimension of symbolism will be rehearsed and repeated in other associated dimensions.[27] Thus we would expect that the monument builder's diacritical symbolism invested in wood and stone will be repeated in horizon alignments that conflates dark Moon with solstice sunsets in a nine-ten year periodic alternation between minor and major standstills.[28] This expectation provides a test for our review of the next two models. We therefore retain from the materiality model that the Stonehenge monument complex unified contemporary mirroring monuments of wood and stone, each of which included cryptic elements of their 'opposite' materiality, while this diacritical combination is expected to be replicated in each monument's landscape location and horizon astronomy, and that in its approach to Stonehenge the Stonehenge Avenue had a substantial timber palisade running close to its last two sections.

---

[23] L. Sims, 'Stonehenge Decoded?', *Culture and Cosmos* 17, no. 2 (2013): pp. 138–42.
[24] North, *Stonehenge*, pp. 458–81 and 358–62.
[25] L. Sims, 'What is a Lunar Standstill III?', *Documenta Praehistorica* XLIII, (2016): pp. 467–78.
[26] North, *Stonehenge*, pp. 365–73.
[27] R. Rappaport, 'The Obvious Aspects of Ritual', in R. Rappaport, ed., *Ecology, Meaning and Religion* (Berkeley: North Atlantic Books, 1979): pp. 173–221.
[28] L. Sims, 'What is the minor standstill of the Moon?', *Journal of Skyscape Archaeology* 2, no. 1 (2013): pp. 67–76, 95–102.

The Landscape Model

Higginbottom's landscape model[29] shows that many megalithic sites in western Scotland, including single standing stones, stone rows and stone circles, have been intentionally located with a careful selection of the horizon even in cases, such as in Mull, where the landscape geomorphology goes against their preferences. Sites are paired such that each display reversed but closely similar horizons, one to the north and the other to the south. When indicated by megalithic structures, distinctive distant northern pronounced horizon features seem to have been chosen to mark the Sun's summer solstice limits and the Moon's major and minor northern standstill limits. The portfolio of features for each site includes combining close and far horizons in approximate cross-cardinal alignments, frequently resulting in each structure nestled into reversed semi-amphitheatre/half-bowl site settings, large bodies of water in the direction of the far horizon, prominent landscape features optimally edging these bodies of water, and together the hill/mountain features and water edges optimally bounding the horizon limits for solstice and major and minor lunar standstill risings and settings. Contrarily reversed megalithic sites with distant horizons to the south indicate pronounced horizon features which mark the winter solstice Sun's horizon limits and the Moon's major and minor southern standstill limits. This dual combined suite encapsulates and combines topography and horizon astronomy which emphasise views of the liminal boundaries between mountain and water, land and sky, celestial risings and settings and lunar-solar cycles. According to Higginbottom, by combining landscape oppositions with celestial transformations alternating between north and south, the monument builders forged a common cultural identity through an appreciation of the enveloping liminal forces that bounded their world. This model presently makes no connection with stone/wood dualism or to screens of any kind that might indicate some connection with Palisade-type structures. Nevertheless its insight into choosing locations which control inter-visibility between paired monument locations we will see is one property of the Stonehenge Palisade.

This landscape model is marked by a strong field survey, virtual modelling and statistical testing methodology which has substantially strengthened and extended earlier work conducted by Ruggles.[30] The

---

[29] Higginbottom, 'The World'.
[30] C.L.N. Ruggles, *Astronomy in Prehistoric Britain and Ireland* (London: Yale, 1999).

constructed elements of the complex, shared with the materiality model, are optimally located to combine central views of a body of water with paired obverse sites which are not inter-visible. However, the interpretation offered by Higginbottom for the data is problematical. Alignments for each site are shown to be grouped within single topographical horizon features so that the Sun and Moon emerge from one distinct mountain and enter another. But instead of interpreting this as a conflation of the Sun and the Moon in the same small horizon region, the alignments are considered separately. Rather than keeping to the field data finding that obverse sites are optimally separate and their alignments hidden from each other, it is claimed that alignments are paired *across* them. A southern standstill lunar alignment in one site is linked to the summer solstice in a companion site orientated to the north and vice-versa northern standstill alignments are linked to a companion site orientated to the south on the winter solstice. But Higginbottom's interpretation is weakened by the admission that there is no statistical support for the summer solstice alignments which are required to justify the choice of a full Moon for the high frequency alignments on the southern minor and southern major lunar standstills.[31] Also, this argument undermines the central principle of the model which emphasises the selection of topographic settings mobilising opposite, alternating and inverse dimensions, not by amalgamating them *across* obverse sites. Instead, the preferred interpretation is in the service of a prior *ad hoc* commitment to synchronising solstice alignments with *full* Moons. This would be equivalent to pairing Stonehenge's axial alignment on winter solstice sunset with Durrington Walls' axial alignment on the northern major standstill moonsets and concluding that the combined monument complex is associated with a full Moon ritual. Rather than 'the monuments clearly highlight[ing] the cosmic order of opposites at the extremes', a separate justification is used for selecting just full Moons – full Moon gives a 'spectacular visual display...unaffected by the position in the lunar standstill cycle'.[32] Yet position in the lunar standstill cycle *is* marked by the horizon positions of both Sun and Moon rising and mainly setting into the same significant horizon features, and this provides an alternative and simpler interpretation of the data. The sidereal Moon for

---

[31] Higginbottom, 'The World', p. 24.
[32] G. Higginbottom, A. Smith and P. Tonner, 'A Re-creation of Visual Engagement and the Revelation of World Views in Bronze Age Scotland', *The Journal of Archaeological Method and Theory*', doi: 10.1007/S10816-013-9182-7, (2013): pp. 28–29.

about two years returns to these horizon positions every 27 or so days in a series of time lapsed reversed phases while the Sun only occupies the same position once a year for a week or so. During one of those weeks the Sun comes to meet the Moon and occupy that same horizon region. Then dark Moon will set with the winter solstice sunset to herald the beginning of the longest darkest night to the south and with the summer solstice to the north. Without the requirement for additional explanations, a dark Moon at each solstice automatically offers another 'spectacular visual display' for observing the greatest possible number of stars. The cognitive impact of 'spectacular displays' cannot discriminate between either of these options whereas the landscape model itself suggests the builders placed their monuments to conflate Moon and Sun when simultaneously present at their limits on a single pronounced horizon feature – dark Moon at a solstice. The full Moon interpretation is doubly undermined by Higginbottom's finding of many southern alignments on the major and minor Moon *and* the winter solstice Sun at these latitudes to be completely or partially obscured below the southern horizon.[33] Why would the monument builders, if they were investing so much thought into carefully selecting horizons, frequently choose exactly the horizon landscape that so dramatically blocks a 'spectacular visual display' of the full Moon? For a culture that believes in a stationary flat earth one interpretation stands out: if the Moon is below the horizon then it must be travelling through the underworld. Since such a cosmology would also understand that when the Moon has disappeared from the sky at dark Moon it must be travelling through the underworld, then the purposeful choosing of sites with interrupted horizon views of the southern standstill moons is consistent with lunar-solar alignments that conflates the Sun's solstice settings with the standstill dark Moon.

Once amending this landscape model by removing an *ad hoc* commitment to full Moon alignments at their standstills, then its finding of mirrored landscape settings is isomorphic with the critiqued materiality model's mirrored monuments in wood and stone. Just as stone and wood structures are each combined with a germ of the other in the materiality model, so each site's landscape horizon is the mimesis of the other. Looking at one site's horizon alignments presents the 'same' horizon shape as its reverse site, and therefore each present the promise of the other. Each horizon is diacritically related to the other. Just as materialities are diacritical in one model, so landscape placement and horizons are in the

---

[33] Higginbottom, 'The World', pp. 35, 36, 45, 50.

other model. The same diacritical principle applies to both models' 'astronomies'. The mirroring of the materiality model is not just that of reversed duplication, but includes that of mirrored fusion in which the hint or germ of one is included in its mirrored opposite. The landscape model also possesses this property in its horizon alignments of lunar-solar conflation between south and north horizons, each generating culminating solstice dark Moons during major and minor standstills. Rather than the complex and critiqued alignments suggested by Parker Pearson, the landscape model offers a simplified arrangement of alignments consistent with its central principle of alternating cycles and topographies. The critiqued landscape model finds a matching alternation between northern and southern horizons with the addition of an accompanying expanse of water over which can be observed the transit of Sun and Moon at the limits of their cycles culminating with a solstice dark Moon, not full Moon, or with the Moon and Sun transiting below the southern horizon. This combination of topographies and astronomy suggests that prominent horizon peaks reaching to the sky are chosen to contrast with expanses of water and lunar-solar transits below the horizon which reach to the underworld. This interpretation is more consistent with Higginbottom's data and deepens her central interpretive concept of paired obverse landscapes by extending the concept of landscape into the watery underworld. Instead of 'the world begins here, the world ends here'[34] the symbolic structure emerging from these two models suggests we need to consider rituals simulating a journey *between* worlds that include an underworld. This exercise in paring cognate models to the point that they can be integrated therefore suggests that other relevant models should be consistent with diacritical materialities, landscapes and astronomies that extend beyond this world.

The Diacritical Model
Sims' model shows that some Neolithic/EBA monuments are designed using asymmetrically paired categories in a dualistic and reversible monument complex. Just as we have seen that at Stonehenge the stone monument has a hint of wood and that the wooden Durrington Walls has a hint of stone, and in Western Scotland each monument horizon is paired with another monument's matching but reversed horizon, so horizon alignments are also combined asymmetrically with their reverse equivalent in paired monuments whose design categories overlap each other. They are

---

[34] Higginbottom, 'The World'.

22 Through the Gloomy Vale: Underworld Alignments at Stonehenge

organised in dualistic diacritical combinations. According to this model these 'astronomical' and material categories are organised in this compelling alternation in the service of a belief in a cosmos sensed in need of repair and management in forestalling stasis partly by conducting ritual journeys through a simulated underworld.[35] Significant common ground exists between this model and the materiality and the landscape models.

Approaching Stonehenge uphill along the Avenue when at the Heel Stone can be seen the winter solstice sunsets between the Grand Trilithon uprights into the Altar Stone, while directly above at an elevation of 4°–5° is an alignment beneath the Grand Trilithon lintel on the southern minor standstill moonsets.[36] Moving clockwise round the Stonehenge outer ditch from the Heel Stone to the southern entrance looking from the Station Stone 92 at 90° to the axial alignment a secondary ray touches the back of stone 16, the stone with a surface rendering of oak bark which is hidden from the Heel Stone behind Grand Trilithon stone 56, and continues on to Station Stone 93 to a horizon alignment on the northern major standstill moonsets (Figs. 2 and 3).[37] But the same alignment on the northern major standstill is also the *axial* alignment through the Durrington Walls southern timber circle where it is conflated with the Sun's summer, not winter, solstice settings.[38] In combination with this axial lunar-solar alignment at Durrington Walls there are again secondary orthogonal and cryptic alignments onto the winter solstice sunset to the south-west from the inner horseshoe of timber posts and by the subsidiary monument alongside of Woodhenge with its few megaliths in its southern quadrant.[39] In all four cases lunar-solar conflated alignments to the south-west are bracketed with stone and those to the north-west with wood. These paired and reversed diacritical alignment combinations do not just overlay an identical syntax of materialities, but also intersect with reversed horizons as predicted by the landscape model. The mainly wooden Durrington Walls has the same dimensions as Stonehenge and therefore presents the 'same' horizon shape as Stonehenge but now the alignments are rotated 90° to look to the north-west rather than the south-west, just as do many stone monuments in Western Scotland. At the Stonehenge monument complex the axis of the paired monuments of mainly stone and mainly wood conflate alignments

---

[35] Sims, 'Entering'.
[36] L. Sims, 'The Solarisation of the Moon: Manipulated Knowledge at Stonehenge', *Cambridge Archaeology Journal* 16, no. 2 (2006): pp. 191–207.
[37] The same arrangement can be found at Avebury – see in discussion below.
[38] North, *Stonehenge*, pp. 347–73.
[39] Pitts, *Hengeworld*, p. 264.

on lunar standstills and the Sun's solstices, while secondary orthogonal cryptic alignments are made on the obverse standstills and solstices. The materiality and landscape models' claims therefore withstand refutation when diacritically reformulated and combined with an archaeoastronomy model of lunar-solar conflation. All three models' separate insights, when parsed against the evidence, culminate in their combination. Stone and wood, south and north horizons, minor and major standstills are asymmetrically combined in reversed proportions across a dualistic monument complex.

The integrated model
We argue that each monument is in a relation of reversible dualism with its doppelganger but in a unity which while split is consistent within this dualism. Both lunar alignment minor and major combinations, while endlessly alternating around a nine-year cycle, are united in aligning on the sidereal Moon which in both cases display reversed time-lapsed lunar phases attenuated over the course of one year and both culminate with dark Moon at their respective solstice sunsets.[40] The cryptic orthogonal alignment within each monument is the germ of the next turning point in the endlessly alternating 9- or 10-year periods of the minor and major standstills[41]. And each dark Moon solstice sunset ritual is finalised by solstice sunrise in orthogonal alignments across both monuments and in the subsidiary monuments Woodhenge and the northern timber circle at Durrington Walls.[42] As the Sun sets and disappears on the western horizon, in a stationary flat earth cosmology, it journeys eastwards past the buried dead through the underworld to be resurrected on the eastern horizon. Just before dark Moon however, the Moon can be seen *rising* on the east horizon as waning crescent Moon shortly before sunrise. During dark Moon it is then lost in the Sun's glare and after one to three days can then be seen resurrected and *setting* on the western horizon after the Sun sets. For about seven to nine days, until it is first quarter Moon, it cannot be seen rising on the eastern horizon because of the light of the already risen Sun. Therefore the Moon's resurrection takes place while dying on the western horizon. The Moon's symbolic repertoire is more suitable for a cosmology that requires contradiction and complexity rather than the one-dimensional Sun. Constant alternation in rituals is built into the monument

---

[40] Sims, 'The Solarisation', p. 199.
[41] Sims, 'What is a Lunar Standstill III?'.
[42] Parker Pearson, *Stonehenge*, p. 89.

24 Through the Gloomy Vale: Underworld Alignments at Stonehenge

design as necessary, managed and controlled in this reversible dualism by the overlapping interplay of each monument's dominant and diacritical aspects. Each ritual can only be completed by moving on to the next. By requiring timed rituals, and so time itself, to keep moving it forestalls any sensed crisis in which time might stop.

The integrated model and the Avebury monument complex
If this model is robust then we would expect the Stonehenge Palisade to be consistent with its key components of diacritically combined stone and wood, north and south horizons, minor and major solarised standstills in the service of simulations of underworld ritual journeys. First we will summarise an earlier partial application of this model, and then move on to the Stonehenge Palisade. In a preliminary study of the Avebury monument

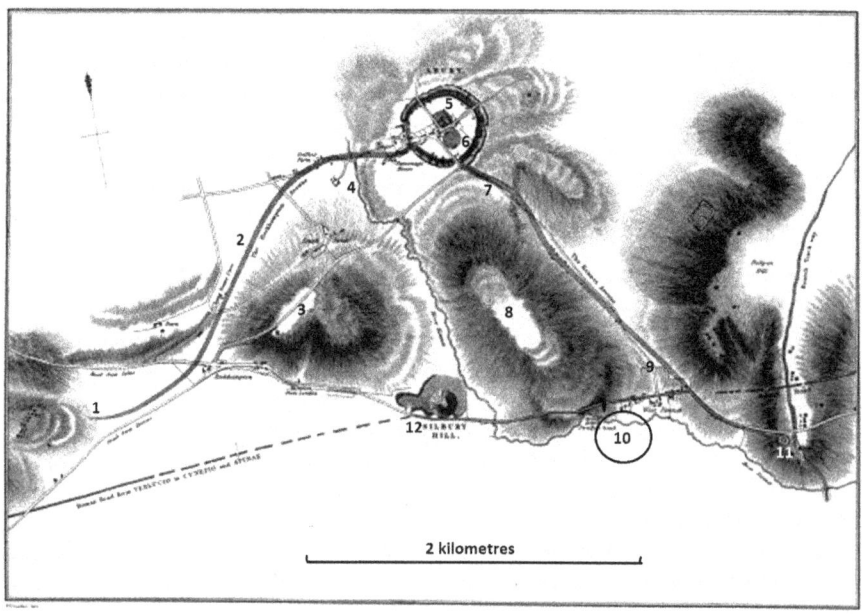

Fig. 4a The Avebury monument complex with features identified in the text. Key: 1 Fox Covert, 2 Beckhampton Avenue, 3 Folly Hill, 4 River Winterbourne, 5 Northern Inner Circle of the Avebury Henge and a double post circle stood approximately where the number 5 is positioned, 6 Southern Inner Circle of the Avebury Henge, 7 West Kennet Avenue, 8 Waden Hill, 9 North Kennet Springs, 10 West Kennet Palisades (see 4b), 11 Sanctuary, 12 Silbury Hill (Adapted from Crocker, 1821).

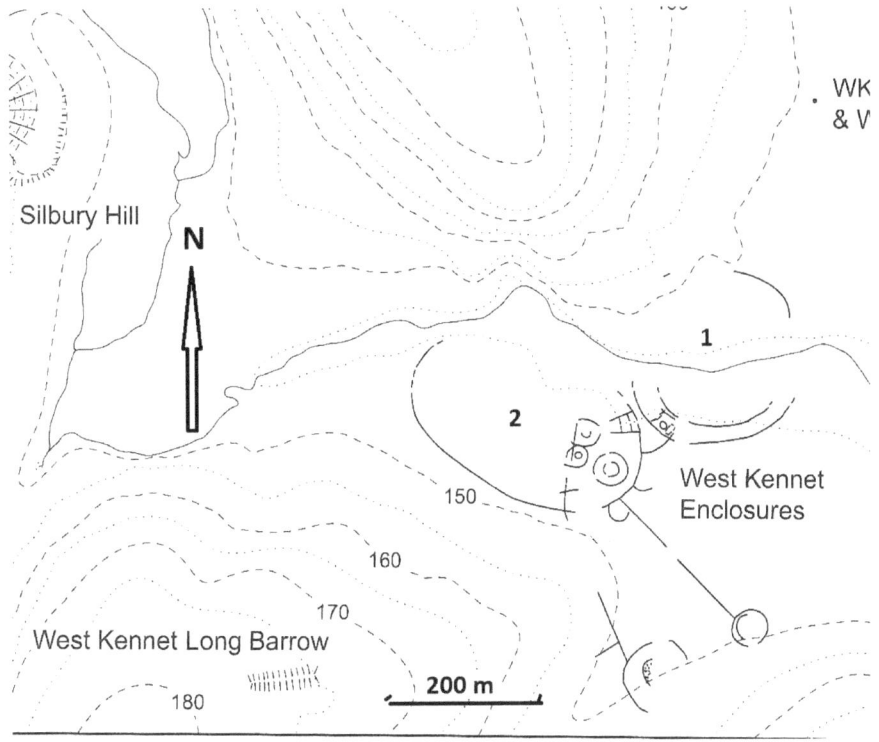

Fig. 4b. Detail of the West Kennet Palisades. Key: 1 Enclosure 1, 2 Enclosure 2 (Adapted from Whittle 1997, 55).

complex[43] the Avebury stone circle and henge and the West Kennet Palisades are shown to be separated and obscured from each other by the intervening Folly Hill and Waden Hill (Fig. 4a). Nevertheless they are linked to each other by the Beckhampton Avenue,[44] the West Kennet Avenue and the North Kennet springs.[45] The prescribed routes along these avenues have been carefully chosen by the monument builders to provide intermittent views not of these two monuments but a third - the centrally located Silbury Hill which the avenues skirt at a relatively constant distance. These inter-visibility locations are from the centre of the southern inner circle of Avebury circle, the central structure of Enclosure 2 of the

---

[43] Sims, 'Entering'.
[44] L. Sims, 'The Logic of Empirical Proof', *Time and Mind* 2, no. 3 (2009): pp. 333–45.
[45] S. Marshall, *Exploring Avebury* (Stroud: The History Press, 2016), pp. 134–37.

West Kennet Palisades (Fig. 4b), from the Fox Covert start of the Beckhampton Avenue and the Sanctuary end of the West Kennet Avenue. The Avebury Circle combines axial alignments on winter solstice sunset with the southern major standstill moonsets, with an orthogonal alignment on summer solstice sunsets that also intersects diacritically with the timber henge alongside the northern inner circle.[46] The West Kennet Palisades axial alignment is on the northern minor moonsets and the summer solstice sunsets, both seen setting into Silbury Hill. Between the two large monuments of mainly stone and mainly wood, the smaller stone *and* wood Sanctuary aligns on the southern minor standstill rises and sets, and provides the diacritical alternation by materials and alignments.[47] Integrated into this complex are a digital series of views of Silbury Hill which act as facsimiles of the crescent Moon before, during and after dark Moon.[48] With its flat summit in-line with the background horizon at its point along the Beckhampton Avenue when it crosses the River Winterbourne and from the Sanctuary, it simulates a Moon that has set and is in the underworld. There is just one place from which an underworld Moon can be observed and that, of course, is the underworld itself. Ritual participants by their own agency when viewing this are led to believe that they must be with the Moon in the underworld. Avenues and monuments serve to simulate a journey through the underworld. This is reinforced within the centre of the southern inner circle to see the cropped chalk top of Silbury Hill as the waxing crescent Moon setting on the south-south-western horizon at 80° south of west, thirty degrees beyond what it ever does in this world and therefore an underworld alignment, journeying through the underworld to its resurrection on the western horizon.[49] For those who participated in rituals along these Avenues seeing the Moon in the underworld and within these monuments they therefore simulated a journey through the underworld according to the rhythms of the lunar-solar cycle.

The integrated model and the Stonehenge Palisade
We have found that diacritically combined materials and alignments predicted by the integrated model are present at both the Stonehenge and

---

[46] North, *Stonehenge*, pp. 271–76.
[47] L. Sims, 'Toads Turning Time: Verifying Visualisations of the Sanctuary (Avebury, Wiltshire)', (submitted).
[48] Sims, 'Entering', p. 401, Fig. 6.
[49] See below.

Avebury monument complexes. However the Avebury evidence for these operating in the service of rituals simulating underworld journeys is reliant on the monument of Silbury Hill, and no such monument exists within the Stonehenge monument complex. The diacritical model predicts that some other arrangement of landscape, structures and their associated 'astronomy' should be present there in the service of the same cosmology.

The materiality model suggests that ritual processions begin at Durrington Walls, descend along a short avenue to the River Avon, then leave the Avon to process along the Stonehenge Avenue in a circuitous route to Stonehenge (see Fig. 1). In this journey from Durrington Walls processionists carry with them its association with wood and main alignment on summer solstice sunsets and the northern major moonsets. The Avenue route leaves the River Avon out of sight of and away from Stonehenge rising up to King Barrow ridge. In the gap between the Old and New King Barrows '…the [Avenue]…seems to have been located at the very point where Robin Hood's Ball comes into view'.[50] This causewayed enclosure was first built in 3600 BCE and was in use until about 1,600 BCE.[51] Built on the side of a south-east facing hill on the far north-west horizon each of the two encircling ditches provided the material placed *uphill* for two chalk banks. Seen from below along the Avenue these two chalk walls merge to present the appearance of a single long shallow mound of chalk sitting on the north-west horizon. The angle of view of Robin Hood's Ball from the King Barrow Ridge is about 47° north of west. Continuing westwards along the Avenue to Stonehenge is a descent into Stonehenge Bottom. Just above the 85 metre contour and before entering the flat and boggy[52] Bottom the eye-height of an adult Neolithic Man[53] lines up with the top of the Stonehenge Palisade and the cropped top of the north-west horizon revealing just the chalk crescent of Robin Hood's Ball at about 50° north of west – the alignment previously seen at Durrington Walls southern circle on the northern major standstill moonsets (Fig. 5a). Over the course of one year abstracting lunar observations to just this time-lapsed horizon limit therefore selects thirteen

---

[50] S. Exon, V. Gaffney, A. Woodward, and R. Yorston, *Stonehenge Landscapes* (Oxford: Archeopress, 2000), p. 75.
[51] A. F. Whittle and A. Bayliss, *Gathering Time*, (Oxford: Oxbow, 2011), pp. 197, 706, 900.
[52] Exon, *Stonehenge*, pp. 40, 52.
[53] D. R. Brothwell and M. L. Blake, 'The Human Remains from Fussel's Lodge Long Barrow: Their Morphology, Discontinuous Traits and Pathology', *Archaelogia* 100 (1966): pp. 48–63.

reverse phased lunations covering the whole lunar cycle and culminating with dark Moon at summer solstice – a lunar-solar and reversed binary 'astronomy' entering into Robin Hood's Ball. Embedded in the late Neolithic/EBA Stonehenge monument complex Robin Hood's Ball is constructed as a facsimile of the sidereal Moon's upper limb at its northern horizon extreme.

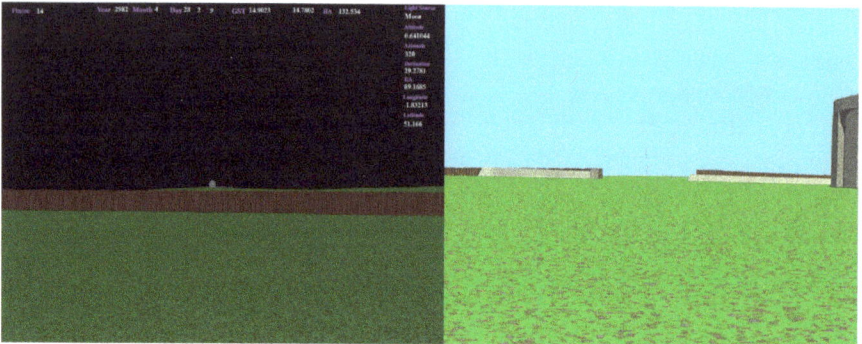

Fig. 5a (left) Virtual model of the view when walking down the Avenue just before dropping below the 85 metre contour, showing the northern major standstill moon setting into Robin Hood's Ball on the north west horizon.
Fig. 5b (right) Virtual model of the view of Robin Hood's Ball from Aubrey Hole 20 by south entrance to Stonehenge. (Dashed central vertical construction line indicates horizon position of Robin Hood's Ball.)

Continuing along the Avenue and into the Stonehenge Bottom not just the archaeoastronomy but landscape phenomenology brackets the sudden loss of sight of Robin Hood's Ball below the level of the Palisade with the high encircling horizons, and '[t]he walker experiences a negative and out-of-worldly state'[54] through a waterlogged landscape.[55] Arriving at the Avenue 'elbow' this 'out-of-worldly' sense is reinforced by coming within two metres from the high Palisade. Rising out of the Bottom the Avenue takes a sharp turn and proceeds uphill at 40° south of west[56] and an altitude of 1° giving an un-obscured ground level view at the centre of Stonehenge to winter solstice sunset. Directly above an upper window raised 4–5° by the Stonehenge lintels captures the southern minor standstill moonsets. The

---

[54] Exon, *Stonehenge*, p. 75.
[55] Exon, *Stonehenge*, pp. 40, 42; Parker Pearson, *Stonehenge*, p. 140.
[56] R.J.C. Atkinson, 'Some new measurements on Stonehenge', *Nature* 275 (1978): pp. 50–52.

Palisade diverges from this last section of the Avenue at 26°, and rising uphill out of Stonehenge Bottom and away from the lowering Palisade emphasises the sense of coming up out of an underworld. Continuing on to the beginning of the 250 m gap in the Palisade about 150 m from the entrance to Stonehenge allows another view of Robin Hood's Ball, although now at 60° north of west (Fig. 5b).[57] Once entering Stonehenge moving to the southern entrance the ditch terminals are marked by a human cremation, wood ash and an cattle skull to the left and a grooved ware sherd and cattle mandible to the right and between them both an unobscured line of sight along the edge and at 90° to the line of the Palisade to Robin Hood's Ball at 63° north of west. From this vantage can now be seen the half-height stone 11 probably with its timber lintel, the otherwise hidden stone 16 whose surface is rendered as oak bark, and next to the edge of the western section of the Palisade post-pipe an Iron-Age adult male burial looks north-west also at 90° to the line of the Palisade. These sixteen signifiers suggest that this was a likely prescribed view.

Coming out of the encircling horizons and Palisade from Stonehenge Bottom and then walking up the final section of the Avenue towards Stonehenge during the week of winter solstice, the southern minor standstill of the Moon will be dark Moon. The Moon was last seen as waning crescent Moon rising on the *eastern* horizon just before sunrise to then emerge after dark Moon as waxing crescent Moon on the *western* horizon. In a stationary flat-earth cosmology the Moon will be perceived as having travelled west through the underworld. Durrington Walls alignment on the northern major standstill of the Moon is repeated by the choice of Avenue route in its sight of Robin Hood's Ball on the same alignment when entering Stonehenge Bottom. Now seeing Robin Hood's Ball again, but at 63° north of west, leaves just one possible interpretation – this is an underworld alignment. Seeing the Moon in its underworld transit therefore reverberates back on the experience of walking westwards through Stonehenge Bottom as a simulation of walking through the underworld in the company of the Moon's westwards underworld transit.

Conclusion
We have seen that the monument builder's careful choice of landscape, monument design and alignments at Stonehenge and Avebury are diacritically organised into asymmetric paired categories in dual monument

---

[57] The yellow shading in Fig. 1 also shows this intermittent visibility of Robin Hood's Ball along the Avenue route.

complexes. Three independently developed dualistic models, once parsed against the evidence, all converge to the same emergent reconstruction suggesting that they are embedded in a cosmology driven to keep time turning by organising dual ceremonies simulating a journey through the underworld according to the alternating rhythms of solarised lunar standstills between major and minor standstills. The locally available landscape at Avebury allowed the building of two avenues' prescribed routes and Silbury Hill to symbolically represent a Moon that enters, travels through and emerges from the underworld. At Stonehenge the Avenue and the discontinuous Palisade served a similar function to construct prescribed views of Robin Hood's Ball as a facsimile of the Moon entering and returning through 'the dark and gloomy vale'[58] of the underworld.

Acknowledgements
We would like to thank the two anonymous peer reviewers and the editor Frank Prendergast for their helpful comments on earlier drafts of this paper.

---

[58] Psalms 23:4, Lutheran translation, (1984).

The Bush Barrow and Clandon Barrow Gold Lozenges and the Upton Lovell Golden Button: A Possible Calendrical Interpretation

Claude Maumené

Abstract: The Bush and Clandon Barrow gold lozenges and the Upton Lovell golden button, discovered in burial grounds near Stonehenge and Mount Pleasant in southern England, were most frequently thought to be ornamental breastplates designed to show the high level political or religious status of the wearers. The author attempts to demonstrate that a purely decorative interpretation must be rejected and proposes a complementary evaluation of these items which all show similar decorations. Counting the lines and interpreting the patterns on both breastplates and button have led to the proposal that these objects were made to facilitate counting, memorisation and transmission of the numbers of days of one or several synodic cycles of Venus, Mars and Jupiter, in agreement with a number of Moon and Solar cycles. In terms of anthropology, the symbolic lozenge shape associated with fertility and fecundity, appeared in many the cultural areas of ancient Europe. Venus, appearing alternately as an evening and morning star, is also an essential symbol of life, death and rebirth. This may be important within the funerary context of the culture of Wessex.

Introduction

Bush Barrow is an early Bronze Age burial site (1900–1700 BCE) situated on the extreme west of the Normanton Down cemetery, close by the Stonehenge stone circle. This burial mound which ranks among the major sites of the Stonehenge complex was excavated in 1808 by William Cunnington.[1] Clandon Barrow which is situated around fifty miles from Stonehenge, west of Dorchester, in the parish of Winterborne St Martin, near the Maiden Castle site and around 4 miles from the archaeological site of Mount Pleasant Henge was first excavated by Edward Cunnington in 1882. Upton Lovell Barrow, also called the Golden Barrow, located less

---

[1] S. Needham, A. J. Lawson and A. Woodward, '"A Noble Group of Barrows": Bush Barrow and the Normanton Down Early Bronze Age Cemetery Two Centuries On', *The Antiquaries Journal* 90 (2010): pp 1–39.

Claude Maumené, 'The Bush Barrow and Clandon Barrow Gold Lozenges and the Upton Lovell Golden Button: A Possible Calendrical Interpretation', *The Marriage of Astronomy and Culture,* a special issue of *Culture and Cosmos*, Vol. 21, nos. 1 and 2, 2017, pp. 31–50.
www.CultureAndCosmos.org

32 The Bush Barrow and Clandon Barrow Gold Lozenges and the Upton Lovell
Golden Button: A Possible Calendrical Interpretation

than twelve miles southwest of Stonehenge, on the banks of the River Wylye, was excavated by William Cunnington in 1803 and again in 1807. The Bush and Clandon Barrow gold lozenges and the Upton Lovell golden button, all golden objects from the Bronze Age (2000–1650 BCE), are held to be archetypal of the 'Wessex culture'.[2] Needham and Woodward consider that the Wessex group does not represent a precise cultural entity, but rather the result of multiple connections to foreign lands.[3] Later Needham et al argue that these burial deposits of the Early Bronze Age Period 3 (1950-1750/1700 BCE), on Normanton Down, can be seen as representing a dynastic succession that controlled access to Stonehenge during this period and presided over the ceremonies therein.[4]

The Bush Barrow lozenge
Being unsure of the exact position of the body found in the barrow, Needham et al have proposed a reconstruction of the grave layout, based on Cunnington's accounts.[5] They suggested that the body would have been crouched on its left side. By the right side of the body was a mace, with the handle being embellished with bone zigzag mounts. Three worked sheet gold objects were also found – a large diamond shaped lozenge resting on the man's chest and a large belt-hook lying by his waist, both finely decorated; as well as another small diamond shaped lozenge, which may have been mounted on the handle of the mace.

The lozenge itself has been the object of a precise description by Anthony Johnson:

> The face carries a series of concentric lozenges engraved at equally spaced intervals, each defined by a series of four precise, closely spaced parallel grooves. The smaller zone carries two pairs of lines that divide the very centre into a pattern of further nine small, equally sized lozenges. The border between the outside and first inner lozenge is filled with a series of single inscribed zigzag lines which create a pattern of nine interfaced triangles on each of the sides.[6]

---

[2] S. Needham and A. Woodward, 'The Clandon Barrow Finery: A Synopsis of Success in an Early Bronze Age World', *Proceedings of the Prehistoric Society* 74 (2008): pp. 1–52.
[3] Needham and Woodward, 'The Clandon Barrow Finery'.
[4] Needham et al., 'A Noble Group of Barrows'.
[5] Needham et al., 'A Noble Group of Barrows'.
[6] A. Johnson, *Solving Stonehenge, The New Key to an Ancient Enigma*, (London: Thames & Hudson, 2008), p. 180.

The second smaller lozenge angle presents a simplified ornamentation, composed of four concentric lozenges. The positioning of the largest Bush Barrow lozenge on the corpse does indeed suggest, at first sight, a decorative function, but this does not exclude a symbolic value. If one accepts the denomination of regalia used by Needham *et al*, as similar to emblems of royalty, a symbolic interpretation can be sought.[7] The relationship between royalty and the calendar deserves mention: '*Describere annum*' is indeed a task which tradition attributes to the religious kings of Rome.[8] By analogy, the idea of the possession of objects of a 'calendrical' nature, in the hands of an elite, also of Indo-European origin, which make it possible to harmonise the relations between nature and society and regulate sacred and profane time, does not seem farfetched.

The Clandon Barrow lozenge
The Clandon Barrow lozenge was associated with other objects and especially with the head of a mace encrusted with five gold bosses. The quality of this golden lozenge, in particular, the precision required for its outline and the sophisticated nature of its various geometrical patterns has been emphasised by John North.[9] The pattern of the Clandon Barrow Lozenge is similar in style to that of the Bush Barrow lozenge, but differs in the lack of zigzags on the rim of the object. An inspection of the golden objects of the Wessex culture by Joan Taylor pointed out the similarity of the cross pattern found on the Upton Lovell button and the Clandon Barrow lozenge.[10] In addition, given the relative proximity of the archaeological sites from which the Bush Barrow and the Clandon Barrow lozenges come, Taylor suggests that both might have been made by the

---

[7] S. Needham, A. Woodward and J. Hunter, 'The Regalia from Wilsford G5, Wiltshire, (Bush barrow)', in A. Woodward *et al.*, *Ritual in Early Bronze Age Grave Goods* (Oxford: Oxbow Books, 2015), p. 235.
[8] P. M. Martin, 'La fonction calendaire du roi à Rome et sa participation à certaines fêtes', *Annales de Bretagne et des pays de l'Ouest* 83, no.2 (1976): pp. 239–44.
[9] J. North, *Stonehenge: A New Interpretation of Prehistoric Man and the Cosmos* (New York: The Free Press, 1996), pp. 511–13.
[10] J. Taylor, 'Early Bronze Age Technology and Trade, the evidence of the Irish Gold', *Expedition Magazine* 21, no. 3 (1979): p. 25. [online] https://www.penn.museum/sites/expedition/early-bronze-age-technology-and-trade/ [accessed April 2017].

same craftsman.¹¹ Elsewhere Johnson has proposed, using computer modelling to find the 'most likely method used' to configure the form of the Bush Barrow and the Clandon Barrow lozenges, that they were respectively based on pure geometric forms: a hexagon and a decagon.¹²

The Upton Lovell button
The Upton Lovell button was found with other objects under a bowl barrow, called 'The Golden Barrow', alongside some burnt human bones.¹³ The cremations were accompanied by very rich burial goods; a necklace of 'about 1000' amber beads with spacer plates, a group of eleven (originally thirteen) gold drum-shaped beads, a large sub-rectangular golden sheet plaque decorated with incised lines, two golden sheet cap-ends or 'boxes', a grape cup, two urns (one lost), a bronze awl and a large conical button of shale.¹⁴

The cone shaped button decorated with zigzags and circles in groups of three or four, forms a set with a disc of the same diameter as that of the base of the button. The latter is decorated with bars assembled in chevrons following a quadripartite pattern, recalling the central part of the Bush Barrow lozenge. The conical part fits back together with the shale core. This one may have been used as a form of design by the goldsmith. On the golden button itself eight circular lines are to be seen, but there were ten lines originally. It does seem that the base of the golden button may have been damaged and the first series of circles might be incomplete. Indeed, ten lines are visible on the shale core.

Previous interpretations
In 1988 Thom, Ker and Burrows suggested seeing, in the zigzag alignments of the Bush Barrow lozenge, indicators of the position of the Sun and the Moon on significant dates in the year.¹⁵ By so doing they incorporated ideas of Alexander Thom, who proposed a prehistoric

---

[11] J. Taylor, *Bronze Age Goldwork of the British Isles*. (Cambridge: Cambridge University Press, 1980), p. 46 ; J. Coles and J. Taylor, 'The Wessex Culture: A Minimal View', *Antiquity* 45 (1971): pp. 6–13.
[12] Johnson, *Solving Stonehenge*, pp. 180–81.
[13] Needham *et al.*, 'Upton Lovell G2e, Wiltshire', p. 220.
[14] Needham *et al.*, 'Upton Lovell G2e, Wiltshire', p. 220–29.
[15] A.S. Thom, J.M.D. Ker and T.R. Burrows, 'The Bush Barrow Gold Lozenge: Is it a Solar and Lunar Calendar for Stonehenge?', *Antiquity* 62 (1988): pp. 492–502.

division of the year into sixteen parts.[16] This calendrical interpretation was later refuted by John North who considered that the size of the object and its fragility could not allow for the determination of precise azimuths.[17] Similarly, Clive Ruggles estimated that the proposed orientations do not correspond to the pattern represented on the lozenge; 'several of the alignments actually fall between the markings, while many of the markings do not fit any of the alignments at all'.[18] He also remarked that if the Bush Barrow lozenge had an astronomical function such as that proposed by Archibald Thom, other lozenges of the same type would exist, saying 'Why should only this one function as a calendrical device?'[19] Instead he preferred to conclude that the function of the object was purely ornamental.[20] Needham had a similar opinion; 'gold is used almost exclusively for ornamentation during the early metal age'.[21] However he mentioned another possibly ritualistic meaning by saying, 'although maces and similar regalia could equally have had more 'collective' religious-cum-ceremonial connotations, as has been suggested elsewhere for lunular or precious cups'.[22]

However the calendrical hypothesis expressed by Archibald Thom has recently been defended by Thomas Gough who indeed supports the thesis that the directions, defined by the regular zigzag pattern, and corresponding to sixteen different epochs of the year, are correct at more or less 0.5°.[23] He stresses in particular the accuracy of the acute 81° angle of the lozenge which represents exactly the angle formed by the solstices between each other on the latitude of Stonehenge.[24] This hypothesis was

---

[16] A. Thom, 1967, *Megalithic Sites in Britain* (Oxford: Oxford University Press, 1967), p.109–12.
[17] North, *Stonehenge*, pp. 508–10.
[18] C.N.L. Ruggles, *Ancient Astronomy: An Encyclopaedia of Cosmologies and Myth* (Santa Barbara: ABC-CLIO, 2005), pp.52–54.
[19] Ruggles, *Ancient Astronomy*, pp. 52–54.
[20] Ruggles, *Ancient Astronomy*, pp. 52–54.
[21] S. Needham, 'Discussion: Reappraising "Wessex"', in A. Woodward and J. Hunter, with D. Bukach, S. Needham and A. Sheridan, *Ritual in Early Bronze Age Grave Goods: An Examination of Ritual and Dress Equipment from Chalcolithic and Early Bronze Age Graves in England* (Oxford: Oxbow books, 2015), p. 255.
[22] Needham, 'Discussion: Reappraising "Wessex"', p. 255.
[23] T. T. Gough, 'Further Evidence for the Existence of Prehistoric Celestial Alignments in Western Scotland: Calendrical Alignments on the Island of Mull', *Mediterranean Archaeology and Archaeometry* 14, no 3 (2014): p. 254.
[24] Gough, 'Further Evidence', pp. 254–55.

also defended by Euan MacKie in 2009. MacKie's argument relies partly on the diversity of evidence and especially on the similarity of the acute angle of the Bush Barrow lozenge with the 82° angles form by the two golden arcs, along the sides of the Nebra disk (Fig. 1).[25] The Nebra sky disk is a bronze disk found in Germany and dated from the Bronze Age (c. 1600 BCE); it is decorated with gold patterns interpreted generally as a sun or full moon, a lunar crescent, and stars. According to both MacKie and Gough, the Bush Barrow lozenge could be a small version of a large wooden lozenge with angles carved on it, to help with setting out long sight-lines marking the main points of the prehistoric solar calendar.[26]

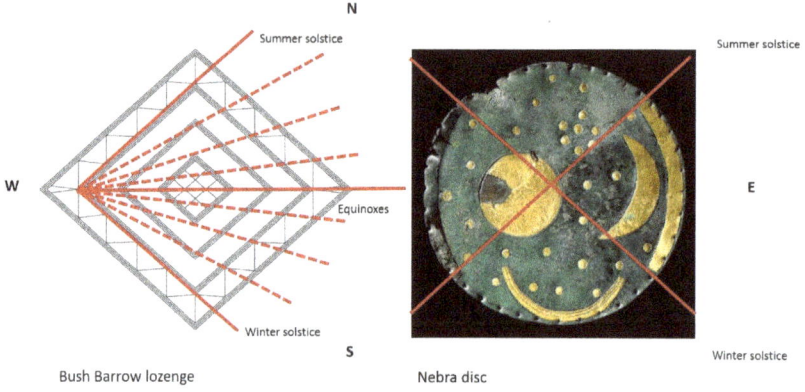

Fig. 1. The Bush Barrow lozenge (left) and Nebra disc (right). The figure illustrates how both artefacts have been interpreted as solar calendars, with height subdivisions in the case of the lozenge. The lines are passing through the inner points of the zig-zag pattern and according to the sunrise direction indicate the different epochs. Photo: D. Bachman.[27]

---

[25] E. MacKie, 'The Prehistoric Solar Calendar: An Out-of-fashion Idea Revisited with New Evidence', *Time and Mind: The Journal of Archaeology, Consciousness and Culture* 2, no. 1 (2009): pp. 9–46.
[26] MacKie, 'The Prehistoric Solar Calendar', p. 32; Gough, 'Further Evidence', p. 254.
[27] D Bachman, 'Nebra Sky Disk', 2006. [online] https://commons.wikimedia.org/wiki/File:Nebra_Scheibe.jpg, [accessed Apr. 2017].

Another purely numerological interpretation based on the value of the lesser and the larger angle of different lozenges was proposed by Keith Critchlow.[28] The acute angles of the small and the large Bush Barrow lozenges (respectively 60° and 80°) represent one-sixth and two-ninths of a circle and the obtuse angle of the Clandon Barrow lozenge (102.75°) two-sevenths, according to Critchlow.[29] This proposal was criticised by John North who considered the switching from the lesser angle in some cases to the larger angle in others, as a sleight of hand, only to obtain the desired fraction of a circle.[30]

Another possible astronomical interpretation
This paper proposes that in the Bush Barrow and Clandon Barrow lozenges and in the Upton Lovell Button (respectively Fig. 2, 4, 5) forms of memorisation or of transmission of the amount of days corresponding to the apparent cycle of the planets, especially that of the planet Venus, as well as those of Mars and Jupiter, can be seen.

The general idea that guided the selection of the three artefacts is that objects with common characteristics (the same materials, the same context of discovery, the same period and same geographical origin) have every chance of being produced by the same culture and with the same intention. On the diagram of form-cum-function relations of the golden objects of the first half of the second millennium of England and Ireland proposed by S. Needham, these three objects are very close to one another and share also the same degree of sophistication.[31] Examples available for other objects that meet the above criteria are rare. So the selection was limited to these three objects for examination.

The artefacts themselves reveal the advanced knowledge of geometrical shapes of those who designed and produced them. Associated closely with the culture of Stonehenge III (c.2000 BCE) their ingenuity is beyond any doubt. The proximity and the contemporaneity of the Bush Barrow, the Upton Lovell mounds and the Stonehenge site, as well as other wood or stone rings, suggest that the designers of these objects possessed sophisticated knowledge, above all astronomical, relating to the movements of the Sun and the Moon as demonstrated convincingly by

---

[28] K. Critchlow, *Time Stands Still: New Light on Megalithic Science* (London: Gordon Fraser, 1979) p. 240.
[29] Critchlow, *Time Stands Still*, p. 240.
[30] North, *Stonehenge*, p. 508.
[31] Needham, 'Discussion: Reappraising "Wessex"', p. 258.

Lionel Sims.[32] The knowledge of the cycles of the planets in Wessex culture, supported by this paper although feasible, has, for the time being, not been confirmed. However the cycle of Venus was known to the Babylonians during the same period of time. The Venus tablet of Ammisaduqa refers to the record of astronomical observations of Venus (heliacal risings and settings) probably compiled around the mid-seventeenth century BCE.[33]

Other evidence of encoded astronomical knowledge
Anthropology also enables us to confirm that objects of a proven calendrical nature have existed within cultures without writing systems. An example is provided by a lunar calendar used by the natives of the former kingdom of Dahomey in Africa, kept in the Musée de l'Homme (Paris).[34] It is a band of cloth stitched with thirty elements (seeds, kernels, hard fruit, pebbles and so on) succeeding one another in length, symbolically illustrating the thirty days of the period.

Archaeology has also discovered other artefacts for which the calendrical function is firmly established. Terracotta objects belonging to the Cycladic civilisation of the beginning of the Bronze Age (Early Helladic period I), whose function is unknown, called 'Frying pan vessels', would, according to Tsikritsis *et al's* recent work, have encoded mathematical and astronomical knowledge.[35] Analysis of the engraved symbols has revealed a symbolic script describing phenomena relating to cycles of the planets, the Sun and the Moon.[36]

Similarly, the Trundhom 'Sun chariot' disc (Denmark, Nordic Bronze Age, ca. 1800–1600 BCE) has been interpreted as a representation of the

---

[32] L. Sims, 'The "Solarization" of the Moon: Manipulated Knowledge at Stonehenge', *Cambridge Archaeological Journal* 16, no. 2 (2006): p. 200–3, p. 191.
[33] E. Reiner and D.E. Pingree, *Babylonian Planetary Omens*, Vol. 3 (Volume 2 of Bibliotheca Mesopotamica Babylonian Planetary Omens) (Leiden, Brill, 1998), p. 1.
[34] G. Ifrah, *Histoire Universelle des chiffres – L'intelligence des nombres racontée par les nombres et le calcul,* (Paris: Robert Laffont, Coll. Bouquins, 1994), p. 61.
[35] M. Tsikritsis, X. Moussas and D. Tsikritsis, 'Astronomical and Mathematical Knowledge and Calendars During the Early Helladic Era in Aegean "Frying Pan" vessels', *Mediterranean Archaeology and Archaeometry* 15, no.1 (2015): pp. 135–49.
[36] Tsikritsis *et al.*, 'Astronomical and mathematical knowledge and calendars', pp. 135–49.

Sun or of the full Moon, whose decorations depict a calendrical count.[37] The ritual conical golden hat known as the 'Berlin Gold Hat' (Swabia or Switzerland, Bronze Age 1000–800 BCE), could represent a lunisolar calendar. A detailed analysis indicates that the decorations (mostly discs of concentric circles, sometimes of wheels) permit determination of the dates for both lunar and solar periods. However, the solution for deciphering them has not yet entirely been found.[38]

According to Joseph Loth, in the Irish laws, there is always question of periods of three nights, of nine nights, not to mention a longer period. According to him, there was a week of nine days and a sidereal month of 27 days divided into three novenas. The use of the novena in the Celtic world and the existence of traces of antique calendars all based on ten months of 36 days, in China, in Armenia, and more near to us in the Etruscan civilization and in ancient Rome, may help to understand why days may have been grouped by nine and 36.[39]

The symbolism of the lozenge shape
With regard to the Bush Barrow lozenge, John North had already highlighted the symbolic value of the lozenge shape, extensively used from Ireland to Persia and even beyond.[40] According to Marija Gimbutas 'The lozenge and triangle with one or more dots are encountered on shrine walls, vases, seals, and typically on the pregnant belly or other parts of the

---

[37] K. Randsborg, 'SPIRALS! Calendars in the Bronze Age in Denmark', *Adoranten* 2009: pp. 1–11 [online] www.rockartscandinavia.com/images/articles/randsborga9.pdf [Accessed Nov. 2016]; A.C. Sparavigna, 'Ancient Bronze Disks, Decorations and Calendars' [online] https://arxiv.org/pdf/1203.2512 [Accessed Nov. 2016].
[38] W. Menghin, 'Der Berliner Goldhut und die goldenen Kalendarien der alteuropäischen Bronzezeit' *Acta Praehistorica et Archaeologica*, Potsdam, Unze, no. 32 (2000): pp. 31–108.
[39] C. Maumené, 'Interprétation de la division de l'espace à Larchant selon le calendrier gaulois', *Acte de la société Belge d'études Celtiques*, Bruxelles, Vol XIV, no. 1 (2010): pp. 59–90 ; L. Weibao, 'New Evidence for the Ten-month Calendar', *Publication of the Yunnan Observatory* (1996–01); L. Weibao, 'Advance in the Research on the Ten-month Solar Calendar', *Progress in Astronomy* (1997–01); L. Weibao and J. Chen, 'Examination of the "Seats of the Five Emperors" in Traditional Chinese Constellations', *Astronomical Research & Technology* (2010–02); see also the importance of the novena in the Irish Celtic culture in J. Loth, 'L'année celtique dans les textes irlandais', *Revue Celtique*, (1904): pp. 134–37.
[40] North, *Stonehenge*, p. 503.

Pregnant Goddess, starting in the 7th millennium BC.'[41] Gimbutas also suggests that the lozenge is a representation of the vulva of these goddesses. Moreover, she proposes associating the point frequently present in the middle of the lozenge with the notion of maternity, or even to the image of semen, with the lozenge representing the field sown.[42] For Johanna Stuckey, given that clay representations in the shape of a vulva have been discovered in Mesopotamia in temples dedicated to Ishtar, and that the same symbol is represented on certain seals in the tomb of that very same goddess; their connection to the goddess is coherent.[43] Besides, Pizzimenti holds that Ishtar is a goddess whose astral correspondence is the female version of the planet Venus.[44] For the Minoans, lozenge series are also present on the apron of the snake goddess, who is sometimes identified as the Phoenician goddess Astarte.[45]

On an anthropological level, a back-apron or a string skirt, decorated with lozenge patterns, was still worn recently in Eastern Europe, by young pubescent girls, showing the community they have reached sexual maturity and may be courted with view to marriage.[46] Nowadays, female Turkish weavers believe the lozenge is a symbol of the feminine sex.[47] Moreover in the Niğde region nomads call the motif *dudak* (the lips), a term familiarly

---

[41] M. Gimbutas, '*Le langage de la déesse*', *Des femmes*, Antoinette Fouque, 2005, pp. 171–72.
[42] Gimbutas, '*Le langage de la déesse*', p. 172.
[43] J. Stuckey, 'Of Omegas and Rhombs: Goddess Symbols in Ancient Mesopotamia and the Levant', *MatriFocus, Cross-Quarterly for the Goddess Woman*, Lammas 5/6 (2006): [online] http://www.matrifocus.com/LAM06/spotlight.htm [accessed Nov. 2016].
[44] S. Pizzimenti, 'The Astral Family in Kassite Kudurrus Reliefs. Iconographical and Iconological Study of Sîn, Šamaš and Ištar astral representation', in *Proceedings of 55 RAI*, Paris 6th–9th July 2009.
[45] G. A. Owens, '"All Religions are One" (William Blake 1757–1827), Astarte/Ishtar/Ishassara/Asasarame: The Great Mother Goddess in Minoan Crete and the Eastern Mediterranean', *Cretan Studies* 5 (1996): pp. 207–18.
[46] E. W. Barber, 'On the Antiquity of East European Bridal Clothing', in Linda Welters, ed., *Folk Dress in Europe and Anatolia: Beliefs about Protection and Fertility* (London: Bloomsbury Academic, 1999), pp. 13–32; E. W. Barber and P. T. Barber, *When They Severed Earth from Sky: How the Human Mind Shapes Myth* (Princeton, NJ: Princeton University Press, 2005), pp. 68–70.
[47] M .A. Gallice and A. Diler, 'Fragments d'un langage oublié – La symbolique dans les kilims', [online] http://www.kilims.fr/brochure_ada-symbolique.pdf [accessed Nov. 2016], p. 8.

applied to the female sex organ.[48] The lozenge also appears in bands on one of the best known bell beakers dating to before 2500 BCE, found in West Kennet Long Barrow, near Avebury.[49] A lozenge divided into four appears in the book of Kells (miniature folio 7), which is believed to have been created c. 800 AD, and is represented as a brooch, on the right-hand shoulder of the most ancient representation of the Virgin in the scriptural art of Western Europe.[50] The most pertinent and closest parallel is probably those lozenges seen in passage grave art. G. Robin, in his work on the rock art of the Neolithic tombs around the Irish Sea, has listed all the signs engraved.[51] The quadrangular signs (lozenge or square) are present on 19% of the engraved slabs, and most often associated with chevrons.[52]

These numerous examples underline the relationship between the lozenge shape and the idea of fertility, fecundity, and maternity, as well as death and rebirth. However, this enumeration does not pretend to demonstrate that the symbolism of the lozenge is entirely transcultural, but in the absence of precise cultural evidence from Wessex, this paper proposes the use of analogy to suggest some possible meaning of the lozenge, as an echo of some other cultures.

Early Counting methods
The methods used by the craftsmen are straightforward and would have been understood by populations ignorant of writing systems. They are based on elementary counting techniques for any set of objects, used here for the recording of days. This practice does not require any knowledge of abstract reckoning; a numeration can easily be made just by matching. George Ifrah refers to the principle of correspondence, unit by unit, which makes it easy to compare two collections of beings or objects with or without the same nature, without using abstract counting.[53] He backs up

---

[48] Gallice and Diler, 'Fragments d'un langage oublié', p. 8.
[49] Wiltshire Museum, 'Bell Beaker', [online] http://www.wiltshiremuseum.org.uk/galleries/index.php?Action=4&obID=121&prevID=&oprevID=, [accessed April 2017].
[50] N. Niamh, 'Brooch or Cross? : The Lozenge on the Shoulder of the Virgin in the Book of Kells', *Archaeology Ireland* 10, No. 1, (1996).
[51] G. Robin, *L'architecture des signes – L'art pariétal des tombeaux néolithiques autour de la mer d'Irlande* (France : Presse Universitaires de Rennes, 2009), p.76, p. 91.
[52] Robin, *L'architecture des signes,* p. 76, p. 91.
[53] G. Ifrah, *Histoire Universelle des chiffres – L'intelligence des nombres racontée par les nombres et le calcul* (Paris: Robert Laffont, Coll. Bouquins, 1994), p. 42.

this assumption by evoking for example the rosary, which accompanies the recitation of litanies.[54] Each bead corresponds to a prayer. In this manner there is no need for abstract enumeration in order to recite the litanies without mistake.

Primary counting systems originally used simple forms such as notches.[55] These types of systems rapidly reach their limits when coming close to large numbers. Faced with this difficulty, human beings had to count by packs, using the idea of replacing ordinary pebbles by differentiable objects or images of various sizes, matching conventional forms; such as, for instance, a stick for the unity, a flat ball for a group of ten, a small ball for a group of a hundred.[56] This principle which relied on simple hardware intermediates, proved to be of great help when a large number of objects needed to be accounted for.[57]

By analogy the present paper proposes to link days to the decorative elements of the various objects in question. Based on the pairing principle the 36 zigzag (triangles) placed on the circumference of the Bush Barrow lozenge will be linked to 36 successive days (Fig. 2). Each side of the lozenge presents nine triangles or changes of direction. Another option is to count the sticks which are included in the broken line, assuming that the double sticks in each of the acute angles have to be counted as one, like the two other angles where they are combined. With regard to the Clandon Barrow lozenge and the Golden Barrow button, since there are no triangles, the successive days have been matched to sticks. The counting of the days can be done by going through the perimeter of the lozenge. At the end of the first rotation 36 days have passed and a new series of 36 days begins.

The pack counting principle leads to linking the nine triangles displayed on each side of the Bush Barrow lozenge, to one of the adjacent line of the same length. On this point, it is worth pointing out that Hunter and Woodward have underlined the perfect accuracy of the layout: the lengths of the nine successive triangles, correspond exactly to the length of the adjacent continuous groove.[58] An uninterrupted unit would be worth nine triangles or sticks, so the complete perimeter of the lozenge would equal 36 triangles (Fig. 2).

---

[54] Ifrah, *Histoire Universelle des chiffres*, pp. 43–44.
[55] Ifrah, *Histoire Universelle des chiffres*, pp. 161–68.
[56] Ifrah, *Histoire Universelle des chiffres*, pp. 69–72.
[57] Ifrah, *Histoire Universelle des chiffres*, pp. 41–61, pp. 69–74.
[58] Needham, 'The Regalia from Wilsford G5', p. 238.

In an additive numeration system, the value proper to each sign is independent of its position. To represent nine days it is enough to juxtapose nine triangles (or sticks), or a long line. To represent 36 days, drawing four lines corresponding to the perimeter of the lozenge is enough. By applying this principle one obtains 144 days (4 x 36) per series of four concentric lozenges (Fig. 3). This grouping by periods of 4 x 36 days, a unit of 144 days, could be explained simply as a means of facilitating numbering. The cycle of Venus (very close to 576 days) then corresponded to 4 x 4 x 4 novenas, or 4 x 4 periods of 36 days or 4 periods of 144 days. It is also recognised that four is the limit which makes it possible to identify at first glance a given number of aligned analogous elements. Reaching beyond all of them blurs our minds and the global vision is no longer of any help, according to Ifrah.[59]

Analysis
1. The Bush Barrow lozenge
Based on the methodological principles described above, each perimeter would come to a total of 4 x 9 = 36 days (Fig. 2 and 3). Otherwise, a series of 36 squares can be observed on a plate made from a bone covered by a sheet of gold, from a group of mounds on the Normanton Down, very close to the Bush Barrow mound.[60] The pattern presents the shape of a grid of 6 x 6 squares. Two squares seem to have been sacrificed after the initial outline, for the benefit of a circular notch made on the top part of the pattern.

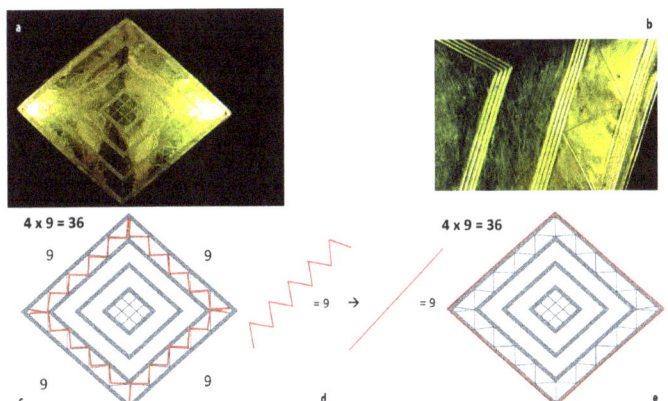

Fig. 2 The Bush Barrow lozenge, Photos: © Wiltshire Museum, Devizes.

---

[59] Ifrah, *Histoire Universelle des chiffres*, p. 33.
[60] Needham, 'Discussion: Reappraising "Wessex"', p.194–95.

44   The Bush Barrow and Clandon Barrow Gold Lozenges and the Upton Lovell
       Golden Button: A Possible Calendrical Interpretation

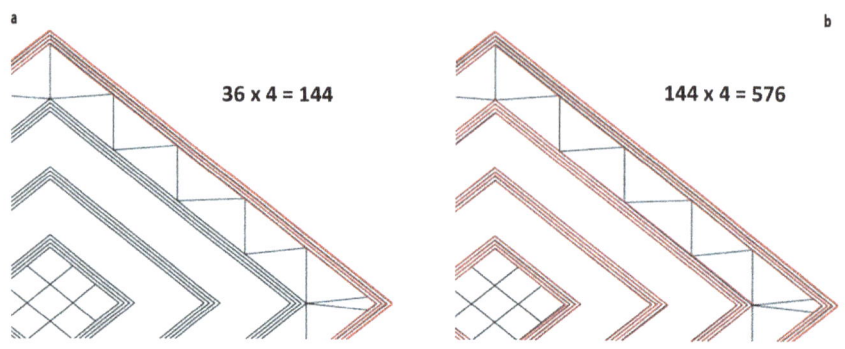

Fig 3 The Bush Barrow lozenge

As said above, on the Bush Barrow lozenge, the concentric lozenges are grouped together in 4's with each series adding up to an overall of 144 days. The fours series of concentric lozenges form a total number of 576 days.

$$[((4 \times 9) \times 4) \times 4] = [((36) \times 4) \times 4] = [144 \times 4] = 576$$

The total of 576 days is very close to the value of the first synodic period of the planet Venus, on average 583.92 days where a synodic period represents the time for a celestial body to return to the same position relative to the Sun, as seen by an observer from Earth.

Using the inferior conjunction of Venus as a marker, the remaining lozenges in the central part of the patterns could represent the eight (or nine days if we include the ninth lozenge in the middle) during which the planet disappears in the west to reappear later at the east. The conjunction of Venus with the Sun occurs when the planet passes between the Sun and the Earth. During this period Venus disappears in the light of the Sun and reappears when it moves far enough away.

$$[((4 \times 9) \times 4) \times 4] + 8 = 584$$

Hence, if this hypothesis is correct, the astronomer-priests of Stonehenge were aware of the cycle of Venus and observed its inferior conjunction which initialises a new cycle. Pursuing this interpretation, it is possible that the small lozenge decorated with four concentric lozenges may have worked as an ensemble along with the large lozenge. It could correspond to the repetition of four cycles of Venus, still according to the principle of the base, with the ultimate aim to reach a higher rank of numbers. With each of the four concentric lozenges representing a cycle of Venus, we would obtain:

$$584 \text{ (or 585)} \times 4 = 2336 \text{ (or 2340 days)}$$

This number of days (2340) would also correspond to 3 synodic periods of Mars, another astronomical fact that might also be illustrated in the Clandon Barrow lozenge.

2. The Clandon Barrow Lozenge

Here each unity is matched to a day. The enumeration is also made, clockwise, following an intuitive progression along the perimeter of the central lozenge, inspired by the natural movement of the Sun or the Moon. The complete perimeter would thus represent 4 x 24 days giving a total of 96 days (Fig. 4 a, b).

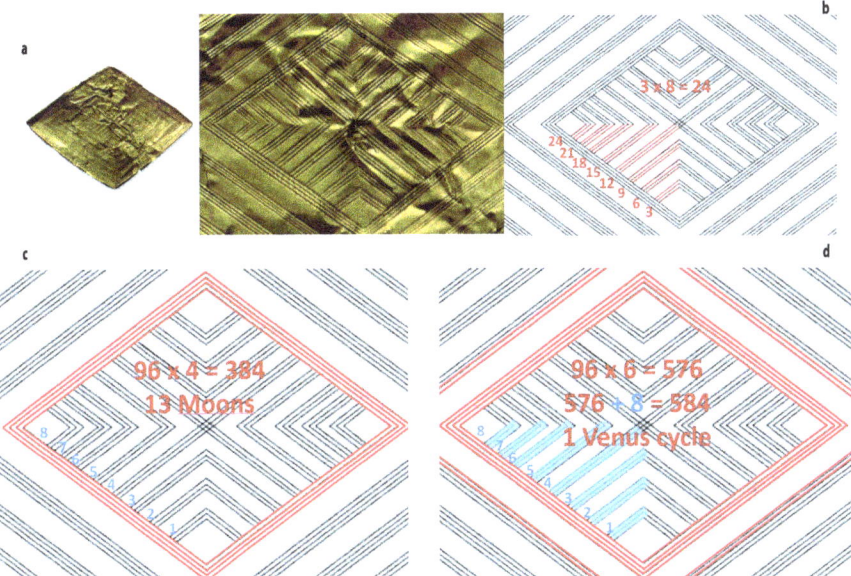

Fig. 4. The Clandon Barrow Lozenge, Photo: Dorset AONB with permission of Dorset AONB.[61]

---

[61] The Clandon Barrow Lozenge [online] http://www.dorsetaonb.org.uk/assets/downloads/South_Dorset_Ridgeway/Resources/Image_bank/Gold_lozenge_from_Clandon_Barrow.jpg [accessed April 2017].

46 The Bush Barrow and Clandon Barrow Gold Lozenges and the Upton Lovell Golden Button: A Possible Calendrical Interpretation

It is to be noted that the central motif is partially asymmetrical. The medians divide the central lozenge into four sectors which are alternately composed of three or four series of chevrons grouped by three. It appears that the goldsmith's intention was to represent 24 sticks on each side of the lozenge, not a number respecting a perfectly symmetrical pattern. If that had been his intention, he would have preferred to consistently represent three sets of chevrons in each sector (or four), but not an alternation of three and four series. This infringement of a perfect symmetry runs counter to a non-functional purely ornamental hypothesis. So the goldsmith's choice, based on the number 24 appears to have been intentional.

Each perimeter would represent 4 x 24 = 96 days. Applying the principle of pack counting, each series of four perimeters would represent 4 x 96 = 384 days. Now, 384 is a number of astronomical interest since it represents a precise estimation of the duration of thirteen Moons (Fig. 4 c).

$$384 = 13 \times 29.5309 + 0.0983$$

The theoretical error is just 2 hours 21 minutes and few seconds for a lapse of more than one year. Note that 384 days for thirteen lunar months represents a better estimation than 354 days for twelve lunar months. It is also to be observed that 384 x 2 = 768, a total close to the span in days of the synodic period of Mars (the synodic period of Mars is observable between two successive oppositions), with an average 780 days with variations from 765 to 800 days. So after 768 days, Mars is potentially opposed to the Sun or is about to be so in the following days.[62] Mars is opposed to the Sun when the planet is visible all night and respectively rises and sets in opposition to the rising and setting of the Sun in the locality. This event can be observed and its computation is straight-forward.[63] The organisation by unit of 768 days (2 x 384) could have potentially increased the interest in this by warning of the next return of Mars to a remarkable phase of its cycle (such, for instance, as its opposition to the Sun).

Also note that 96 x 6 = 576. An average of eight days, are missing to complete a synodic period of Venus. To put it in another way, 576 days,

---

[62] M.J. Powell, 'The Naked Eye Planet in the Night Sky (and How to Identify them)' [online] http://www.nakedeyeplanets.com/ [accessed October 2016].
[63] R.D. Purrington, 'Heliacal Rising and Setting: Quantitative Aspects', *Archaeoastronomy, Journal for the History of Astronomy* 12, no. xix (1988): pp. 72–84.

from its heliacal rising in the east (its first appearance in the morning sky), Venus appears for the last time in the west in the evening sky, before returning after an average of eight days in the morning sky. These eight missing days could be materialised by (and could be the reason for) the eight series of three unities depicted on each side of the central lozenge (576 + 8 = 584) (Fig. 4 d). So the overall pattern would respond to the formula given below and would come to a total of 584 x 4 = 2336 days i.e., four cycles of Venus.

$$[((24 \times 4) \times 6) + 8] \times 4 = [576+8] \times 4 = 584 \times 4 = 2336$$

This complete total itself is very close to three cycles of Mars, since,

$$2336 = 2340 - 4 = (780 \times 3) - 4$$

3. The Upton Lovell button

Should our analysis seem, at first sight, to be highly speculative, it can be tested with reference to a third object selected according to the criteria proposed above: the Upton Lovell button. The latter conical shaped object matches a disc of the same diameter, forming the base of the button, decorated with bars assembled in chevrons, following a similar structure to that of the central part of the Bush Barrow lozenge. The conical part is decorated by ten circular lines, grouped following the series: 4 + 3 + 3. The number retained is that of the matrix. At the edges of the button, on its base, is a succession of zigzags incised between the first two series of circles, corresponding to a series of forty sticks or triangles (Fig. 5 a, b, c).

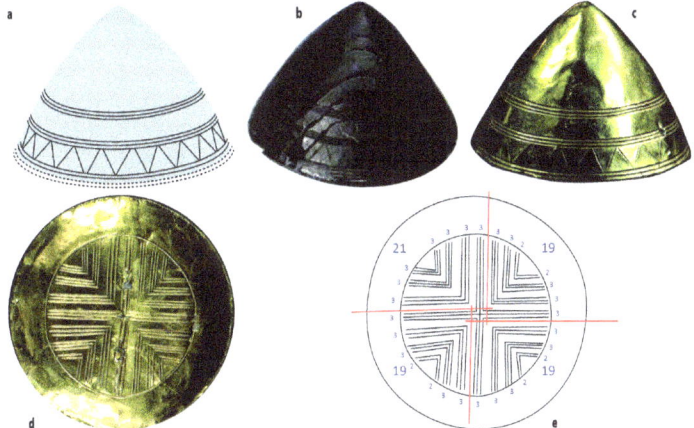

Fig. 5 The Upton Lovell button (a,c), its shale matrix (b) and its base (d,e), Photos: © Wiltshire Museum, Devizes.

48   The Bush Barrow and Clandon Barrow Gold Lozenges and the Upton Lovell
      Golden Button: A Possible Calendrical Interpretation

On the disc forming the base of the button, the number of bars making up the chevrons is respectively, for the four areas: 19 + 19 + 19 + 21, coming to a total of 78, obtained by crossing the perimeter of the circle delimiting the motif. It is important to note that the chevrons are in groups of two or three. For three of the four sections, the total of 19 is obtained by adding: 3+3+2+2+3+3+3 = 19. The total of 21, corresponding to the last section is obtained by the sum of 3+3+3+3+3+3+3 = 21 (Fig. 5 d, e).

Here again the principle of symmetry is not totally respected. The sector comprising 21 unities is deliberately made up of three chevron series only. Therefore, this suggests that the goldsmith's intention was to depict 78 unities and not the 76 or 84 which scrupulous respect of the symmetry would have called for. Applying the rule of counting in series retained in the Barrow lozenge case, allows us to combine the bars on the base of the button with the round line patterns of the button itself. The total leads to a precise approach of the span of the synodic period of Mars, that is to say 780 days, obtained simply by multiplying the ten circles of the button by the 78 unities on its base:

$$10 \times 78 = 780$$

Moreover, and in the same way the chevrons of the Bush Barrow lozenge were combined with the lines forming the sides of the concentric lozenges, we can match the forty triangles decorating the outside part of the button with the ten adjacent, concentric lines. This brings us to a total of 400, within one day of the synodic period of Jupiter. The precise synodic period of Jupiter amounts to 398.9 days.

$$398.9 = 40 \times 10 - 1.1 = 400 - 1.1$$

Analysis of the conical Upton Lovell button tends to consolidate the calendar hypothesis, since it seems to encode the durations of the synodic periods of Mars and of Jupiter.

Discussion

To support this hypothesis, it needs to be emphasised first of all that these objects are all apparently symmetrical, but that they are sufficiently well designed so that a slight asymmetry can be seriously considered. Mouchet suggests that symmetry can be sought to facilitate memorisation as well as transmission.[64] But in the present case, it is the irregularities of the patterns examined, the infringements of a perfect symmetry, which reveal a functional intent. Indeed, the Bush Barrow lozenge, like the Upton Lovell button, both present irregularities betraying the will of their creators to materialise one number rather than another, one structure rather than another; a rupture with perfect symmetry which also represents a complex phenomenon, in the simplest way.

For the Bush Barrow lozenge, an arrangement of 4 x 24 bars in groups of three has been retained to graduate each side of the central lozenge, making a four-part, entirely symmetric division, based on the medians, impossible. In the case of the Upton Lovell button or more precisely of the disc at its base, the four-party arrangement corresponds to a total of 78 unities, using the 19+19+19+21 series. Here the intention to total 78 and not 76 (4 x 19) or 80 (4 x 20) is clearer. So the hypothesis which assigns a purely decorative function to these artefacts must be complemented by an additionally plausible hypothesis of a 'scientific' aim, to use a term from Mouchet.[65]

To defend this functional hypothesis this paper suggests reliance on the standards proposed by Frank Ventura to determine the strong as well as the weak points of an iconographic interpretation.[66] These standards are: i) the fitness within the whole context of what is known about the culture, what Ventura calls 'internal evidence'; ii) preference for interpretation implying concrete processes rather than the use of higher order concepts and abstract thinking; iii) external evidence (where internal evidence is lacking), such as well-established interpretations of analogous artefacts from other cultures and iv) interpretations that lead to testable hypothesis.[67]

If we apply these standards to our interpretation we can say: i) there is no reason to believe that the men who built Stonehenge were not able to observe the courses of the wandering stars and to record them; ii) this record is based on primitive techniques, term-to-term correspondence and the principle of the basis or grouping; iii) from an ethnographic point of view, other objects have been described or interpreted as having an

---

[64] A. Mouchet, *L'élégante efficacité des symétries* (Dunod, 2013), pp. 13–21.
[65] Mouchet, *L'élégante efficacité des symétries*, pp. 13–21.
[66] F. Ventura, 'Reading Messages from the Past: Interpreting Symbols of Possible Archaeoastronomical Significance in Malta', in *The Materiality of the Sky*, Proceedings of the 22nd Annual SAC Conference, (Ceredigion, Wales: Sophia Centre Press, 2016), pp. 3–19.
[67] Ventura, 'Reading Messages from the Past', pp. 3–19.

astronomical function; iv) the proposed model is tested successfully on several similar objects.

Conclusion
The robustness of this paper's proposition rests on the fact that three objects, selected according to criteria as objective and independent as possible, respond to the same astronomical model. These different artefacts, which demonstrate encoded planetary cycles, strengthen the functional and astronomical hypothesis. To further test the theory, one of the few similar objects known, the gold rectangular Upton Lovell plate could be analysed in the future. Although incomplete and damaged, there is a potential cycle of 360 days and 584 days.

So at the end of this journey, a new hypothesis is put forward relating to the function of these artefacts. The intention of their creators, by calling on simplified numerical representations, could have been to facilitate the memorisation and the transmission of knowledge relating to the planetary cycles, among which Venus may have occupied a preeminent place. Beyond the 'scientific' connotation assigned to the objects, they may also have been used in rituals or for the establishment of ceremonial dates. If associated with Venus, it is reasonable to consider that these rituals may have been connected to the cult of fertility, of fecundity, but also, given the context in which they were discovered, to death and to rebirth.

The hypothesis put forward may lead to new findings concerning other objects of the same workmanship and cultural origin. The repetition of results might enable the progressive rejection of the idea of a purely aesthetic function for these objects and promote more attention to the counting of patterns which seem *a priori* to be purely ornamental. To extend the corpus, the criterion of integrity and good state of conservation that we had implicitly retained may have to be abandoned. Other selection criteria could be modified to extend the study to an even longer series, even at the risk of confusing objects of different origin and function.

Acknowledgements
We thank the two anonymous reviewers for their constructive comments. Special thanks to Liz Henty for her patient and precious help, and to Marianna Riedderstad and Roslyn Frank for their kind support. The author would also like to thank, Lisa Brown, Curator of the Wiltshire Museum for providing most of the pictures used in this paper.

Clava Cairns, Midwinter Sunset and the Minor Lunar Limit

J. Anna Estaroth

Abstract: This is a study of the Early Bronze Age cairns at Balnuaran of Clava, near Inverness, Scottish Central Highlands, from the perspective of skyscape archaeology. The site's solar orientations link it with Maeshowe in Orkney. The role of lunar limits is discussed with special reference to Balnuaran of Clava Central and its rays, which link the site to Stonehenge in Wessex. New alignments are investigated, in order to verify skyscape phenomena and integrate them with previous academic research. The importance of seasonal alterations in light and darkness leads to a suggestion that Clava monument-types may be qualitatively different. Passage-graves appear to be associated with midwinter sunset, the nearest new Moon, minor lunar limits and darkness, although light is still significant. Conversely ring-cairns seem to be associated with sunrise, the midsummer full Moon, major lunar limits and light, yet they have dark connotations, leading to conclusions that the role of light and darkness is paramount in understanding these monuments.

Introduction

This paper explores three Clava cairn monuments (one ring-cairn and two passage-graves) from a skyscape archaeology perspective, touching on their geographical, cultural, environmental and chronological context. The cairns at Balnuaran of Clava (henceforth Balnuaran) are the best preserved Early Bronze Age (EBA) monuments of the Clava tradition with two exhibiting a midwinter sunset alignment. Antiquarian interest is noted and archaeological research considered, so that their social significance can be comprehended.

Skyscape archaeology methodology is outlined and fieldwork started with winter solstice measurements, when the setting Sun shone into the two passage-graves. Clava cairns are described, using Balnuaran Northeast as an example, to clarify the integration of cairn and standing stone circle, and the importance of their southwest orientations is discussed. The study set out to explore the rays, apparently unique to Balnuaran Central ring-cairn, which are low rubble causeways connecting the cairn peristalith to four of the surrounding circle orthostats. These rays have several alignments, and comparisons were made with Stonehenge, which exhibits orientation to midwinter sunset and the minor lunar limit. The double alignment (midwinter and minor limit) has implications for the wider context of Orkney and Wessex, which were locations of cultural dominance in prehistory.

Richard Bradley noted that the three EBA monuments at Balnuaran are not in a straight line; the central ring-cairn appears offset, which he realised permitted midsummer sunrise to light up the back of Balnuaran Northeast, Balnuaran Central and Mains of Clava Northwest (the last is in the neighbouring field).[1] Considering this layout generated the main research question as to whether two theoretical groupings could be relevant for this site: the longest darkest night of the year (midwinter), minor lunar limits, new Moons and ceremonies of darkness as opposed to the longest day of the year (midsummer), major lunar limits, full Moons and ceremonies of light. The unique nature of Balnuaran generated further research questions regarding the significance of all Clava ring-cairns and whether they have greater meaning as life-affirming structures, rather than being purely burial monuments. Finally the significance of seasonal dark and light will be considered, asking whether passage-graves and ring-cairns may have had either binary or separate functions, based on their unique relationship with the Sun and Moon.

Clava cairns in their environment
The Clava area had intermittent use, primarily burial, over more than four thousand years. Bradley's radiocarbon dates (calibrated to two sigma confidence level) proved there was evidence of Mesolithic activities between 5474–5242 BCE at Balnuaran Northeast and Central.[2] Bradley suggested that the monuments were constructed during the EBA, when people were largely pastoralists.[3] There was extensive reuse during the Late Bronze Age 1100–800 BCE.[4] Moving into the early medieval period there was a Pictish cremation (653–856 CE) and the onsite notice at Milton of Clava South, less than 800 metres away, said it was converted into a Chapel with graveyard around 1100 CE.[5] This may have lasted until the

---

[1] Richard Bradley, *The Good Stones: A New Investigation of the Clava Cairns* (Edinburgh and Kings Stanley: Society of Antiquaries of Scotland Monograph Series No. 17, 2000), p. 126.
[2] Bradley, *Good Stones*, p. 115.
[3] Bradley, *Good Stones*, p. 157.
[4] Bradley, *Good Stones*, p. 119.
[5] Bradley, *Good Stones*, pp. 115, 114–19; Historic Scotland onsite notice board.

Reformation during the Early Modern Period. George and Peter Anderson reported that unbaptized children were 'still buried' at this disused chapel and Cosmo Innes reported finding a secret burial of 'unchristianed children', in a ruined chapel, in the mid 1800's, suggesting this practice was reasonably widespread, as unbaptized children could not be buried in normal graveyards.[6] While this would be for Christian reasons, it indicates the long-term use of the valley for burial/religious purposes. Bradley found recent depositions of crystals and coins during his 1990's excavations.[7]

The name Clava refers to a district along the valley of the river Nairn near Inverness which has eleven monuments within 0.2 square kilometres. Five are at Balnuaran, (two of which are late Bronze Age additions); one at Culdoich, three at Mains of Clava and two further monuments are at Milton of Clava.[8] Milton of Clava North is an accepted cairn, while Milton of Clava South is disputed, although Niall Sharples discussed a nearby mound as yet another possible Clava cairn.[9] The cairns this paper explores are the three main EBA cairns at Balnuaran.

Clava cairns are a distinct group of EBA monuments found in the Central Highlands of Scotland of which there are two styles: ring-cairns and passage-graves. They are particular to the Central Highlands, mainly concentrated around the Inverness area, which functions as a trading hub today, much as it may have done in prehistory. This region has good water-transport links: V. Gordon Childe noted that 'you could see Orkney' on a clear day from the Moray Firth, and Stuart Needham reasoned the location was 'pivotal' in early Bronze Age trade links with Ireland via Loch Ness.[10]

---

[6] George Anderson and Peter Anderson, *Guide to the Highland and Islands of Scotland, Including Orkney and Zetland, description of their scenery, statistics, antiquities and natural history with numerous historical notes and a complete map* 3rd edition (Edinburgh: G. Parker, 1851), p. 370; Cosmo Innes, 'Notice of a tomb on the hill of Roseisle, Morayshire recently opened; also of the chambered cairns and stone circles of Clava, in Nairnshire', *Proceedings of the Society of Antiquaries Scotland* 3 (1857–60): pp. 47–50, p.48.
[7] Bradley, *Good Stones*, p. 231.
[8] Bradley, *Good Stones*, p. 15.
[9] Niall M Sharples, 'Excavations at Milton of Clava, Inverness-shire', *Glasgow Archaeological Journal* 8 (1995): pp. 1–9, p. 2.
[10] V. Gordon Childe, *Scotland Before the Scots being the Rhind Lectures for 1944* (London: Methuen, 1946), p. 20; Stuart Needham, 'Migdale-Marnoch: Sunburst of Scottish Metallurgy', in Ian Shepherd and Gordon J. Barclay, eds., *Scotland in Ancient Europe: The Neolithic and Early Bronze Age of Scotland in their European Context* (Edinburgh: Society of Antiquaries of Scotland, 2004), pp. 217–45, p. 241.

Alan McKirdy stated the hills were 'deeply scarred' by ice-age erosion, which flowed NE until approximately '11.700 years ago', but the overall topography is largely unchanged for the past 5,000 years.[11]

Childe's 1946 map is adapted to show the cairns in their cultural context, with the Clava cairns largely grouped at the head of Loch Ness and the entrance to the Moray Firth (Fig.1). It also helps to put the cairns in their archaeological context. The circles represent Clava cairns and the triangles represent recumbent stone circles (RSCs) a neighbouring cultural group, for which Richard Bradley identified various similarities, such as lunar orientation, though he found many distinctive differences, such as their 'landscape setting'.[12] More Clava cairns have been identified since Childe's day; Bradley lists fifty.[13]

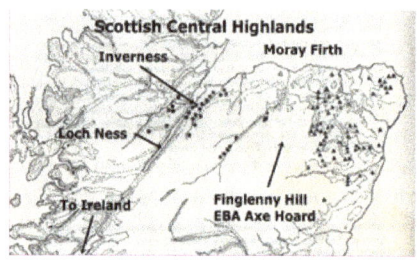

Fig. 1 Scottish Central Highlands. Clava cairns (circles) largely cluster around the mouth of river Ness at Inverness (indicated). The RSCs (triangles) are to the east. The Finglenny Hill Axe Hoard is located between the two groups. Map adapted from Gordon Childe.[14]

Without radiocarbon dating Childe was unaware that Clava cairns were contemporaneous with the later stages of both RSCs and Stonehenge. The most recent archaeological survey of Clava cairns was undertaken by Bradley in the 1990's and he firmly placed these monuments in the EBA, with construction dates between '2340 to 2030 BCE'.[15] Liz Henty, who researched Tomnaverie RSC, provided construction dates between '2580

---

[11] Alan McKirdy, *Set in Stone: The Geology and Landscapes of Scotland* (Edinburgh: Birlinn, 2015), pp. 36, 67.
[12] Richard Bradley, *The Moon and the Bonfire: An Investigation of Three Stone Circles in North-east Scotland* (Edinburgh: Society of Antiquaries of Scotland, 2005), p. 114.
[13] Bradley, *Good Stones*, p. 172.
[14] Childe, *Scotland Before the Scots*, p. 104.
[15] Bradley, *Good Stones*, p. 157.

BCE to 2220 BCE' based on Bradley's excavation data.[16] Martyn Barber noted that one of the earliest bronze axes manufactured in Britain comes from a hoard buried on the boundary between these two cultural groups at Finglenny Hill (marked with an arrow in Fig.1).[17] The location of Clava cairns was considered significant by Needham, in his study of early metallurgy, because they lay at the head of Loch Ness, a position which controlled trade routes for both raw materials from southern Ireland and prestige finished goods, the manufacture of which 'centered around Buchan'.[18] Other Neolithic cultural monuments such as the Orkney-Cromarty-Hebridean chambered tombs lie to the northwest and the Clyde cairns are to the south.

Antiquarians have shown interest in these monuments; Mrs. Campbell's initial excavations in 1820 uncovered some remains now lost at Balnuaran Southwest and George Bain described the lack of an 'authentic account' for the 1850 opening of Balnuaran Northeast.[19] James Fraser's 1884 'ground plan' indicated which stones were restored, enabling a degree of assurance about using certain stones for measuring alignments today, although poor site maintenance irritated George Browne and images of his visit (around 1920) can be seen online at the Canmore website, which stores thousands of records for Historic Environment Scotland, the government body responsible for maintaining these monuments.[20]

At these latitudes (north of 57°) and with the curvature of the earth, light and dark are diurnal, but more significantly they are seasonal. In summer the Sun never fully sets, it lies just beneath the horizon creating twilight; for which the Scottish term is the gloaming. After sunset a thin layer of lighter blue sky highlights the northern horizon, dimming stars to an altitude of five to ten degrees, but the rest of the night sky is visible as

---

[16] Liz Henty, 'The Archaeoastronomy of Tomnaverie Recumbent Stone Circle: A Comparison of Methodologies', *Papers from the Institute of Archaeology* 24, no. 1 (2014): pp. 1–15, p. 3.

[17] Martyn Barber, *Bronze and the Bronze Age Metalwork and Society in Britain c2500–800 BC* (Stroud: Tempus Publ., 2003), p. 50.

[18] Needham, 'Migdale-Marnoch: Sunburst of Scottish Metallurgy', p. 236.

[19] George Bain, *History of Nairnshire* (1893; Nairn: Telegraph Office, 1928), p. 6.

[20] James Fraser, 'Descriptive Notes of the Stone Circles of Strathnairn and Neighbourhood of Inverness', *Proceedings of the Society of Antiquaries of Scotland* 18 (1884): pp. 328–62, p. 342; George Forrest Browne, *On Some Antiquities in the Neighbourhood Of Dunecht House Aberdeenshire* (Cambridge: Cambridge Press, 1921), Pl XXXVI, p.192; https://canmore.org.uk/collection/121933 [accessed 10 Feb. 2015].

normal. John Lister-Kaye, who lives near Clava, commented that 'night recedes to only ninety minutes in twenty-four hours at midsummer' compared to eighteen hours of darkness at midwinter.[21] Conversely Margaret and Gerald Ponting stated the winter full Moon during major limit years lights up the night sky for 'twenty two hours' at Calanais.[22] Scotland is a place for extremes of light and dark.

Methodology
Using a clinometer, a GPS locator and good quality compass the methodology used was that advocated by Fabio Silva, in his study of Portuguese tombs, where he defined a 'window of visibility' as the 'region of the horizon, given the structure's corridor and entrance geometry, which can be seen from within the chamber'.[23] Although passages at Balnuaran are sufficiently straight as to merit theodolite measurement, this was part of a larger study, where the bulk of cairns were considerably more ruinous. No straight line existed and many were in inconvenient locations, hence this selection of measuring devices. Measurements were taken from the edge of the right-sided orthostat at the chamber end of the passage diagonally opposite to the left-sided orthostat at the passage entrance and from the left-sided chamber stone to the right-sided entrance stone, generating a range of azimuths and declinations, rather than just one figure. Similarly when measuring from the kerb-stone edges to the most substantial circle orthostat, two measurements were taken outlining the area of sky emphasised by that stone's location; always from the right side of one stone to the left side of the other and again from the left side of one stone to the right side of the other. Measurements were also taken in reverse to check for magnetic anomalies. The largest SW circle orthostat was selected because the largest kerb-stones and their stone circle counterparts were matched by the builders. This was to identify an area of sky, which might be regarded as significant to the builders of the monument. Stellarium planetarium software was used to ascertain solar/lunar positions in 2000 BCE and Photoscape software was used to annotate diagrams.[24]

---

[21] John Lister-Kaye, *Song of the Rolling Earth* (London: Abacus, 2004), p. 148.
[22] Gerald and Margaret Ponting, *Clach an Tursa, Carloway Isle of Lewis* (Stornoway: Ponting, 1981), p. 4.
[23] Fabio Silva, '"A Tomb with a View": New Methods for bridging the gap between land and sky in megalithic archaeology', *Advances in Archaeological Practice* 2, no. 1 (2014): pp. 24–37, p. 27.
[24] http://www.stellarium.org [accessed on 4 Jan. 2017];

Skyscape archaeology – the Balnuaran Enclosure

A common feature of all Clava cairns is their SW focus: passages are SW orientated and the largest orthostat for each of the cairns is always positioned towards the SW. Thus the monuments' potential relationship with the SW horizon was considered most important. Bradley's contour map of the Balnuaran enclosure area, shown from above, indicates the three main cairns, and is adapted to illustrate alignments (Fig. 2).

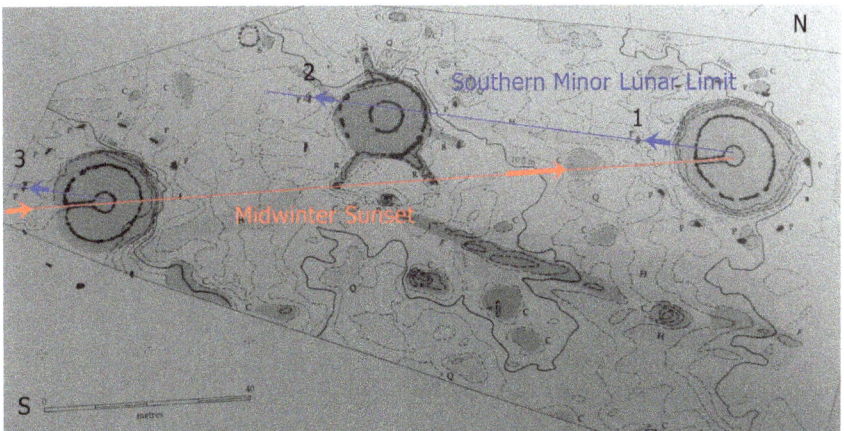

Fig. 2 Balnuaran Enclosure: The grey areas represent the cairns' platforms and the darker marks are the orthostats which constitute the chamber boundaries, the central ring, the peristaliths and the standing stones encircling each cairn. The grey contour lines indicate bedrock. The red line indicates midwinter sunset showing how it connects the passage-graves with two of Central's circle-stones. The two blue lines show kerb and circle-orthostats which align with the southern minor moonset. Numbers indicate the tallest circle-orthostats. Diagram adapted from Richard Bradley.[25]

The right-hand cairn is Balnuaran Northeast, (57° N 28' 24.25" 004° W 04' 23.63" elevation 98 m), the middle cairn is Balnuaran Central (57° N 20' 22.76" 004° W 04' 26.79" elevation 99 m) and the left cairn is Balnuaran Southwest (57° N 28' 21.01" 004° W 04' 28.25" elevation 96 m). The largest circle-orthostat for each cairn is in the SW, indicated by the numbers 1, 2 and 3.

The midwinter Sun sets along Balnuaran Southwest's passage (3) connects with the two stones which are part of Balnuaran Central's circle

---

http://www.photoscape.org., [Accessed on 13th January 2015].
[25] Bradley, *Good Stones*, p.16.

58 Clava Cairns, Midwinter Sunset and the Minor Lunar Limit

(2) and along Balnuaran Northeast's passage (1) to the back wall; indicated by the red line in Fig. 2. Balnuaran Central ring-cairn's rays appear to deliberately integrate its circle with the two passage-graves. This midwinter sunset alignment was first discovered by Boyle Somerville in 1910 and Alexander Thom expanded upon how the ray-marked stones of Balnuaran Central were picked out by the midwinter sunset.[26] Clive Ruggles commented that many Clava cairns are 'orientated with respect to the Moon' particularly the major limit midsummer full Moon.[27] Additionally Ruggles noted that the Balnuaran group encapsulated 'symbolism not found elsewhere' such as solar alignments, and incorporated 'coloured stonework', making this group exceptional.[28] Bradley looked at lunar alignments along the passages of Balnuaran Southwest and Northeast noting they were 'less precise' than solar, which is the case, although he did not explore Somerville's historical orientations regarding the largest circle-orthostat.[29]

This research found that the largest circle orthostats (the stones corresponding to numbers 1, 2 and 3 in Fig. 2) align with the southern minor limit moonset, for which Ruggles suggested a declination of -19.6° for 2000 BCE, as indicated by two blue lines in Fig. 2.[30] Ruggles explained that declinations 'gradually change' over time, so any declination needs to be checked against its estimated prehistorical date.[31] Balnuaran Southwest's significant orthostat, (opposite the largest peristalith stone), is broken and much smaller than the other two. The declinations and corresponding Azimuths are as follows:

Balnuaran Northeast -19.33° to -22.75° (Az. 224°-232°, Alt. 0°),
Balnuaran Central -17.49° to -22.75° (Az. 224°-236°, Alt 0°),
Balnuaran Southwest -12.12 to -22.75° (Az. 224°-247°, Alt 0°).

This alignment occurs when measured from their kerb counterparts and in the case of Balnuaran Central and Balnuaran Northeast they also align with

---

[26] H. B. Somerville, 'Instances of Orientations in Prehistoric Monuments in the British Isles.' *Archaeologia* 73 (Second Series) (Jan. 1923): pp. 193–224, p. 200; Alexander Thom, 'Megalithic Astronomy: Indications in Standing Stones', *Vistas in Astronomy* 7 (1966): pp. 1–57, p. 19.
[27] Clive Ruggles, *Astronomy in Prehistoric Britain and Ireland* (New Haven: Yale University Press, 1999), p. 246.
[28] Ruggles, *Prehistoric Britain*, pp.157, 246.
[29] Bradley, *Good Stones*, p. 26; Personal Communication.
[30] Ruggles, *Prehistoric Britain*, p. 57.
[31] Ruggles, *Prehistoric Britain*, p. 57.

Culture and Cosmos

each other and the centre of Balnuaran Northeast's chamber, making five separate points of connecting rays. Fraser informed us that all three significant circle-orthostats had been 'shifted' in antiquity, which may explain why the NE orthostat is only close to -19.6°(-19.33 to -22.75), although the largest kerb-orthostats remain unaltered.[32] According to Henshall and Ritchie the fallen largest SW orthostat for Balnuaran Northeast was reset about 6 feet (1.8m) further 'southwest from the kerb' prior to Fraser's visit, roughly in line with its original position.[33] The greatest problem is posed by how far it was misplaced sideways of the original position, within the circle. Henshall and Ritchie estimated the neighbouring circle stones (which were also reset in antiquity) were 'close to the original position'; as the largest circle-orthostat appears half way between these stones, it is probably near its original position, but this location is inaccurate.[34] These alignments hint at intentionality for lunar orientation from the unaltered kerb-orthostat positions but, given the repositioning in antiquity of the largest circle-stone at Balnuaran Northeast, it is insecure.

Somerville, who measured every circle stone for the three monuments, provided diagrams which show the 'winter solstitial sunset' alignment along Balnuaran Northeast's passage.[35] This indicates a similarity to Maeshowe in Orkney, where Aubrey Burl described the midwinter sunset glow could 'reach the central chamber', although Maeshowe is Neolithic and retains its roof, which these EBA monuments do not.[36] Bradley described the sunset alignments as linking Clava, Maeshowe, 'Cornish entrance-graves' and Irish Wedge-shaped tombs.[37] Orkney's Maeshowe was utilised, according to Gordon Noble, from '2,800 to 2,200 BCE', which suggests that the monument fell into disuse just as Clava cairns

---

[32] Fraser, 'Descriptive Notes of the Stone Circles', p. 342.
[33] Audrey Shore Henshall and James Neil Graham Ritchie, *The Chambered Cairns of the Central Highlands: An inventory of the Structures and their Contents* (Bath: Edinburgh University Press, 1988), p. 204.
[34] Henshall and Ritchie, *Chambered Cairns*, p. 205.
[35] H. B. Somerville, *Royal Anthropological Institute* MS 267, f. 159.
[36] Aubrey Burl, *Prehistoric Astronomy and Ritual* (1983; repr. Aylesbury: Shire Archaeology, 1997), p. 26.
[37] Richard Bradley, 'After the Great Stone Circles' in Richard Bradley and Nimura Courtney, eds., *The Use and Reuse of Stone Circles: Fieldwork at Five Scottish Monuments and its Implications* (Oxford: Oxbow Books, 2016), pp. 112–21, p.113.

began to be built.³⁸ The visual proximity of the Moray coast and convenience of Clava as a trading centre could have made it attractive to sky-conscious Orcadians. As both monuments align with midwinter sunset, this suggests significant social links, or a common cosmology, hinting at a cosmology of considerable antiquity. Henshall noted that only Maeshowe and Clava are decorated with stone markings.³⁹ If Clava cairns had a secure combination of midwinter sunset and minor lunar limit, it would indicate strong links with Stonehenge (for which Mike Parker-Pearson suggested the 'last phase of stone construction' dated to '2300–1900' BCE, with the structure being abandoned approximately '1640–1520' BCE), suggesting the possibility that sky knowledge was either shared between Wessex, Orkney and Clava or that they accessed a common cosmology from another source.⁴⁰ This would reinforce Bradley's argument that 'political dominance' was transferred from 'Northern Britain back to Wessex'.⁴¹

Skyscape archaeology – Balnuaran's two Passage-graves
Jim Knowles' photograph is adapted to show Balnuaran Northeast which is a good example of a passage-grave, containing a central chamber with a passage, (Fig. 3). The kerb-stones of the cairn's peristalith are much larger than the abundant cairn material. The cairn itself is surrounded by a grass-covered rubble platform, which is then surrounded by a free-standing stone circle. The defining Clava feature is that all orthostats in both the kerb and circle are smallest at the NE and gradually increase in height, with the tallest in the SW. The overall size of the monuments is such that an extended family group could easily build them. Henshall provided these measurements for Balnuaran Northeast circle: the NE stone was 1.6m high while the SW stone was 2.74 m.⁴² Significantly, many circle orthostats were matched with their opposing kerb-stones, either by shape, colour, size or type of stone, as can be seen in Fig. 3.

---

³⁸ Gordon Noble, *Neolithic Scotland Timber, Stone, Earth and Fire* (Edinburgh: Edinburgh University Press, 2006), p. 107.
³⁹ Audrey Shore Henshall, *The Chambered Tombs of Scotland Vol. 2* (Aberdeen: Edinburgh University Press, 1972), p. 284, [Hereafter Henshall, Vol. 2].
⁴⁰ Michael Parker-Pearson, *Bronze Age Britain* (London: B. T. Batsford, 1993), p. 93.
⁴¹ Richard Bradley, *The Social Foundations of Prehistoric Britain: Themes and Variations in the Archaeology of Power*, Longman Archaeology Series (London and New York: Longman Group, 1984), p. 65.
⁴² Audrey Shore Henshall, *The Chambered Tombs of Scotland* Vol. 1 (Aberdeen: Edinburgh University Press, 1963), p. 364, [Hereafter, Henshall Vol. 1].

J. Anna Estaroth 61

Fig. 3 Balnuaran of Clava Northeast passage-grave. The tallest circle stone is at the SW and the smallest in the NE. The binary nature of the monument's design is subtly indicated by matching circle-stones and kerb-stones indicated by the orange dotted lines. Photograph Jim Knowles.[43]

The two circle stones marked 1 and 2 match with their kerb-counterparts, with circle stone 1 being square and white, as is its kerb counterpart and circle stone 2 being curved and dark, as is the kerb-stone opposite. This matching indicated that the circle and cairn were coetaneous constructions and Bradley found radiocarbon evidence that platform material was 'contemporaneous with the cairn and circle'.[44]

Bradley discovered that the platform was added about fifty years after creation of the cairn – sealing the entrance to anything except light; it only half filled the passage so that 'sunlight could still penetrate to the interior at midwinter'.[45] There is evidence of fire in the platform area leading up to the passage and although they were dealing with an earlier Irish culture, Chris Fowler and Vicki Cummings noted evidence of the ritual use of fires

---

[43] Jim Knowles, West Lothian Archaeology Group, http://www.armadale.org.uk/balnuaran.htm [Accessed on 2 Aug. 2016].
[44] Richard Bradley, *The Significance of Monuments on the Shaping of Human Experience in Neolithic and Bronze Age Europe* (London & New York: Routledge, 1998), p. 113.
[45] Richard Bradley, 'The Land, the Sky and the Scottish Stone Circle', in Chris Scarre, ed., *Monuments and landscape in Atlantic Europe: Perceptions and Society During the Neolithic and Early Bronze Age* (London and New York: Routledge, 2002), pp. 122–38, p. 130.

in front of passages, 'even after entrances were blocked'.[46] Fires have practical as well as ritual uses, fending off predators; McKirdy cites bone evidence of bear, wolf, and lynx found at Inchnadamph (Sutherland), for which Rick Schulting and Michael Richards provided radiocarbon dates '3640–2940 BCE'.[47] Janice Short stated that Royal Acts were passed in the 1400's by King James I for the 'destruction of wolves' and traditionally the last wolf killing was on the 'River Findhorn', near Clava, in '1743'.[48] It appears that predators were in the Clava area during the construction of these cairns.

Checking midwinter sunset required a non-overcast sky and photographs taken of midwinter sunset 2014 confirmed this alignment still occurs. Bradley, plus Henshall and Ritchie, referred to tests with tarpaulin which proved that the Sun could reach the far wall of Balnuaran Northeast's chamber.[49] Douglas Scott photographed the event in 1999 and his image can be viewed on the Highland Environment Record website.[50] Scott's records were referred to by Bradley, regarding 'midsummer sunrise' at Mains of Clava Northeast.[51] Scott also found that circle orthostats, among the Clava group as a whole, marked the 'rising and setting Sun every 45 days'.[52] Ruggles' suggested midwinter sunset declination for 2000 BCE is -23.95°, and Balnuaran Northeast's passage measurements provided the window of visibility from declination -22.63° to -25.74° (Az. 207-217° Alt.3°) and Balnuaran Southwest's passage

---

[46] Chris Fowler and Vicki Cummings, 'Places of Transformation: Building from Water and Stone Monuments in the Neolithic of the Irish Sea', *The Journal of the Royal Anthropological Institute* 9, no. 1 (March 1, 2003): pp. 1–20, p. 13.

[47] McKirdy, *Set in Stone*, p. 71; Rick J. Schulting and Michael P. Richards, 'The Wet, the Wild and the Domesticated: The Mesolithic–Neolithic Transition on the West Coast of Scotland', *European Journal of Archaeology* 5, no. 2 (01 August 2002): pp. 147–89, p. 164.

[48] Janice Short, History of wolves in Scotland, http://www.wolvesandhumans.org/wolves/history_of_wolves_in_scotland.htm [accessed on 22 Oct. 2016], the website of the wolves and humans foundation.

[49] Bradley, *Good Stones*, p. 122; Henshall and Ritchie, *Chambered Cairns*, p. 121.

[50] http://her.highland.gov.uk/FullImage.aspx?imageid=183409&uid=MHG3013 [accessed on 10 Aug. 2016].

[51] Douglas Scott, 'An Astronomical Assessment of the Clava Cairns' unpublished paper held by the RCAHMS, 1991; Bradley, *Good Stones*, p. 125.

[52] Douglas Scott, 'The Solar-Lunar Orientations of the Orkney-Cromarty and Clava Cairns', *Journal of Skyscape Archaeology* 2, no. 1 (2016): pp. 45–66, p. 61.

measurements were -22.63 to 26.27 (Az. 205-217° Alt 3°) confirming the alignment at both.[53]

Balnuaran Southwest passage-grave has a prominent platform; the passage aligns with midwinter sunset and the orthostat to the right of the passage entrance aligns with the southern minor limit moonset. Interestingly the orthostat to the left of the passage entrance generated a false-positive alignment with the southern major limit moonset. While the alignment seems correct and appears to sight upon a hill feature, the event itself could not have been seen because the midsummer full Moon at major limit is exceptionally low in the sky and the hill height intervenes, as will be discussed later. Photographs of midwinter sunset in 2014 showed trees currently obscure the Sun's journey along the passage.

Skyscape Archaeology – Balnuaran Central Ring-cairn
The ring-cairn is different from the passage-graves and Henshall and Richie commented that there were 'as yet no parallels elsewhere' for this style of monument.[54] Henshall noticed ring-cairn centres were 'much larger' than passage-grave chambers, noting the lack of post holes, the charcoal-covered floor, stating they were 'too wide to corbel over'.[55] It was therefore never roofed-over and may be close to the original cairn height. Bradley says Balnuaran Central (Fig. 4) was damaged, some southerly peristalith stones had been removed, but he noted several circle-stones align with the inner kerb-stones by sharing the same 'colour, shape or raw materials'.[56] These combinations generate multiple alignments and exploring their meaning could be a potentially rewarding area of research. Some colour-sharing also occurs along three of the four rays.[57] Apart from Balnuaran, all of the study ring-cairns were located where hills and trees cannot obscure their light open aspect to the sky, a common design feature, but in stark contrast to the dark beehive-shaped passage-graves.

---

[53] Bradley, *Good Stones*, p. 157; Ruggles, *Prehistoric Britain*, p. 57.
[54] Henshall and Ritchie, *Chambered Cairns*, p. 17.
[55] Henshall, Vol. 2, p. 271; Henshall Vol. 1, p. 26.
[56] Bradley, *Good Stones*, p. 23.
[57] Bradley, *Good Stones*, p. 23.

64 Clava Cairns, Midwinter Sunset and the Minor Lunar Limit

Fig. 4 Balnuaran of Clava Central ring-cairn: looking southwards with Balnuaran Southwest passage-grave in the top left corner. The red line indicates the beam of midwinter sunset, with black dotted line arrows showing the position of the midwinter sunset's shadows cast by the ring-cairn's circle-orthostats. The blue lines delineate the four rays, 1 W, 2 NE, 3 E, 4 SE. Photograph John Knowles.[58]

This picture by Knowles includes members of the West Lothian Archeology Group to provide scale for the monument.[59] The rays are of interest; these are the low rubble rays running from the peristalith to a circle stone. This ring cairn has four, marked here with blue outlines and numbered. Henshall provided measurements for the rays as being 'over two metres wide, but only thirty centimetres tall'.[60] Fig. 4 shows the line of midwinter sunset picking out two ray-marked circle-stones on the left, which Henshall and Richie considered were possibly placed to link Balnuaran Central with the passage-graves, in order to generate a 'unitary design' for the monument group.[61] It is these two stones which generate distinctively extra-long shadows at the solstice, due to the exceptionally low Sun. Bradley discovered the ray reaching to the broken stone at the foot of the picture (ray 2); another is obscured by the trees top right (ray 1).[62] The two rays not associated with midwinter sunset are both aligned

[58] Knowles, West Lothian Archaeology Group, [accessed on 2 Aug. 2016].
[59] Knowles, West Lothian Archaeology Group, [accessed on 2 Aug. 2016].
[60] Henshall Vol. 1, p. 362.
[61] Henshall and Ritchie, *Chambered Cairns*, p. 116.
[62] Bradley, *Good Stones*, pp. 20–21.

Culture and Cosmos

with the northern minor limit – moonrise and moonset. Ruggles provided this equation for the lunar centre + (ε - 1) - P, and for 2000 BCE his suggested declination figure was +17.95°.[63] My figures are declinations +8.96° to +21.08° (Az. 282°-306°Alt. 3°) for the westerly ray (Fig. 4, ray 1) and declinations +8.11° to +22.45° (Az. 49°-78°Alt. 2°) for the damaged orthostat ray (Fig. 4, ray 2). Because the rays have never been interfered with, this is a safe alignment and definitely suggests that this monument is connected to the minor lunar limit.

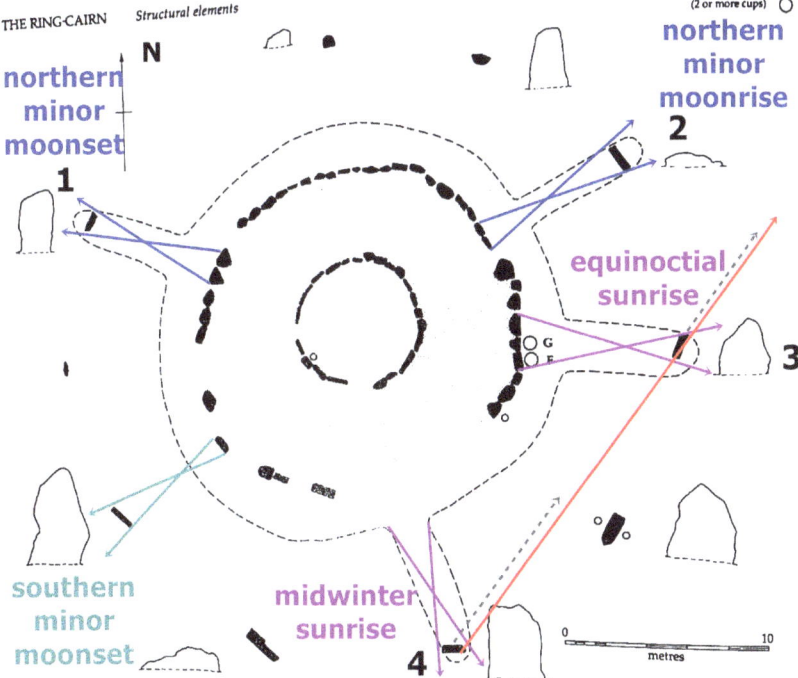

Fig. 5 Balnuaran of Clava Central ring-cairn alignments: looking northwards a reverse image of Fig. 4. The red line shows the midwinter sunset beam of light, touching two circle-orthostats connected to the cairn by rays, with grey dotted lines showing the direction of shadows. The purple lines indicate measurements of these rays: E (3) aligns with equinoctial sunrise and SE (4) with midwinter sunrise. The blue lines indicate measurements taken of the other two rays: W (1) aligns with northern minor moonset and NE (2) with northern minor moonrise. The light blue lines indicate measurements taken of the largest SW circle-orthostat which aligns with southern minor moonset. Diagram is adapted from Richard Bradley.[64]

---

[63] Ruggles, *Prehistoric Britain*, pp. 37–57.
[64] Bradley, *Good Stones*, p.20.

To avoid bias all ray orthostats were measured from the peristalith of Balnuaran Central. The other two rays are not aligned to lunar horizonal activity, but to solar: one is orientated on equinoctial sunrise declinations -3.54° to +8.20° (Az. 81°-103° Alt. 4°) (Fig. 4, ray 3), and the other orientated on midwinter sunrise declinations -23.01° to -27.00° (Az. 147°-163° Alt 4°) (Fig. 4, ray 4). They link the cairn with the pair of circle-stones which are connected by the shadows cast by midwinter sunset. This resembles, the stone pairs that Olwen Pritchard described as being deliberately 'arranged to create shadow phenomena' at the solstices.[65] The above shadows would have been visible for participants outside the passage-graves. Essentially this monument is complex in nature having both multiple solar and lunar alignments, unlike any other Clava cairn; it is also the only known ring-cairn to have rays. Bradley's diagram was adapted to clarify this monument's alignments (Fig. 5).

Discussion
The main focus of this study was horizonal events along the southwest. Lionel Sims gave this definition of a lunar standstill or limit: they are the 'declination measure of the Moon's geocentric extremes', allowing for parallax and refraction.[66] Because all Clava Cairns are orientated towards the southwest, my focus was on the triple phenomena of the southern moonset at its most southerly extreme of the 18.61 year nodal cycle (southern major limit moonset), the midwinter sunset (an annual event occurring on the horizon halfway between the two lunar events) and the southern moonset at its least southerly extreme of the nodal cycle (southern minor limit moonset). The minor limit peak occurs 9.3 years after the major limit, which in turn follows 9.3 years after the previous minor limit in an unending cyclical basis. Sims cogently explained how the new Moon closest to midwinter during minor limit years, generated the 'longest and darkest' period of time for the whole nodal cycle, making it suitable for rituals of darkness.[67] Sims used the term 'phase-locked' to describe how

---

[65] Olwen Pritchard, 'Shadows, Stones and Solstices', *Journal of Skyscape Archaeology* 2, no. 2 (2016): pp. 145–64, p. 160.
[66] Lionel D. Sims, 'What is a Lunar Standstill? Problems of Accuracy and Validity in 'The Thom paradigm', *Mediterranean Archaeology & Archaeometry* 6, no. 3 (2006): pp. 157–63, p.157.
[67] Lionel Duke Sims, 'The "Solarization" of the Moon: Manipulated Knowledge at Stonehenge', *Cambridge Archaeological Journal* 16, no. 2 (Jun. 2006): pp. 191–207, p. 203.

the builders of Stonehenge deliberately integrated both solar and lunar alignments, as the builders of these monuments appear to have done.[68] J. McKim Malville refuted Sims' arguments regarding dark Moon, minor lunar limits and midwinter sunset darkness ceremonies but, citing ethnographic evidence, argued instead for a 'three-day death/rebirth' festival at midwinter, with the purpose of regenerating the waning Moon, rather than focusing on darkness.[69] Sims responded to this by arguing that 'rituals follow a time-resistant syntax' allowing for multiple adjustments from hunter-gathering and then pastoralism to agricultural economies and that the dark Moon was essential to these transformations.[70] Whichever meaning was attached to the lunar limit and midwinter combinations, they appear to have been important for the Clava cairn builders.

Bradley has recently contrasted night and human remains with 'sunrise and the domestic world' and this theme of darkness and light is reflected in the seasonal changes Scotland experiences.[71] Significantly, the full Moon, falling closest to midsummer on major limit years, generally provides light for twenty-four hours. There are other combinations, such as minor limit dark moon and midsummer, or major limit full Moon and midwinter. However the combinations considered in this paper are the new Moon nearest midwinter on minor limit years connected with darkness, and the full Moon nearest midsummer on major limit years connected with light, because these lunar limits are associated with the most extensive periods of darkness and light, which is particularly relevant for these latitudes.

Looking at the combined three cairns (Fig. 2) it is noticeable that they were not built in a straight line. The central cairn had no great height and it would not have blocked midwinter sunset arriving at both passage-graves. Something prevented the builders placing them in one line. Sims provided an excellent diagram of solar and lunar risings along the entire horizon for Stonehenge, which has midwinter sunset largely opposite midsummer

---

[68] Sims, 'What is a Lunar Standstill?', p.160.
[69] J. McKim Malville, 'The Enigma of Lunar Standstills', *Journal of Skyscape Archaeology* 2, no. 1 (2016): pp. 85–94, p. 93.
[70] Lionel Sims, 'Forum Reply', *Journal of Skyscape Archaeology* 2, no. 1 (2016): pp. 96–99, p. 97.
[71] Richard Bradley, 'The Dark Side of the Sky: The Orientations of Earlier Prehistoric Monuments in Britain and Ireland', in Marion Dowd and Robert Hensey, eds., *The Archaeology of Darkness* (Oxford: Oxbow books, 2016), pp. 51–61, p. 59.

sunrise, because it is appreciably flatter than in the Nairn valley.[72] John North commented that even small changes in altitude have a 'very marked effect': elevating the horizon by a quarter of a degree may move the rising point by a 'complete solar diameter'.[73] Unlike Stonehenge, at Clava the hills cause the visible midsummer sunrise to occur further to the east; midsummer sunrise is therefore not opposite midwinter sunset. Bradley produced a diagram which illustrated these solar alignments showing the two passage-graves align with midwinter sunset, while midsummer sunrise picks out the back of Balnuaran Northeast, Balnuaran Central and Mains of Clava Northeast.[74] This is not an alignment based on Balnuaran Central's individual circle-stones and Scott's 2014 survey indicated that the cairn aligns with sunrise on '20th May and 20th August'.[75] Conversely my survey of the rays indicates equinoctial sunrise and midwinter sunrise for Balnuaran Central, all of which suggests it is a sunrise cairn. Bradley also noted how red sandstone was used to face the setting Sun at Balnuaran Northeast, but those stones facing the rising Sun 'contain inclusions of quartz'.[76] Frank Prendergast explained in this year's SEAC public lecture, that quartz was usually employed at the front of prehistoric buildings, but this exceptional position made sense, if the quartz was there to reflect the midsummer sunrise.[77] Ruggles however connects quartz with ceremonies associated with 'movements of the moon'.[78]

If the builders had wanted a largely opposite midsummer/midwinter aligned monument, they could have done so at nearby Newtown of Petty which is only 4.7 kilometres north of this site, has an open outlook over the Moray Firth and 'radiocarbon dates' indicate it was constructed contemporaneously.[79] This suggests that the central cairn could have been

---

[72] Lionel D. Sims, 'What is a Minor Standstill of the Moon', *Journal of Skyscape Archaeology* 2, no. 1 (2016): pp. 67–76, p. 68.
[73] J.D. North, *Stonehenge: Neolithic Man and the Cosmos* (London: HarperCollins, 1996), p. 227.
[74] Bradley, *Good Stones*, p. 126.
[75] Scott, 'Solar-Lunar Orientations', p. 52.
[76] Bradley, *Good Stones*, p. 126; Bradley, 'Dark Side of the Sky', p. 54.
[77] Frank Prendergast, Exploring the Monuments and the Cosmology of the Boyne Valley, https://www.youtube.com/watch?v=Jpqvno7kDqU [accessed 26 Sept. 2016].
[78] Ruggles, *Prehistoric Britain*, p.124.
[79] Richard Bradley and Margaret Mathews, 'The Clava Ring-cairn at Newton of Petty, Excavations by John Thawley 1975–77', in Richard Bradley (ed.) *The Good Stones a New Investigation of the Clava Cairns* (Edinburgh and Kings Stanley: Society of Antiquaries of Scotland Monograph Series No. 17, 2000), pp. 131–59, p. 158.

intentionally offset in order to catch midsummer sunrise. Balnuaran Central combines midwinter sunset and the southern minor lunar limit, plus midsummer sunrise and the northern minor lunar limit, both of which have dark Moon connotations. Potentially the passage-graves were associated with midwinter sunset, dark and the transformation of the dead, while the ring-cairn was a cairn of midsummer sunrise, light and the transformation of the living.

On major lunar limit years the Sun is just below the northern horizon and the full Moon nearest midsummer is just above the southerly horizon. The full Moon is effectively blocked by the southerly hills which at Balnuaran have an altitude of five degrees. It varies, but Stellarium planetarium software shows that the midsummer full Moon usually achieves a maximum altitude around 2°50'.[80] The Moon is so low that, not only does it not appear above the hills on major limit years, but it has insufficient height to raise above them on years either side of the major limit. Occasionally the Moon sits on the hilltop, briefly slides or skims along the hill then sets, which could be visually dramatic. For up to five years of the 18.61 lunar nodal cycles, the midsummer full Moon at Clava does not visibly rise and set normally; where normal means the upper lunar rim rises first, the moon gains some height and then the upper rim sets last. On major lunar limit years the Clava valley is cast in shadow when the rest of the district is bathed in moonlight. The Moon could be thought of as connecting with the underworld or hiding in the earth. It is significant that at Balnuaran no orthostat actually orientates to the major lunar limit, the orientations are towards minor lunar limits and solar orientations, primarily midwinter, but midsummer and equinox as well.

Conclusions
This paper considered the three main cairns at Balnuaran, describing them in their geographical and cultural context as well as considering the relevant academic research. They were examined from a skyscape archaeology perspective as part of a larger study comparing hilltop with riverside cairns. The results confirmed solar passage-grave alignments, but also established new lunar and solar alignments. The study found that Balnuaran Central has rays aligned with northern moonrise and moonset during minor limit years and an orthostat which potentially aligns with the southern minor limit moonset. In addition two other rays align with equinoctial sunrise and midwinter sunrise, while marking out two circle-

---

[80] http://www.stellarium.org [accessed on 4 Jan. 2017].

orthostats which generate shadows during midwinter sunset. The danger of antiquarian re-structuring was recognised, which activated other methods of ascertaining whether lunar alignments existed. The methodology enabled confirmation of Somerville's and Thom's work, while incorporating some of Bradley's archaeological discoveries.

The specific topography at Balnuaran is why, on major lunar limit years, these cairns are in darkness while everywhere else is bathed in moonlight. This posed a number of questions about this location, leading to further comparative research across the Clava region relating to the major lunar limit. The size of many ring-cairns visited by the author suggests an inclusive summer event was possible. The builders might have also gathered, nine or ten years later, for new Moon nearest midwinter during minor limit years; the passage-graves being off-limits. For at least thirteen years of the lunar nodal cycle the midsummer full Moon would bathe the central ring-cairn in light, making it suitable for gatherings. From this it could be concluded that ring-cairns might have functioned as social gathering-centres and potentially as star-schools, particularly during major limit years, when the Moon was invisible and its light did not mask starlight. These unusual circumstances may have led them to build so many monuments at this location. Achieving dark during cosmic light was quite exceptional. To add weight to this conclusion further research at the next non-overcast, major limit midsummer full Moon would be desirable.

Both solar and lunar alignments were significant for the builders of Clava cairns and it is possible that they learnt their sky-observation skills from Orkney practitioners. The shared archaeoastronomical evidence suggests that sky-watching skills were evident at Maeshowe in Orkney, at Stonehenge in Wessex and at Balnuaran and these could have been dispersed by trade and interchange or been seeded from an earlier common cultural heritage. Balnuaran and the Clava region's pivotal position regarding prestigious early bronze axe manufacture and the exceptionally low midsummer full Moon, which was hidden by the hill on major lunar limit years, may have made it a magnet for celestial expertise. This conclusion indicates the need for integrating Clava cairns into any exploration of the wider archaeoastronomical, socio-political environment in EBA Britain, particularly with Maeshowe, Stonehenge and Aberdeenshire's RSCs.

The Balnuaran complex exhibits the winter solstice sunset alignment, which might be considered sufficient to explain their purpose. However the ring-cairn's alignments are with the southern minor moonset and, through the rays, with the northern minor moonrise and set, with equinoctial sunrise, midwinter sunrise and by shadow with midwinter sunset. It would have been lit up by summer solstice sunrise. It is open to the sky and appears to have been designed as an open air, day lit monument, only being dark during new Moons and during the period when the midsummer full Moon on major limit years was invisible, a factor which may have been significant.

Given the theme of darkness and light it appears the first half of the theoretical balance has been met, the role of the dark new Moon nearest midwinter solstice appears associated with the minor lunar limit, the womb-like passage-graves and the transformation of the ancestral dead. However the field data does not support a combination of midsummer, major lunar limit and light, due to the southern hill hiding the midsummer full Moon. The open ring-cairn is normally well lit and may have functioned as a centre for the living community. However the disappearance of the normally bright full Moon nearest midsummer solstice brings darkness when there would usually be twenty-four hours of light. The full Moon appears to visit the underworld. It suggests a different kind of combination to those posited above: just as the midwinter sunset casts light within the passage-grave's dark chamber, during the darkest part of the year; the midsummer full Moon, being hidden by the hill generates darkness within the normally light ring-cairn's centre, during the lightest part of the year.

Orientations of Late Neolithic to Bronze Age and Iron Age Long Cairns in Coastal Finland

Marianna P. Ridderstad

Abstract: In this study, the orientations of c. 138 long cairns located in coastal Finland were measured and examined, along with other properties of the cairns. The length of the cairns varies from a few metres to almost 50 m. The dominant color of the stones in most of the cairns is red, and they were usually built on locally elevated terrain, e.g. on ridges, rocky outcrops or small islets on the ancient shore. It was found that in the category of long cairns there were several different types of elongated cairns: the 'simple' and curved long cairns, some of which were attached to round cairns; the rectangular cairns with one or more central chambers; the very large rectangular cairns; and two different types of ship-formed cairns, Type 1 and Type 2, the latter of which was a previously an unrecognised type of Late Neolithic/Early Bronze Age long cairn. The comparison of the orientations of the cairns of different types and locations suggest that there was some cultural continuity between the Neolithic and the Early Bronze Age cultures on the western coast of Finland. However, based on the present analysis, this continuity does not seem to have extended beyond the Middle Bronze Age. It is also suggested that the appearance of the Type 2 ship-formed cairn in the Ostrobothnia region in the Late Neolithic may have resulted from outside cultural influences, perhaps from the earliest contacts with the central ideologies of the Nordic Bronze Age.

According to present knowledge, cairn building in Finland started in Northern Ostrobothnia during the Early and Middle Neolithic, ca. 5100–2500 BCE.[1] The Middle and Late Neolithic were a period of cultural upheaval in the region and elsewhere on the Ostrobothnian coast. This included the building of large villages of wooden semi-subterranean houses, terraced houses as long as 100 m, and the so-called Giants' Churches (hereinafter GCs), which were huge rectangular stone enclosures

---

[1] See Jari Okkonen, *Jättiläisen hautoja ja hirveitä kiviröykkiöitä – Pohjanmaan muinaisten kivirakennelmien arkeologiaa*, Acta Universitatis Ouluensis B 52 (PhD Thesis, University of Oulu, 2003); Teemu Mökkönen, 'Stone Setting Filled with Red Ochre from the *Keelaharju Site,* Northernmost Baltic Sea Region: A Stone Age Grave in the Context of North European Burial Traditions', *Fennoscandia Archaeologica* 30 (2013): pp. 13–36.

Marianna P. Ridderstad, 'Orientations of Late Neolithic to Bronze Age and Iron Age Long Cairns in Coastal Finland', *The Marriage of Astronomy and Culture,* a special issue of *Culture and Cosmos,* Vol. 21, nos. 1 and 2, 2017, pp. 73–86.
www.CultureAndCosmos.org

likely constructed to serve as some type of communal spaces used, for example, for ritual gatherings.[2] The culture relied on marine resources and the villagers carried out extensive foreign trade, which brought outside influences with it. The orientations of the GCs that are arguably connected to astronomical events could have resulted from the cultural transmissions via such trade connections.[3]

The earliest built cairns were at first round and torus-shaped, i.e., round with a large central pit. There is no direct evidence that the cairns were used as graves, but it is usually assumed that this was the case or that they were used for other ritual activities. There were other types of cairn too, with apparent profane usage purposes: huge piles of burnt, fire-cracked stones have been found around the GCs; it is believed those had been used to boil water in the processing of seal blubber.[4]

During the peak of the GC culture in Ostrobothnia, Southern Finland was inhabited by the Corded Ware (Battle Axe) Culture, which lacked built cairns or other large monuments of stone. By the end of the Neolithic, c. 1800 BCE, the GCs seem to have been abandoned and new types of cairns, long cairns, started to be built on the Ostrobothnian shores. Long cairns, as well as cairns with other shapes, were then built on the coast of Finland ever since until the end of the Iron Age c. 1000 CE.

In the early Bronze Age, starting c. 1500–1300 BCE in Finland, the cultural differences between Ostrobothnia and Southern Finland disappeared and similar style cairns were built all along the coast of Finland – with the exception of Northern Ostrobothnia, where the culture there seems to have suffered a permanent cultural decline, probably due to the increasingly unfavourable climatic conditions.[5] The cairns were mostly huge and circular or oval, sometimes with a large central pit or 'chambers'; later in the same period they became smaller. Sometimes, the Bronze Age cairns were triangular, irregular or attached together so that they formed cross-like or cloverleaf-shaped formations. In the Iron Age, rectangular

---

[2] See Okkonen, *Jättiläisen hautoja ja hirveitä kiviröykkiöitä*, pp. 241–42.
[3] Marianna Ridderstad and Jari Okkonen, 'Orientations of the Giant's Churches in Ostrobothnia, Finland', in M. Rappenglück, B. Rappenglück, and N. Campion, eds., *From Alexandria to Al-Iskandariya: Astronomy and Culture in the Ancient Mediterranean and Beyond, Proceedings of SEAC 2009*, British Archaeological Reports International Series, in press.
[4] Okkonen, *Jättiläisen hautoja ja hirveitä kiviröykkiöitä*, p. 195.
[5] See Reijo Solantie, 'Aspects of Some Prehistoric Cultures in Relation to Climate in Southwestern Finland', *Fennoscandia Archaeologica* 22 (2005): pp. 28–42.

and triangular stone settings increased in popularity relative to other types of cairns, although the most common cairn type was always the round or oval small cairn. Long cairns were always built along with other types, and a special form of long cairn from the Bronze Age onwards was the ship setting, where the stones were placed in a formation resembling a boat or a ship.

Fig. 1. Examples of long cairns in this study: (a) Type 1 M simple long cairn; (b, c) Type 2 M 'curved' long cairn; (d) large rectangular cairn; (e) chambered cairn; (f) Type 1 Ship setting; (g) Type 2 Ship setting. The small inset schematic drawings present the basic possible shapes of all six types of cairns; see the text for further information on the different types. All images are by the author.

The present study only concerns the cairns of coastal Finland (Fig. 1). At least since the Bronze Age, this region has formed a cultural sphere distinct from inland Finland, with the coastal zone showing the central features of the cultural sphere of the Nordic Bronze Age, while the (rather poorly known) inland cultures were probably hunter-gatherers. Cairns were also built in inland Finland, but the inland Bronze Age and Early Iron Age cultures have not yet been properly investigated.

Methodology
During the project, 138 cairns along the coastal Finland from Kotka in the south to Keminmaa in the north were examined and their orientation measured. The cairns in the study area are not evenly distributed along the coast, but are mostly concentrated in Ostrobothnia and Western Satakunta; the largest group of Bronze Age ship settings is in Kotka.

The fieldwork was carried out in 2009–2013. The axial orientations of the cairns were measured with a magnetic compass (Suunto compass with 360° scale), in addition to other possible axes for evidence of triangular and more complex construction style. The final values for the axial orientations were calculated from the averages of the reciprocal measured axial orientations.

The morphology of the Finnish cairns, especially the long cairns, is a subject not extensively studied. Therefore, the shapes of the cairns were recorded for comparison purposes. Then, to group the measured cairns for comparison with each other by type, the shapes and sizes of the cairns could be taken into account. Other features recorded were the general placements of the cairns relative to the ridges and geology on which they had been built, as well as the sizes, shapes and colors of their constituent stones.

Most of the cairns have only been dated using the shore displacement method, i.e., by using their height relative to the present sea level. The post-glacial rebound of the bedrock in Finland proceeds at a known rate and this can be used to estimate when a given site was on the shoreline in prehistoric times.[6]

Since the sites are now located in forestry due to the post-glacial rebound, the original horizon heights had to be calculated from a digital terrain model. For practical reasons, the horizon line for each individual cairn was not

---

[6] See, M. Eronen, 'Land Uplift: Virgin Land from the Sea', in M. Seppälä, ed., *The Physical Geography of Fennoscandia* (Oxford: Oxford University Press, 2005), pp. 17–34.

modelled, but a generalised horizon height model was used. The model was based on the observation that the cairns had usually been built on elevated locations in the coastal zone with good visibility towards their surroundings in all directions and that towards the west and/or the south, the horizon line was often the sea horizon. The astronomical declinations were calculated for the cairns using horizon altitudes from zero to one degree, and the resulting differences in these results were taken into account and regarded as uncertainties in interpreting the results of the analysis.

Results
The length of the cairns examined varies from c. 3 m to 48 m; their width varies from c. 1.5 m to over 15 m. The sizes and shapes of the stones that had been used to build the cairns vary from round beach stones of c. 10 cm to large sharp-edged stones of over 60 cm in diameter to occasional large boulders of over 1 m in size that had been placed in the middle or end of a cairn. The dominating color of the stones used to build the cairns is red; especially among the Bronze Age cairns, all of the stones in a cairn are often red or pink granite and/or sandstone.

The cairns were most often been placed on locally elevated terrain on the ancient shores, e.g., the top of a ridge, a rocky outcrop or a small islet. These locations then offered naturally good, open views towards the surroundings and the sea horizon. However, the cairns themselves have a low profile and could not have been seen from afar, e.g., from a boat approaching the coastline, suggesting that the argument that the cairns could have been used as some sort of territorial markers does not seem to be valid.

It was observed early during the field work that the monuments known as 'long cairns' do in fact consist of several different types of elongated cairns. Even the 'simple' long cairns have considerable differences in their outer appearance. This was expected for cairns from different regions and periods, but cairns belonging to the same region and period as dated by the shoreline displacement method turned out to be of varying types.

The cairns investigated in this study are divided into several different types. The basic 'simple' elongated long cairn was termed as Type M; this type was built in all locations and all periods, but the majority of this type of cairn in this study is from Ostrobothnia and the western coast of Finland in general. In Northern Ostrobothnia, all of the cairns belonging to this type are Late Neolithic; in Central and Southern Ostrobothnia and Satakunta, the use of this type extended to the early Bronze Age. A special sub-type of the 'simple' Type 1 M long cairn is the 'curved M', Type 2 M (Figs. 1 and 2).

78   Orientations of Late Neolithic to Bronze Age and Iron Age Long Cairns in Coastal Finland

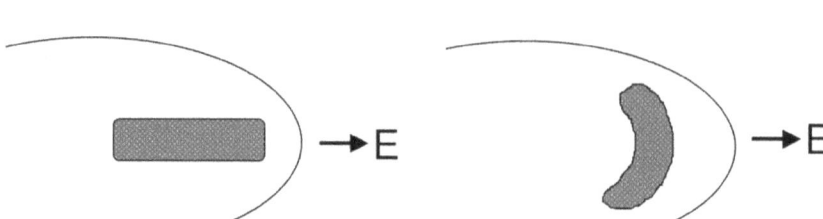

Fig. 2. Observed placements of Type 1 M and 'curved' Type 2 M long cairns relative to the edge of a ridge on which they were built.

It was observed during the study that while the basic M Type 1 cairns had most often been placed along the long axis of a ridge or rocky formation, the 'curved' Type 2 M had been built along the edges of a ridge (usually the eastern edge, see Fig. 2). Both the basic Type 1 M and the curved Type 2 M cairns are occasionally attached to round or torus-shaped cairns usually placed at the ends of the long cairn. This type of construction may indicate that one way of building a long cairn was to combine two round cairns with an elongated setting of stones. Most of the cairns of the curved Type 2 M are located on the Late Neolithic shorelines of Central and Southern Ostrobothnia.

Among the Bronze Age cairns investigated in this study, two special types of large cairns were observed: the rectangular or oval cairns with one or more rectangular central chambers with walls constructed out of stones, and the very large rectangular cairns, the largest of which are almost 50 m long. The smallest of the chambered cairns are c. 10 m and the largest over 20 m in length. All of the cairns belonging to these two types are located on the western coast of Finland, from Central Ostrobothnia to Satakunta.

Two different types of ship-formed cairns, Type 1 Ship and Type 2 Ship, were observed in the sample. The former, Type 1 Ship setting is a rather small, ship-formed setting of stones with carefully laid out edges and usually a triangular stone at the 'front' of the 'ship'. Sometimes, at the back of the 'ship' a large boulder had been placed. These cairns are usually built of mainly red stones. Type 1 Ship settings are mainly from the Middle and Late Bronze Age, but one dated to the Iron Age. The Iron Age ship is remarkably similar to the older Bronze Age Type 1 Ships. All excavated Type 1 Ships have turned out to be graves. They are usually located near the former waterline, along ancient waterways, on rocky beaches and beach cliffs.

The latter type, Type 2 Ship setting, is a previously unrecognised sub-type of the Late Neolithic and early Bronze Age long cairn. They are 10 m to over 20 m long, rather narrow and are a shallow setting of stones located on the very edge, or on the western slope, of a bare rocky hill or outcrop on the ancient shore. They are always positioned perpendicular to the cliff or outcrop edge, with good views towards the ancient sea in the west, thus seeming to be 'sailing' towards the west. They have a triangular red stone marking the front; sometimes a smaller one in the back of the 'ship' and, sometimes, the back has a large boulder similar to Type 1 Ship settings. These ships are always built of mainly red stones. An interesting detail is that, often, another triangular red stone was placed a few metres away from the ship setting; similar triangular stones can be observed near many GCs and these could be recent. All of the Type 2 ship-formed cairns are located on the ancient coastline from Central Ostrobothnia to Satakunta, while the Type 1 ship settings are encountered from Southern Ostrobothnia to Kotka in the SE coast of Finland, c. 115 km eastwards from Helsinki.

The orientation distribution of all 138 cairns (not presented here due to limitations of space) closely resembles a random distribution, which is to be expected since the cairns span several thousands of years and more than 1000 km of coastline. In addition, the orientation distribution of all Type 1 M cairns (not presented here) is close to a random distribution. However, when the declination distributions of the axial directions of the Northern Ostrobothnian and the Central and Southern Ostrobothnian Type 1 M cairns are examined separately, they turn out to be mutually different and non-random (Figs. 3a and 3b).

80 Orientations of Late Neolithic to Bronze Age and Iron Age Long Cairns in Coastal Finland

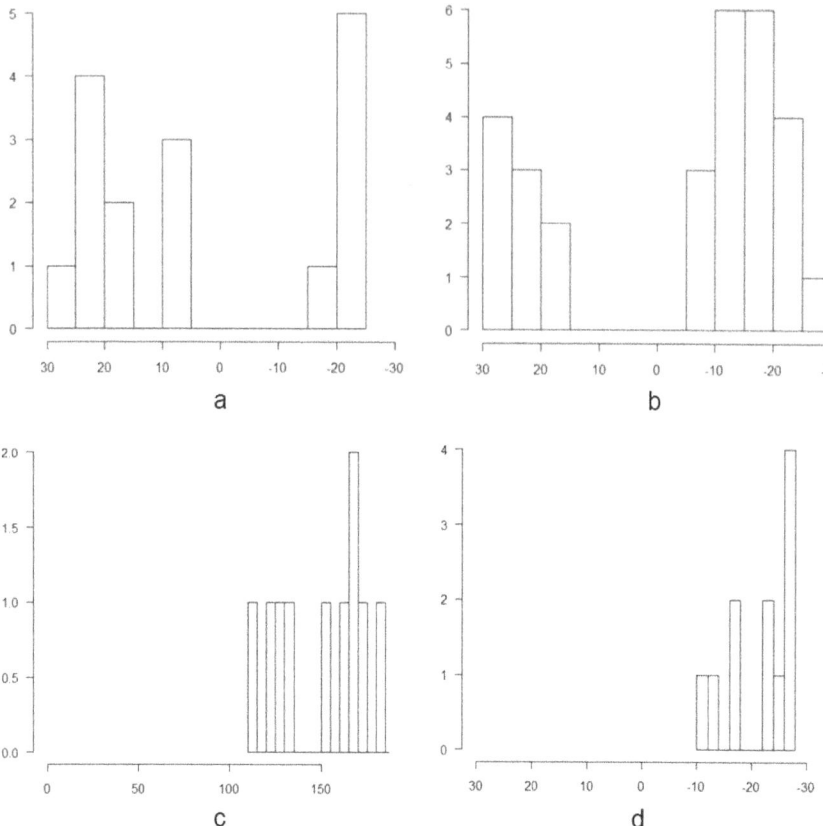

Fig. 3. (a) Astronomical declination distribution towards the eastern horizon for the orientations of the long axes of long cairns of Type 1 M in Northern Ostrobothnia (N=16); (b) declination distribution for Type 1 M in Central and Southern Ostrobothnia (N=29); (c) azimuth and (d) declination distributions towards the eastern horizon for the orientations of Type 2 M 'curved' long cairns (N=11). In the Figures, the horizontal axes are in degree units and the vertical axes are frequencies.

The Northern Ostrobothnian Type 1 M cairn orientations are concentrated around the declinations of +20° and -20°, and +10° (in the eastern horizon; note that due to the axial symmetry of these cairns, the declination distribution towards the western direction is similar with opposite signs for the declination values), and the Central and Southern Ostrobothnian the Type 1 M cairn orientations trend along the NW-SE and

N-S directions, avoiding the +10° orientation (again in the eastern horizon; the orientations towards the west were of opposite sign). The Northern Ostrobothnian orientations are similar to those seen in the GCs and the large house pits in Ostrobothnia.[7]

As seen above, the Late Neolithic cairns of the Type 2 M 'curved' long cairn were probably intended to 'look' towards the centre of the 'curve'. In Figs. 3c and 3d it can be seen that they looked towards the SE and S directions, mostly directly towards the south, suggesting possible interest towards the autumnal and winter sunrises, with individual orientations also to winter solstice.

The azimuth and declination distributions for the axial orientations of the Bronze Age chambered cairns towards the eastern horizon, shown in Figs. 4a and 4b, do not reveal any interesting clustering features but are rather evenly distributed along the horizon; this may be due to the small size of the sample. It is suggested that these types of cairns should be further investigated, not only because of their rather interesting outer appearance but also because some of the GCs (e.g., Tallbackharju, N of Pedersöre) have chambered cairns nearby, or incorporate those as parts of their wall structures. Moreover, this type of cairn is encountered also in the only known early Bronze Age enclosure of Hednatemplet in Jepua, Uusikaarlepyy in Southern Ostrobothnia, and could thus provide clues to the possible cultic continuation of GC-like enclosures into the Bronze Age.

The azimuth and declination distributions for the axial orientations of the very large rectangular Bronze Age cairns towards the eastern horizon, shown in Figs. 4c and 4d, show orientations close to the directions of the solstitial and the equinoctial sunrises (and, correspondingly, the solstitial and equinoctial sunsets in the west, taking the effect of the horizon line into account). Even though the sample size is very small, the strong concentration of the orientations towards these directions tempts one to suggest that these orientations were of some ritual significance.

---

[7] See Marianna P. Ridderstad, 'Orientations and Other Features of the Neolithic 'Giants' Churches' of Finland from On-site and Lidar Observations', *Journal of Astronomical History and Heritage* 18, no. 2 (2015): pp. 135–48; Marianna Ridderstad, 'Orientations and Placement of the Middle and Late Neolithic housepits of Ostrobothnia: A First Investigation Based on On-site and Lidar Observations', *Suomen Museo* 2015 (2017): pp. 5–74.

82 Orientations of Late Neolithic to Bronze Age and Iron Age Long Cairns in Coastal Finland

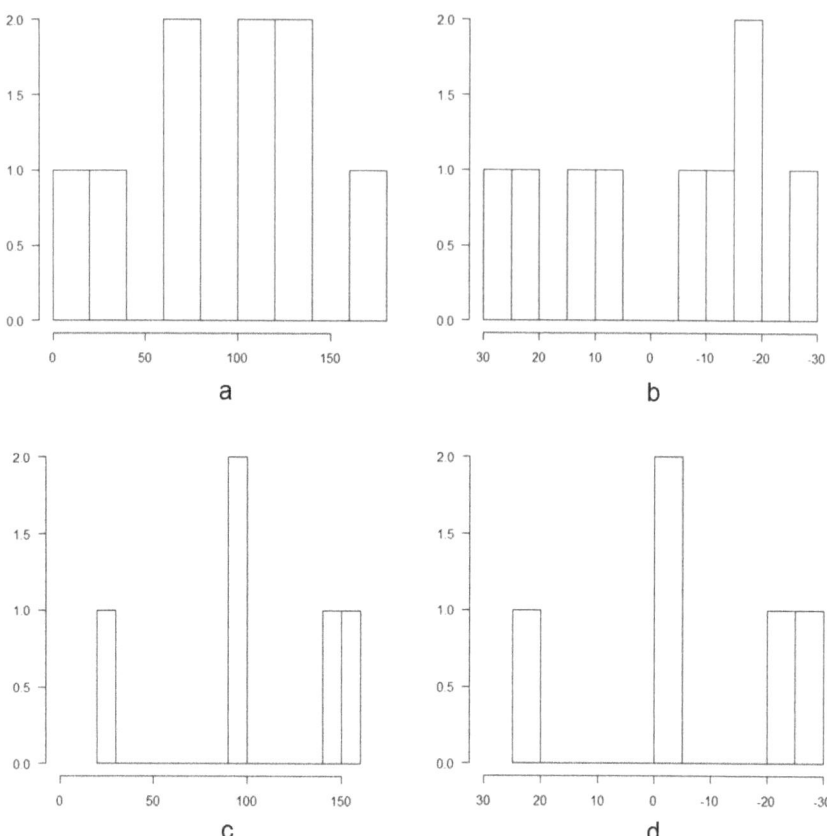

Fig. 4 (a, b) Azimuth and declination distributions towards the eastern horizon for the axial orientations of chambered cairns (N=9); (c, d) azimuth and declination distributions towards the eastern horizon for the axial orientations of large rectangular cairns (N=5). In the Figures, the horizontal axes are in degree units and the vertical axes are frequencies.

Figs. 5c and 5d show the azimuth and declination distributions for the orientations of the Bronze and Iron Age ship settings (Type 1 Ships), and the Late Neolithic and early Bronze Age ship-formed long cairns (Type 2 Ships). These cairns are not symmetrical in shape and, therefore, their orientation was taken to be towards the front of the 'ship'. The orientations and declinations of the former group (shown in Figs. 5a and 5b) are mainly along, or close to, the cardinal directions, with two towards the declinations of +10° and -10°, while the orientations and declinations of the latter group (shown in Figs. 5c and 5d) are mainly indicative of declinations +20° and -20° and +10°. The orientations of the Type 2 Ships thus resemble the orientations of the Northern Ostrobothnian Type 1 M cairns, as well as the orientations of the GCs and the large Neolithic house pits.

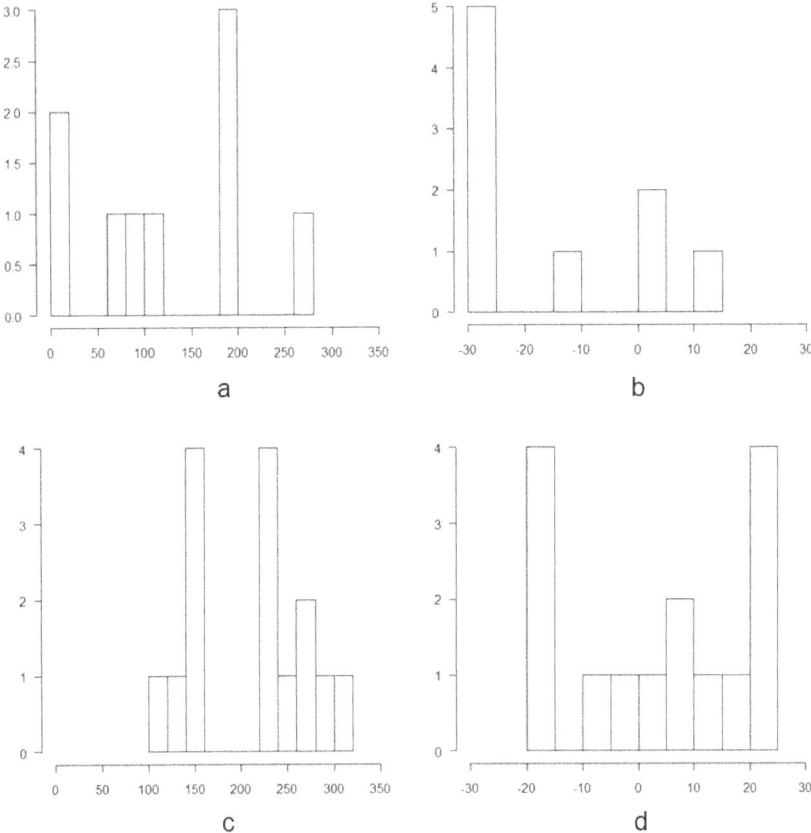

Fig. 5 (a, b) Azimuth and declination distributions of Type 1 Ship settings (N=9); (c, d) azimuth and declination distributions of Type 2 ship-formed cairns (N=15). In each case, the orientation was taken to be towards the front of the 'ship'. In the Figures, the horizontal axes are in degree units and the vertical axes are frequencies.

Discussion

Based on their orientations, it seems clear that the Type 1 M Late Neolithic long cairns of Northern Ostrobothnia were an expression of some kind of cultural continuity from the GC culture to the long cairn-building culture in the region. In Central and Southern Ostrobothnia, on the other hand, the orientations of the Type 1 M long cairns seem only partly similar. However, the cairns of the 'curved' Type 2 M cairns, most of which are located in Central and Southern Ostrobothnia, are looking towards the same segment of the horizon as the GCs, which could suggest cultural continuation. Further investigation could also reveal local variations and partial cultural continuation in the orientations of the Type 1 M cairns in the region.

Based on the results of this study, it is not possible to deduce with certainty to what extent there was a cultural continuation between the Late Neolithic and Bronze Age cultures on the western coast of Finland. As orientations to the main solar events, the solstices and equinoxes, are an idea universally encountered in many cultures, the orientations of the very large rectangular Bronze Age cairns to, or closely towards, the solar astronomical declinations of the solstices and equinoxes could be the result of cultural continuity from the Neolithic as well as a new ideology. The orientations of the Type 2 ship-formed cairns, however, may indicate the existence of some cultural continuation from the Neolithic to the Early or even Middle Bronze Age. On the other hand, the appearance of the new type of ship settings, Type 1 in the Bronze Age, and its different orientations relative to Type 2 suggest that there was little or no continuation from the Late Neolithic to the Middle and Late Bronze Age.

If the ship-formed cairns of Type 2 really are as early as suggested by the shore displacement dating, this could have a range of interpretations. While the orientations of these ship settings followed the orientations of Neolithic monuments in Northern Ostrobothnia, their ship-like appearance seems to be an innovation possibly reflecting new external ideas and influences. These new cultural influences could belong to the earliest echoes of the Nordic Bronze Age arriving from Sweden, or the development could have been indigenous and connected to the many representations of boats in the rock paintings and carvings in Finland, Lapland and Karelia during the Bronze Age.[8] Alternatively, the Bronze

---

[8] See, for example, Antti Lahelma, *A Touch of Red: Archaeological and Ethnographic Approaches to Interpreting Finnish Rock Paintings*, Iskos 15 (PhD

Age ideology, where the ship was strongly connected to solar symbolism, the afterlife and the prestige of earthly power, could have arrived in Finland both from the west and the east, as so many other cultural features have during the millennia. Naturally, this new ideology fused with the central importance of boats already existent in the local cultures that had their roots in the Neolithic and Mesolithic hunter-gatherer way of life.

The idea that a burial cairn could and should be ship-formed can be connected to the role of ships and boats as psychopomps, i.e., the astral or solar ship which carried the dead to the otherworld. This is connected to the belief that at sunset, especially during the extreme horizontal movements of the sun on the solstices and other possible solar keys days, the border between the world of the living and the netherworld of the dead could be transcended. The Iron Age Finns, for example, seem to have believed that at the faraway sea horizon, where the sky and the sea met and the sun moved around, one could physically crawl under the sky vault and enter the other realms. Thus, a ship setting that was built on the shoreline with its front oriented to the west was likely meant to 'sail' towards the western sea horizon and the realm of the dead, carrying the soul of its owner(s) with it.

Conclusions

In this study, the orientations of 138 long cairns located in coastal Finland were measured and examined along with other properties of the cairns. The length of the cairns varied from c. 3 m to 48 m. The dominant color of the stones in most of the cairns was red, and they were usually built on locally elevated terrain, e.g., on ridges, rocky outcrops or small islets on the ancient shore. It was found that in the category of long cairns there were several different types of elongated cairns: the 'simple' and curved long cairns, some of which were attached to round cairns; the rectangular cairns with one or more central chambers; the very large rectangular cairns; and two different types of ship-formed cairns, Type 1 and Type 2, the latter of which was a previously unrecognised type of the Late Neolithic and early Bronze Age long cairn. The comparison of the orientations of the cairns of different types, and their locations, suggest that there was some cultural continuity between the Neolithic and the early Bronze Age cultures located on the western coast of Finland. However, based on the present analysis here, this continuity does not seem to have extended beyond the Middle

---

Thesis, University of Helsinki, 2008), pp. 25–26; Eero Ojanen, *Karjalan kalliopiirrokset* (Keuruu: Otava, 1981).

Bronze Age. It is also suggested that the appearance of the Type 2 ship-formed cairn in the Ostrobothnia region in the Late Neolithic may have resulted from outside cultural influences, perhaps from the earliest contacts with the central ideologies of the Nordic Bronze Age.

Acknowledgements
I wish to thank the editor and the anonymous referees for many useful suggestions.

Megalithic Skyscapes in Galicia

A. César González García, Felipe Criado-Boado and Benito Vilas Estévez

Abstract: We present the results of our analysis of two singular Neolithic monuments and two prominent megalithic groups in Galicia. The two singular monuments are the dolmen of Dombate (Baio, Coruña county), perhaps the largest megalithic chamber in Galicia (or at least the most investigated and well-known) that houses an elaborate decorative program with engravings and paintings, and Forno dos Mouros (Bocelo mountains, Coruña county), also housing paintings and belonging to a bigger group aligned along an historical path following the mountain ridge. Both chambers house interesting illumination effects. The group analysis concerns the Barbanza (Coruña county) and Leboreiro, (Ourense county and borderland with Portugal) necropoleis. There, we find that apart from chamber orientation, location and spatial relations of the monuments within the landscape, the monuments incorporate skyscape associations that complemented and dialogued with that of the chamber orientations. Besides, if the particular directions that we find are related to the movements of the sun and/or moon they may indicate the appropriate ritual time for the dead. Of course, skyscape is not the only or the main factor to explain the location of the mounds within the necropolis but are part of a complex system of relations making those monuments part of a cultural landscape. When taking all factors into consideration a complex picture emerges where we can envisage the ways of construction of social time and space in the megalithic period.

Megalithic Astronomy has for a long time focused on analyzing the null hypothesis – do the megalithic monuments of a given area share a similar orientation pattern?[1] This approach was needed to break the resistance existing among the archaeological community and it showed that the orientation of megalithic monuments in particular regions and belonging to a similar culture tend to be coherent. In several instances, the simplest

---

[1] See for example, S. Iwaniszewski, 'Por una astronomía cutural renovada', *Complutum* 20, no. 2 (2009): pp. 23–38.

A. César González García, Felipe Criado-Boado, and Benito Vilas, 'Megalithic Skyscapes in Galicia', *The Marriage of Astronomy and Culture,* a special issue of *Culture and Cosmos,* Vol. 21, nos. 1 and 2, 2017, pp. 87–103.
www.CultureAndCosmos.org

explanation for those trends is the orientation towards a heavenly body, perhaps in connection with particular topographic features.[2]

However, these data-driven analyses are restricted in nature, as they do not ask about the intent of megalith builders in using such orientations. In order to answer this question, we must contextualize those results within a number of other cultural and social elements of the megalithic builders.[3]

A number of previous works, both from the field of landscape archaeology and from archaeoastronomy, have highlighted the different relations of the megaliths with either moving strategies and/or visibility[4] or the search for regions where prominent topographic features are spotted and interesting astronomical alignments occur.[5] More recent works try to complement these early works to examine the connections between the landscape and the sky.[6]

---

[2] See for instance the now classic works by C. Ruggles, *Astronomy in Prehistoric Britain and Ireland* (New Haven, CT: Yale University Press, 1999) and M. Hoskin, *Tombs, Temples and their Orientations* (Bognor Regis: Ocarina Books, 2001).

[3] For a recent critique on these themes see, for example, Fabio Silva, 'The Role and Importance of the Sky in Archaeology: An Introduction' in F. Silva and N. Campion, eds., *Skyscapes* (Oxford: Oxbow Books, 2015).

[4] C. Tilley, *A Phenomenology of Landscape: Place, Paths and Monuments* (Oxford/Providence: Berg, 1994); M.O. Baldia, 'A Spatial Analysis of Megalithic Tombs', (Ph.D dissertation Southern Methodist University, Dallas, 1995); F. Criado-Boado and V. Villoch Vázquez, 'Monumentalizing Landscape: From Present Perceptions to the Original Meanings of Galician Megalithism (NW Iberian peninsula)', *Trabajos de Prehistoria* 55, no. 1 (1998): pp. 63–80; Q. Bourgeois, 'Monuments on the Horizon: The Formation of the Barrow Landscape through the 3rd and 2nd Millenium BC' (Leiden: Sidestone Press, 2013); J. A. Lozano, G. Ruiz-Puertas, M. Hódar-Correa, F. Pérez-Varela, A. Morgado, 'Prehistoric Engineering and Astronomy of the Great Menga Dolmen (Málaga, Spain). A Geometric and Geoarchaeological Analysis', *Journal of Archaeological Science* 41 (2014): pp. 759–71.

[5] C.L.N. Ruggles and R.D. Martlew, 'The North Mull Project (3): Prominent Hill Summits and their Astronomical Potential', *Archaeoastronomy* 17 (JHA, xxiii, 1992): pp. S1–S13; Ruggles, *Astronomy*, pp. 112–24; J.A. Belmonte and M. Hoskin, *Reflejo del Cosmos* (Madrid: Equipo Sirius, 2002), pp. 69–72.

[6] See F. Prendergast, 'Interpreting Megalithic Tomb Orientation and Siting within Broader Cultural Contexts', *Journal of Physics: Conference Series* 685 (2016) 012004; Gail Higgingbottom and Roger Clay, 'Origins of Standing Stone Astronomy in Britain: New Quantitative Techniques for the Study of Archaeoastronomy', *Journal of Archaeological Science: Reports* 9 (2016): pp. 249–58 and references therein.

Our methodology in the last years has been to systematically analyze the visual features of megalithic mounds together with the scenery effects and features including the skyscape.[7] To do so we try to embed the possible astronomical relations within other spatial analysis including not only the orientation of the chambers but the visibility of one monument from the others and the orientation towards those defining the skyline horizon or towards prominent topographic features. This also includes the illumination effects at particular times of the year occurring both inside and outside of the chambers, perhaps indicating a moment for the dead.

Galicia houses a large number of megalithic monuments, being one of the hot spots for megalithism in the Iberian Peninsula. The most frequent megalithic monument is the dolmen or passage grave.[8] Very similar to the Portuguese seven-stone antas, most of them are composed by a hepta-orthostatic chamber with a corridor that leads to the chamber. In general, a tumulus ('mamoa' in Galician) supports the megalithic structure, without covering it completely. Although weathering and erosion must have decreased the height somehow archaeologists are confident that in most investigated cases such erosion was not severe and hence the megalithic structure was visible above the mound. Therefore, although this erosion might affect the visibility of these mounds, in the cases investigated here, such factor is not crucial. Finally, the outer part of the mound display a cover of stones, usually quartz. This cover is usually more prominent next to the entrance to the corridor.

Dating of these monuments varies, although the range of dates indicates a period of use of the megalithic chambers in Galicia starting in the late fifth millennium BC until the end of the fourth millennium BC, in the regional Middle and Late Neolithic.[9]

---

[7] F. Silva 'The Role ...' indicates that such term tries to comprehend the different skies 'seen' by different cultures under the same sky. However, here we do not know or we do not have direct access to the meaning of the sky for the megalithic builders. Instead by skyscape we refer to the conbination of landscape and sky, understanding by such the social construction behind such terms.

[8] R. Garrido Peña, M.A. Rojo Guerra, C. Tejedor Rodríguez, I. García Martínez de Lagrán, 'Las máscaras de la muerte: ritos funerarios en el Neolítico de la Península Ibérica', in M.A. Rojo Guerra, R. Garrido Peña, I. García Martínez de Lagrán, eds., *El Neolítico en la Península Ibérica y su Contexto Europeo* (Madrid: Cátedra, 2012), pp. 143–71.

[9] M.P. Prieto Martínez, P. Mañana Borrazás, M. Costa Casais, *et al.*, 'Galicia', in, Rojo Guerra, Garrido Peña, García Martínez de Lagrán, *El Neolítico en la Península Ibérica y su Contexto Europeo*, pp. 216–53.

90 Megalithic Skyscapes in Galicia

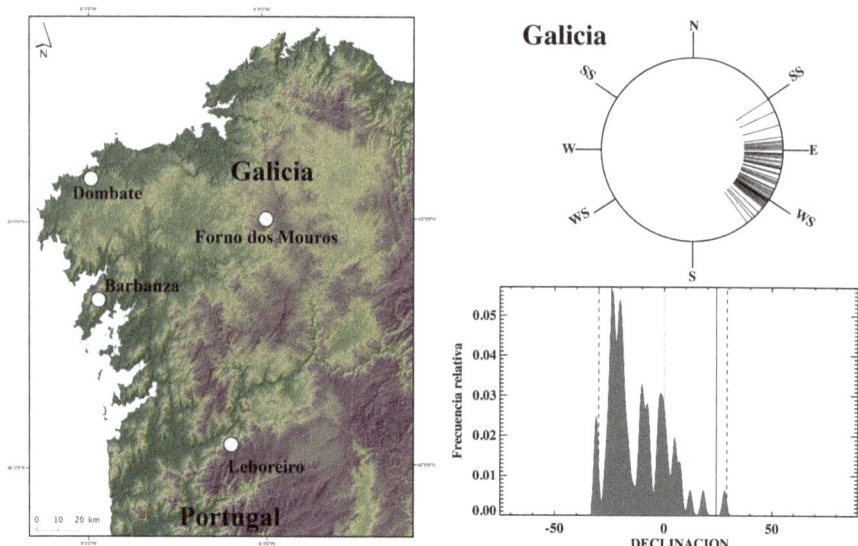

Fig. 1. Left, map with the different locations mentioned in the text. Right, orientation diagram (top) and declination histogram of the 62 megalithic chambers measured so far in Galicia. For details, see text.

Michael Hoskin and his collaborators measured the orientation of up to 32 passage graves in Galicia.[10] In the recent years the members of our team have expanded such number, and we have verified and corrected a number of them, which will be published elsewhere.[11] Fig. 1 displays the orientation diagram and declination histogram for the 64 dolmens measured so far in Galicia.

All dolmens display orientations towards the eastern half of the horizon and, once the altitude of the horizon is considered, all measured dolmens so far are within the luni-solar limits, with several clear concentrations. We would like to highlight the most frequent towards declination -24° with a secondary maximum close to -20° and which could be connected to winter solstice sunrise, and the next two, towards -3° that we would name an equinoctial orientation and at a declination of approximately -12°. We will come back to these numbers later.

---

[10] M. Hoskin et al., 'Studies in Iberian Arcaheoastronomy (5) Orientations of Megalithic Tombs of Northern and Western Iberia', *Journal for the History of Astronomy*, Archaeoastronomy Supplement, 29 (1998): p. S39–S88.

[11] A.C. González-García, B. Vilas-Estévez, F. Criado Boado, 'Skyscape of Galician Megalithism: The Celestial Dimensión of the Landscape', in preparation.

Culture and Cosmos

Figure 1 also indicates that the probability to find an orientation near the winter solstice is six times higher that any other, indicating that such pattern is not random and therefore that intentionality is highly probable.

In the following sections we will analyze four cases, starting with singular chambers where the orientation follows the most common group of orientations for Galicia, i.e., they are oriented towards winter solstice sunrise, and where interesting illumination events occur. Then we will complement our approach by studying two necropolises to investigate if the location of the monuments within the landscape is also relevant from the point of view of archaeoastronomy.

1. Methodology

The methodology employed in the four cases is necessarily different but tries to go beyond the mere statistical treatment used until now. In all the four cases we explored how the location and orientation of chambers and mounds correlates with sunrise/sunset and conversely with moonrise/moonset. To do so we use the inspection of each site's landscape in search for the orientation of the megalithic chamber (when this was measurable) and the possible relation with other monuments (mounds) defining the local skyline. In such cases we will also measure the direction and horizon altitude of such mounds that define the skyline.

We employed two Suunto professional compass/clinometer tandems to perform these measurements. The nominal accuracy of each measurement in azimuth with this instrument is of ¼°, while the accuracy in the vertical is of ½°. Azimuth readings are magnetic ones and were corrected either using triangulation of known topographic features or employing a magnetic declination model.[12] The actual errors in our measurements take into account also the variations introduced by the state of preservation, and the width of the window of visibility, when required, and the amplitude of horizon occupied by the mounds.

When curvigrams are used, we employ a Gaussian kernel for each individual measurement, with a bin-width of twice the estimated error of our measurement.[13]

---

[12] http://www.ign.es/web/ign/portal/gmt-declinacion-magnetica
[13] For a detailed description of the procedure see, e.g., A.C. González-García, I. Sprajc, 'Astronomical Significance of Architectural Orientations in the Maya Lowlands: A Statistical Approach', *Journal of Archaeological Science: Reports* 9 (2016): pp. 191–202.

The hypothesis to test is if, apart from the orientation being possibly related with some astronomical phenomena, the location of these monuments in a particular spot in the landscape might also have something to do with how astronomical events are viewed from there.

We will finally interpret our results following the scheme proposed by Criado-Boado.[14] This scheme proposes to go from a specific finding into a generalization, following several steps. Thus, the first case studied will be Dombate, arguably the best known megalithic monument in Galicia and one that has been excavated and therefore offers a wide range of possibilities to contrast our potential findings with the material culture. Dombate, especially the small chamber (see below) has offered a very early dating (c. 3900 cal BC) followed by, a few centuries later, the construction of the large monument (c. 3400 cal BC) setting the period of interest for our study in the middle Neolithic in the area in the fourth millennium BC.[15] A second case studied is Forno dos Mouros, a monument similar to Dombate, where we may find a parallel to contrast our findings in a different monument more than 50 km away from the first one and where the actual landscape is possibly closer to the Neolithic one than in Dombate that has been largely altered, especially in the recent past due to its musealization. In this way, we may complement the findings in Dombate by including the relation to the environment. Finally, the Barbanza necropolis may allow us to expand to what level our finding at the micro scale of the mound could be generalized to the macro scale of the necropolis. The Leboreiro necropolis will allow us to see if any finds in Barbanza could be generalized, and how, to other necropoleis in Galicia.

2. Dombate

Dombate (Cabana de Bergantiños, A Coruña) is one of the most noteworthy Neolithic monuments in the Iberian Peninsula. Not only is it one of the most outstanding examples of Iberian passage graves but also one that illustrates the main structural, socioeconomic, territorial and symbolic dimensions of the Neolithic and Early Bronze Age megalithic phenomenon across the European Atlantic façade.[16]

---

[14] F. Criado Boado *Arqueológicas: La razón Perdida* (Barcelona: Bellaterra. 2012).
[15] Rojo Guerra, Garrido Peña, García Martínez de Lagrán, *El Neolítico en la Península Ibérica y su Contexto Europeo*, p. 586.
[16] F. Cebrián del Moral & J. Yáñez Rodríguez, eds., *El dolmen de Dombate: arqueología, arquitectura y conservación* (A Coruña, Diputación de A Coruña: 2011).

A classical mound as those described above covers the large megalithic chamber. But the mound not only partially covered this monument but also an earlier one, making it invisible. Later research has confirmed that this phenomenon of *integration* was a quite recurrent trend of the funerary monument building tradition in this[17] and in other European regions.[18]

All orthostats in the chamber and corridor of the passage grave are decorated with paintings – black and red linear motifs and dots over a white background – and engravings. These decorations seem to be contemporaneous with the date of construction of the monument (first half of the fourth millenium BC), as the preparation layers for the paintings and some ^{14}C contexts suggest.[19]

Recent excavations have completed our vision of the monument's history and meaning. A most striking result of this recent work has been the identification of a series of idols placed in sockets especially made to locate them standing up at the entrance of the inclined passageway leading into the megalithic corridor, indicating the special nature of this area within the whole monument. A closing stone that presumably was open when needed blocked the entrance to the corridor.

Finally, we must note the identification of a series of hearths and ditches in close proximity to the monument.[20] While *ditch 2* and *ditch 3* might be contemporary to the Neolithic occupations of Dombate, *ditch 1* (ca. 1 km of estimated perimeter) seems to be later in date. The northeastern limit of this ditch is located just 4 m south to the monument, and probably delimits a Bronze Age settlement area.

---

[17] Patricia Mañana-Borrazás, 'Túmulo 5 de Forno dos Mouros (Ortigueira, A Coruña). Primeiros resultados', *Cuadernos de Estudios Gallegos* 52, no. 118 (2005): pp. 39–79.

[18] R. Bradley, R. Williams, and H. Williams, eds., *The Past in the Past: The Reuse of Ancient Monuments*, special issue of *World Archaeology* 30, no. 1 (1998).

[19] F. Carrera Ramírez, 'El arte prehistórico y su conservación. Pinturas y grabados en Dombate', in Cebrian del Moral and Yáñez Rodríguez, *El dolmen de Dombate*, pp. 229–66.

[20] M. Lestón Gómez, 'Las excavaciones arqueológicas' in Cebrián del Moral and Yáñez Rodríguez, *El dolmen de Dombate*, pp. 139–266.

94 Megalithic Skyscapes in Galicia

Fig. 2. The dolmen of Dombate. Top: notice the prominence of the stone crust next to the entrance, the anthropomorphic idols next to this area and the closing stone of the corridor. Bottom, reproduction of the illumination event on Winter Solstice sunrise as simulated with a 3D model of Dombate and the Stellarium software. Note that the illuminated part never reaches further up than the red and white paintings.

Dombate presents an orientation of 126° and a horizon altitude of 2½° that translates into a declination of -23.8°, compatible with the measurements given by Hoskin.[21] Thus Dombate is oriented towards winter solstice sunrise at the time of construction. The result of this orientation is that the inner parts of the chamber are only lit by direct sunlight during the few days the sun rises close to its winter solstice position.

Given the orientation, the width of the entrance allows the direct illumination of the backstone and its paintings for only a small number of days around WS (Fig. 2). This effect is now precluded by the position of the closing stone at the entrance of the corridor. To verify the effect we built a 3D model of the megalith and uploaded it to the Stellarium planetarium software. A reconstruction of the effect is given in Fig. 2. It is interesting to note that, given the shape and structure of the corridor, the illuminated part never reaches beyond the painted red geometrical pattern

---

[21] Hoskin, 'Tombs', p. 235.

of the back stone. After a first flash of light projecting from the entrance frame onto the red and white paintings, the trapezoidal frame shifts downwards to the right of the stone as the sun gains altitude above the horizon, following which the backstone returns to shadows after a while.

Fig. 3. Illumination of the Forno dos Mouros dolmen at the winter solstice sunrise. Note the projection of the shadow on the large boulder (Pena Moura) right behind the megalith.

3. Forno dos Mouros

A similar phenomenon can be observed at Forno dos Mouros (Toques, A Coruña). This site is located at 713 metres above sea level in the northern area of the Bocelo mountain range, at a small plateau next to the Pena Moura hill. Located next to a traditional route (part of the ancient Road to Santiago) it has a wide visibility with the distant horizon towards S and SE. Forno is part of a small necropolis of another four tumuli and a number of rocky outcrops, where the one known as Pena Moura is quite prominent and is located a mere 50 m behind Forno.

The chamber is of a very similar type, but slightly smaller in size, than that of Dombate. It is again located inside a tumulus that has been recently excavated and restored.[22] The mound, with 20 m of diameter and 1.75 m of

---

[22] Roberto Aboal and Yolanda Porto, eds., *Intervencións de conservación e*

height, although a bit eroded, mainly by the formation of the traditional route to its north side, would not have been significantly much larger than it is today. The chamber again houses a profuse decorative program with paintings in zigzag in red, white and black.

On winter solstice sunrise 2015 we went to Forno to witness the sunrise together with a Spanish national TV team. To our amazement we could verify the impressive light and shadow effects inside the chamber, although in this case the restoration works covered the chamber paintings with soil, and we were not able to verify if a Dombate-like effect happened with respect to the paintings. In this case, we witnessed a different effect: the projection of the shadow of the mound and especially the chamber over the large boulder of Pena Moura (see Fig. 3). It must be stressed that this short period of time around the solstice is the only time of the year when the megalith will overshadow this boulder. Pena Moura is a massive granite (ortogneis) formation of nearly five meters high and as much in diameter, weighing nearly 180 tons, so it seems highly probable that it was there when the monument was built although no archaeological or geological study has yet investigated this question.

4. Barbanza

The Barbanza peninsula is located at the Atlantic coast between two of the 'Rías', the sea arms that enter land following the river valleys and forming a characteristic part of the Galician seascape. An abrupt mountain range also named Barbanza crosses the peninsula. The sierra, with a highest altitude of 680 metres above sea level, has a flat plateau near the top at around 550 metres above sea level of nearly 5 km^2 with a characteristic vegetation, quite similar to that present 5000 years ago when the inhabitants of the peninsula built in this area over 30 mounds.[23]

Based on the typology of the mounds, it has been argued that there were two moments of construction within the necropolis and it has been proposed that they were deliberately positioned in order to indicate the best route to cross the sierra.[24] Finally, it should be noted that the location of the

---

recuperación no xacemento de Forno dos Mouros (Toques, A Coruña) (Santiago: CSIC, 2012).
[23] For a recent review of the ideas schetched here see F. Criado-Boado, P. Mañana-Borrazás, C. Gianotti, 'A paisaxe monumental (4500–2500 a. C.)', in F. Criado Boado, C. Parcero Oubiña, C. Otero Vilariño, E. Cabrejas, eds., *Atlas Arqueolóxico da Paisaxe Galega* (Vigo: Xerais, 2016), pp. 99–144.
[24] F. Criado-Boado and V. Villoch-Vázquez, 'La Monumentalización del Paisaje: Percepción y Sentido Original en el megalitismo de la Sierra de Barbanza

mounds is such that the horizon is closer and higher towards West and North while it is lower and far towards East and South.[25]
It is interesting to note that all the chambers that could be measured within the sierra are of the winter solstice sunrise type.[26] It is also noteworthy that a variable number of mamoas is always visible when one is located next to one of them, with only a few actually defining the skyline.

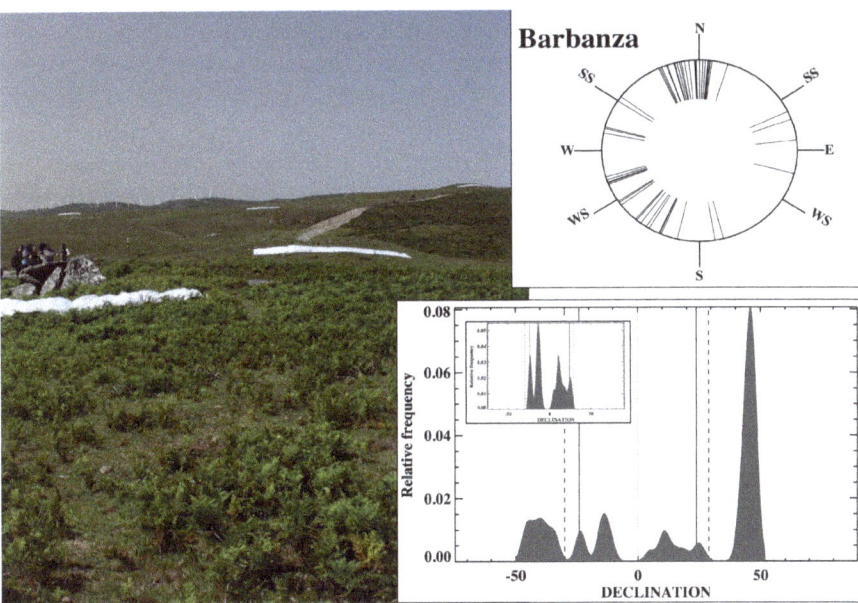

Fig. 4. Left, section of the Barbanza plateau with a number of mounds highlighted with white linen. Note the presence of one mound that defines the skyline as seen from this location. Right, orientation diagram and histogram of the prominent tumuli in Barbanza. For details, see text.

Thus, we visited all mamoas in the Sierra and inspected the visibility to all the rest, measuring the orientation towards those defining the skyline, as

---

(Galicia)', *Trabajos de Prehistoria* 55, no. 1 (1998): pp. 63–80; It is interesting that at the level of the whole peninsula the mounds and chambers are located in close relationship with the easiest routes to cross the peninsula, Marcos Llobera, 'Working the Digital: Some Thoughts from Landscape Archaeology', in R. Chapman and A. Wylie, eds., *Material Evidence: Learning from Archaeological Practice* (London: Routledge, 2015), pp. 173–88.
[25] Criado-Boado and Villoch-Vázquez, 'La Monumentalización del Paisaje'.
[26] González-García, Vilas Estévez, Criado Boado, in preparation.

explained above. To perform such measurements we located ourselves both at the top, bottom and sides of each mound. The readings vary, especially for those cases when the distance to the skyline mound is short, and then the error considered is accordingly larger. We have estimated a conservative error of 1.5° in azimuth. We do not give here a full account of the data, since it is described elsewhere,[27] but the main results are indicated in Figure 4.

We can see that most of the orientations indicate an area towards north, probably following the orientation of the sierra. However, a number of mounds, especially those located at junctions between different paths are located such that on the western part they define relevant astronomical positions, in particular close to declination -24°, winter solstice sunset, and at declination -12°, the same astronomical phenomena we had associated with the orientation of the chambers but now towards the setting part of the horizon.

5. Leboreiro

We wanted to see if the findings that emerged at Barbanza applied elsewhere in Galicia, so we began a similar research project at the Leboreiro necropolis. Leboreiro is located at an altitude of nearly 1200 metres above sea level and at the frontier between Spain and Portugal houses well over a hundred mounds in less than 20 km^2 being one of the densest concentrations of mounds in Iberia.

Given the sheer proportions, the strategy here has been different to that followed in Barbanza so far: instead of assessing the intervisibility between each mound we selected the most prominent, the mound called Mota Grande, and we checked whether it was intervisible with other monuments in the area. We checked for intervisibility by using LIDAR data with a density of 0.5 points/m and an altimetric precission of 20 cm. These data are freely available from the Spanish National Geographic Institute.[28] We used GIS software to work these data, namely QGIS.[29] Finally we verified that map work with a subsequent field survey.

Mota Grande is intervisible with 42 out of the 120 mounds identified within its vicinity. Thirty-eight of those monuments have Mota Grande on their skyline. This is a full 90% of the smaller intervisible sample. We again measured the orientation and altitude of the horizon of Mota Grande

---

[27] González-García, Vilas Estévez, Criado Boado, in preparation.
[28] http://pnoa.ign.es/
[29] http://www.qgis.org/

as seen from each of those thirty-eight mounds. The orientation is in most cases towards the western part of the horizon (Fig. 5).

It should be noted that in most cases, the monuments surrounding Mota Grande appear to use it as a horizon feature in order to align to either summer solstice sunset (or the major northern lunistice) or a setting at a declination between 10° to 12°. In other words, it is almost a mirror image of the most frequent orientations for the chambers in Galicia.

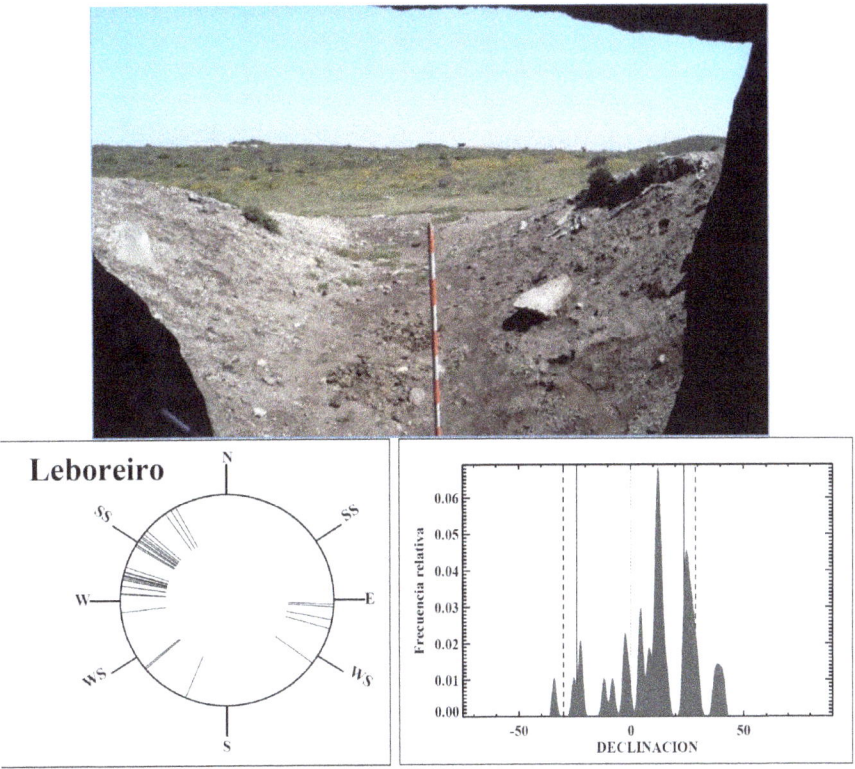

Fig. 5. Top, at the Leboreiro necropolis we measured the orientations towards the Mota Grande mound, when this is defining the skyline. Bottom, orientation diagram and declination histogram. For details, see text.

5. Discussion

The four cases described above present different ways on how the sky might have been incorporated in the building of the megalithic chambers and mounds apart from just their orientation. These include the use of light and shadow effects inside and outside the chambers and how the rise and/or set of heavenly bodies might be seen with respect to other contemporary monuments in the vicinity.

In the cases of Dombate and Forno, it is clear that, if the orientations were connected to winter solstice sunrise, the inner parts of the chamber would be illuminated at that time of the year. It is difficult to know if the construction of the monuments with such orientation was meant to house such illumination events, as it seems clear in other cases within megalithic chambers.[30] However, we argue that the location of paintings and engravings is such that the light and shadow event seems part of the decorative program.

Further, previously existing features of the local landscape are integrated either by incorporating previous mounds within new ones or by including physiographic features such as prominent rocks where shafts of light or shadow might highlight the importance of the moment, in this case winter solstice. Furthermore, such integration of previous elements is done at the level of the whole necropolis as it is indicated by the results in Barbanza and Leboreiro. Whether this indicates that a new monument is located taking into account where the previous ones were or if larger schemes are at play is out of the scope of the present paper.

We do not pretend that the astronomical events are the only factor explaining the location of mounds within the necropolis. Mounds are located such that they follow the easiest route to cross the peninsula,[31] and the horizons seen from them put at the large (macro) level something observed at smaller (micro) stages.

Their location along this route has an effect on the horizons one may find at the site of the mounds.[32] There are clear differences between the horizons towards the east and south, open and distant, in contrast to those towards the north and west (that are closer and restricted). This dichotomy in spatial relation is also observed at the size of the mound, where the area close to the entrance is more elaborated, including a higher prominence of the stone cover and the placement of anthropomorphic idols next to the corridor entrance in Dombate. It is also observed within the megalithic

---

[30] R. Bradley, 'Darkness and Light in the Design of Megalithic Tombs', *Oxford Journal of Archaeology* 8, no. 3 (1989): pp. 251–59; D. Trevarthen, 'Illuminating the Monuments: Observation and Speculation on the Structure and Function of the Cairns at Balnuaran of Clava', *Cambridge Archaeological Journal* 10, no. 2 (2000): pp. 295–315.
[31] Llobera 'Working the Digital'.
[32] Criado-Boado and Villoch-Vázquez, 'La Monumentalización del Paisaje'.

structure itself, where there are differences in the location of the orthostats and the quality of them.[33]

All these points seem to indicate a particular way to integrate the megalithic monuments within the landscape. Finally, when the orientations are considered and a connection with risings and settings of celestial bodies is done, they may also indicate a temporal fingerprint, a moment of importance for the megalith builders and users.

The importance of winter solstice in the region is indicated by the prominence of these orientations of the chambers and the directions indicated by the alignments. However, other concentrations appear towards equinox, taken in a broad sense,[34] as well as that elusive -12° of declination.

Concerning winter solstice, it is interesting to point that the orientation of the funerary monuments facing sunrise at the shortest days of the year is clearly an evocative image possibly symbolising rebirth in an area devoted to the dead. It is even more suggestive if we consider that possibly the chambers could have been open at those moments to perform whatever rituals and then the illumination of the inner parts of the chamber and the paintings could highlight that moment, as a time devoted to the dead or the ancestors.[35] The projection of shadows towards large boulders would also have highlighted such moments.

In this sense, this could be linked with the division of space at the macroscopic scale, where the illumination of boulders or the mound-to-mound alignments with astronomical events may occur, connecting it with

---

[33] F. Criado-Boado, El espacio en el Túmulo', in *La Arqueología en la gasificación de Galicia 3: Excavación del Túmulo n° 3 del Alto de San Cosme*, C. Parcero Oubiña, ed., *Trabajos en Arqueología del Paisaje* 5 (1997): pp. 17–19; F. Criado-Boado, C. Gianotti García, P. Mañana-Borrazás, 'Before the Barrows: Forms of Monumentality and Forms of Complexity in Iberia and Uruguay', in L. Smejda, ed., *Archaeology of Burial Mounds* (Plzen: University of West Bohemia, 2006). Something similar has recently been proposed for the Dolmen of Menga: Lozano *et al.*, 'Prehistoric Engineering and Astronomy of the Great Menga Dolmen (Málaga, Spain)', see note 4.

[34] C.L.N. Ruggles, 'Whose Equinox?', *Journal for the History of Astronomy, Archaeoastronomy Supplement* 28 (1997): pp. 45–50.

[35] L. Goodison, 'From Tholos Tomb to Throne Room: Perceptions of the Sun in Minoan Ritual', in R. Laffineur and R. Högg, eds., *POTNIA. Deities and Religion in the Aegean Bronze Age. The 8th international Aegean Conference* (Austin: University of Texas at Austin, 2001), pp. 77–88; Bradley, 'Darkness and Light in the Design of Megalithic Tombs'.

the meso and microscopic scale, where the orientation of the chambers and illumination inside the chambers happen.

In Dombate, but also at the Barbanza necropolis,[36] the areas connected with settlements are located towards the southeast of the monument. This sector of the landscape linked with the living is also connected with the distant and lower horizons, so the area connected with the sunrise, in particular winter solstice sunrise is the area connected with the living, while the opposite direction, would be connected with restricted horizons, the wild part of the landscape, sunset and, by opposition, with the dead.

In this sense, the projection of light inside Forno dos Mouros at winter solstice could be an evocative image of light entering the area of the dead within the chamber, perhaps bringing life to it or, alternatively entering the realm of the dead. But at the same time the projection of shadows towards the rock of Pena Moura could reinforce such image by projecting darkness into the wild uncultivated part of the landscape.

Although the location of mounds within the landscape, both in Barbanza and in Leboreiro, seems similar, the situation in both cases is not exactly the same. Clearly their locations must comply with the local topography but the spatial concepts and alignments in both cases are not identical. While in both cases there seems to be a preference to locate prominent mounds towards the west and in connection with astronomical referents, the details are different between the two locations. In Barbanza it seems that the setting positions correspond to the same objects seen rising in connection with the orientation of the chambers. However, in Leboreiro, Mota Grande could be indicating either summer solstice sunset or conversely the setting of the winter solstice full moon (if we consider this season as more relevant than others).

At this moment, it could be of interest a discussion on the secondary concentration towards -12° in the orientation of chambers and c. 12° in the alignment towards Mota Grande.

Until recent times the villages next to Leboreiro housed one of the later examples of seasonal house movement in the Iberian Peninsula.[37] At a time near mid-spring (i.e. end of April beginning of May) the peoples of the local parishes moved their houses towards the upper areas of the Sierra together with their flocks in order to secure the pastures for them in the heat of summer. The concentrations around -12° could have been

---

[36] Criado Boado *et al.*, 'A paisaxe…', p. 125.
[37] Orlando Ribeiro, 'Brandas e Inverneiras em Castro Laboreiro', *Revista da Faculdade de Letras* 6 (1939): pp. 297–302.

connected with sunrise around mid February or the end of October, while those towards 12° could be connected with sunset by the end of April and mid August. In this sense it is again of note that a double alignment could be happening in this last area, where one could be witnessing the full moon rise and sunset on the same spot and moment.

Here, at 1200 meters of altitude it would be difficult to imagine a relevant ritual at the harsh conditions of winter and the dates around the end of April and mid August should be preferred. It is clear that the situation has changed a lot in 6000 years, but this gives us a hint that possibly such orientations and alignments could be connected with the economy, social time and rituals of the ancient inhabitants of each area.

Acknowledgments. ACGG is a Ramón y Cajal Fellow of the Spanish MINECO

Different Approaches to Cosmology in Archaeology and Their Application to Maltese Prehistory

Tore Lomsdalen

Abstract: This paper outlines different ways archaeologists have engaged with notions of cosmology and worldview. It begins by highlighting the confusion over definitions of cosmology, particularly within archaeology and archaeoastronomy, and proposes a working definition that tries to be all-encompassing. This definition is then used to illustrate how different bodies of archaeological theory have touched upon the topic of cosmology either explicitly or implicitly. The list provided is not exhaustive but provides an overview of the range of scales (from artefact to landscape) as well as a level of engagement with cosmology (from implicit to explicit). Each of these instances is firstly described in general, followed by a more specific application to Maltese prehistory and its Temple Period. The paper, therefore, highlights the ways in which archaeologists have engaged with cosmology that can complement work done by archaeoastronomers and cultural astronomers. By drawing attention to the holistic and all-encompassing nature of cosmology it is hoped that further steps towards bridging the gaps between archaeology and archaeoastronomy can be taken.

1. Introduction

This paper outlines different ways archaeologists have engaged with notions of cosmology and worldview or, at least, some of its elements. Archaeology has moved on from the view that subjects such as prehistoric ritual, belief and worldviews, being further up Hawkes' 'ladder of inference', are inaccessible to us and, therefore that it is imprudent for archaeology to speculate on these areas.[1] More recently, thanks to the advent of post-processualism spearheaded by Hodder, followed by Shanks and Tilley, archaeology seeks to 'transcend the tired divide between subjective and objective approaches'.[2] Topics such as ritual, belief and

---

[1] Christopher Hawkes, 'Archeological Theory and Method: Some Suggestions from the Old World', *American Anthropologist* 56, no. 2 (1954): pp. 155–68.
[2] Ian Hodder, ed., *Symbolic and Structural Archaeology*, New Directions in Archaeology (Cambridge: Cambridge University Press, 1982); Michael Shanks

worldviews, are now routinely studied by archaeologists, such as Parker Pearson and Insoll.[3] Today archaeology departments feature scholars and research projects dealing with topics as varied as experimental archaeology, cult and religion, anthropology, phenomenology, monumentality, perception of the environment and taskscape, archaeology of the senses, cosmology, land and seascape as well as skyscape archaeology. However, not all scholars explicitly engage with the concept of cosmology.

One possible reason is that the term 'cosmology' takes many definitions, the understanding of which changes from scholar to scholar. It is used scientifically in the fields of astronomy and astrophysics to relate to how the universe functions; it has religious connotations in theology; it relates to belief and worldview in anthropology; to the creation of cosmic order in Classical Greek philosophy; to the spiritual and supernatural in secular or 'new' religions; and often to a three-tiered worldview composed of humans, earth and cosmos. There is often a confusion or, at least, a lack of clarification when using the term 'cosmology' even within the same field and especially in archaeology. According to Silva and Brown the sky is half the cosmos/world of any culture and yet it is rarely mentioned in archaeology.[4] It is therefore necessary firstly to establish a working definition of cosmology, based on both previous literature and the author's own thoughts, which is done in section 2. Section 3 will give a brief and overall overview of Maltese Prehistory, mainly concentrating on the Temple Period. In section 4, some areas of archaeological theory are introduced and discussed in the ambit of cosmology. It will explore how, and to what extent, the term cosmology is being employed and approached within areas of such bodies of theory as site catchment analysis, fragmentation, access analysis, external symbolic storage, archaeology of death, landscape archaeology. More recent holistic approaches that bring

---

and Christopher Tilley, *Re-Constructing Archaeology: Theory and Practice*, Second edition (London: Routledge, 1992).
[3] Mike Parker Pearson, 'Death, Being, and Time: The Historical Context of World Religions', in Timothy Insoll, ed., *Archaeology and World Religion* (New York: Routledge, 2001), p. 204; Timothy Insoll, 'Introduction: The Archaeology of World Religion', in Insoll, *Archaeology and World Religion*, p. 3.
[4] Fabio Silva, 'Cosmology in Transition', (paper presented at TAG Manchester, UK, 16 December 2014); Daniel Brown, 'The Experience of Watching: Place Defined by the Trinity of Land-, Sea-, and Skyscape', *Culture and Cosmos* 17, no. 2 (2013): p. 22.

together land-, sea-, island-, task- and skyscape within a cosmological apprehension will be discussed in section 5.

2. Cosmology

The concept of cosmology can be approached from the perspective of its use in archaeology; however its use is not always consistent. Darvill defines cosmology as 'The world view and belief system of a community based upon their understanding of order in the universe'.[5] Parker Pearson and Richards go one step further by specifying an order of morality, social relations, space, time and the cosmos, that may also be reflected, expressed, and mediated in architectural space, which they document with a wide array of ethnographic examples.[6] Grima, Malone and Stoddart bring the perception of worldview, belief system, temporality and ancestral correlation related to monumentality into the notion of cosmology.[7] Campion states that cosmology recognises that we as human beings, our behaviour, our belief systems and our environments are as much a part of the cosmos, as the sky and stars.[8] The word 'cosmology' as used above implies that it can be studied anthropologically, and may bring in a religious perspective through the total human experience, action and thought in time and space towards maintaining harmony between the way the universe is, and the way human beings behave as suggested by Andrén, Bourdieu and Sims.[9] Mathews claims cosmology is the real world, but also

---

[5] Timothy Darvill, *Oxford Concise Dictionary of Archaeology* (Oxford: Oxford University Press, 2008), p. 111.
[6] Michael Parker Pearson and Colin Richards, eds., *Architecture & Order: Approaches to Social Space* (London: Routledge, 1994), pp. 10–15.
[7] Reuben Grima, 'An Iconography of Insularity: A Cosmological Interpretation of Some Images and Spaces in the Late Neolithic Temples of Malta', *Institute of Archeology* 12 (2001): pp. 55–56; Caroline Malone and Simon Stoddart, 'Conclusions', in *Mortuary Customs in Prehistoric Malta: Excavations at the Brochtorff Cicle at Xaghra (1987–94)*, ed. Simon Stoddart, Caroline Malone, Anthony Bonanno and David Trump, with Tancred Gouder and Anthony Pace (Cambridge: McDonald Institute Monographs, 2009), p. 376.
[8] Nicholas Campion, *A History of Western Astrology: The Medieval and Modern Worlds*, Vol. II (New York: Continuum, 2009), p. 1.
[9] Anders Andrén, *Tracing Old Norse Cosmology: The World Tree, Middle Earth, and the Sun from Archaeological Perspectives* (Lund, Sweden: Nordic Academic Press, 2014), p. 12; Pierre Bourdieu, 'The Berber House or the World Reversed', *Social Science Information* 9 (1970); Lionel Sims, 'Coves, Cosmology and Cultural Astronomy', in Nicholas Campion, ed., *Cosmologies* (Ceredigion, Wales: Sophia Centre Press, 2009), p. 4.

forces, fields, minds, spirits, even deities, as these bodies are capable of being actual, of constituting an actual world.[10]

All scholars mentioned above have addressed what cosmology encompasses, quite independently of each other. However, each one offers a slightly different perspective on the significance and nuances of the term. Nevertheless, it can be argued that one common characteristic all these definitions have, is encapsulated by Darvill's succinct definition, that cosmology is related to how humans or societies create order in the world they live. Consequently, cosmology may be understood as a holistic worldview, a belief system that encompasses all aspects of human society, their environments and the observable, but intangible sky. This is the definition of cosmology that will be applied throughout this paper (see Fig. 1).

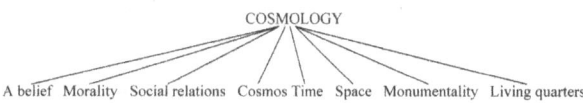

Fig. 1 Various elements that form part of a society's cosmology.

3. A Brief Overview of Maltese Prehistory

The earliest evidence of human presence on the Maltese islands goes back to the early Neolithic (late sixth century BCE), and the archaeological record indicates the population arrived from Sicily.[11] Around a thousand years later, according to Mayrhofer, 'these religious people started with the erection of megalithic temples which at that time were as unique as they are today'.[12] Originally there may have been approximately forty temples on the islands of which about twenty remain today in various conditions.[13] Some excavations were undertaken in the late nineteenth century, but most

---

[10] Freya Mathews, *The Ecological Self* (London: Routledge, 1991), pp. 3–4.
[11] David H. Trump, *Malta: Prehistory and Temples*, with photography by Daniel Cilia (Malta: Midsea Books, 2002), p. 23.
[12] Karl Ing. Mayrhofer, *The Prehistoric Temples of Malta and Gozo: A Description by Prof. Sir Themistocles Zammit* (Ing. Karl Mayrhofer, 1995), p. 5.
[13] Rowland Parker and Michael Rubinstein, *Malta's Ancient Temples and Ruts* (London: The Institute for Cultural Research, 1988), p. 2.

archaeological work was conducted in the twentieth century, with some buildings being reconstructed or rebuilt.[14]

The Maltese temples do not appear as isolated monuments but are frequently found in groups, often paired or even clumped together.[15] What is generally known as the Maltese Temple Period ranges from 4,100 to 2,500 BCE, though the core Temple Period is divided into the Ġgantija Period (3,600–3,000 BCE) and the Tarxien Period (3,000–2,500 BCE).[16] However Trump theorised that indications of temple and religious rituals may go back some centuries to the Red Skorba Phase (4,400–4,100 BCE) that precedes the Temple Period.[17] The period that immediately follows the Temple Period is known as the Bronze Age Period and is chronologically divided into three periods; Tarxien Cemetery (2,400–1,500 BCE) and Borġ in-Nadur (1,400–800 BCE) intervened by a short Baħrija Period (900–800 BCE).[18] From then on Malta goes into its historical periods starting with the Phoenician (700–500 BCE), then the Punic (500–300 BCE) and finally the Roman (200 BCE–1 CE).[19] According to Malone and Stoddart the monuments were constructed and maintained primarily for ritual purposes modelled on a corresponding three-level cosmology consisting of the worlds of the dead (hypogea below the ground), the living (the temples above the ground) and the ancestors (the sky and stars).[20] As temple construction reached its highest flourishing, it went into a sudden and unexplainable decline around 2,500 BCE.[21] Both the sudden start and the mysterious end of the complex giant megalith construction period raises many questions. After the Temple Period, Malta seems to have been

---

[14] John Cox, 'Observations of Far-Southerly Moonrise from Hagar Qim, Ta' Hagrat and Ggantija Temples from May 2005 to June 2007', *Cosmology Across Cultures, ASP Conference Series* 409 (2009), p. 344.

[15] A. Bonanno et al., 'Monuments in an Island Society: The Maltese Context', *World Archaeology* 22, no. 2 (1990): p. 193.

[16] David H. Trump, 'Dating Malta's Prehistory', in Daniel Cilia, ed., *Malta before History* (Malta: Miranda Pubishers, 2004), p. 230.

[17] David H. Trump, *Skorba: Excavations Carried out on Behalf of the National Museum of Malta 1961–1963* (Oxford: Oxford University Press, 1966), p. 10.

[18] Trump, 'Dating Malta's Prehistory', p. 230.

[19] Trump, 'Dating Malta's Prehistory', p. 230.

[20] Caroline Malone and Simon Stoddart, 'Maltese Prehistroic Religion', in Tim Insoll, ed., *The Oxford Handbook of the Archaeology of Ritual and Religion* (Oxford: Oxford University Press, 2011).

[21] Caroline Malone and Simon Stoddart, 'Ritual Failure and the Temple Collapse of Prehistoric Malta', in Vasiliki G. Koutratouri and Jeff Sanders, eds., *Ritual Failure: Archaeological Perspectives* (Leiden: Sidestone Press, 2013).

110 Different Approaches to Cosmology in Archaeology and Their Application
to Maltese Prehistory

colonised by a new Bronze Age civilisation which possessed none of the skills in masonry or the architectural ability of their predecessors.[22] This was also presumably the period the less elaborate megalithic structures named 'dolmens' were erected; however the usage and actual construction periods of the Maltese dolmens remain, according to Sciberras, unidentified.[23]

4. Some Approaches to Cosmology in Archaeology
This section introduces types of archaeological theory which touch upon cosmology either explicitly or implicitly. These range from the purely economic view of cosmology, through various aspects of religion, ritual and symbolism, to more holistic approaches that look at landscapes, waterscapes and skyscapes. The ones presented here were chosen as they provide a good overview of the range of scales (from artefact to landscape) as well as the range of relation to cosmology (from purely implicit to fully explicit). Each of these instances is firstly described in general, followed by a more specific application to Maltese prehistory and its Temple Period.

4.1 Site Catchment Analysis
Vita-Finzi and Higgs define Site Catchment Analysis (SCA) as 'the study of the relationships between technology and those natural resources lying within economic range of individual sites'.[24] According to Chisholm the original ideas of SCA developed and were spelled out by von Thünen in his major work, *The Isolated State*, with the theory that on any specified piece of land, the enterprise that yields the highest net return will be conducted and competing enterprises will be regulated to other land plots.[25]

---

[22] Giulio Magli, *Mysteries and Discoveries of Archaeoastronomy from Giza to Easter Island* (New York: Copernicus, 2009), pp. 48–49.
[23] Danil Sciberras, 'The Maltese Dolmens', in Anton Mifsud and Charles Savona-Ventura, eds., *Facets of Maltese Prehistory* (Malta: The Prehistoric Society of Malta, 1999), p. 106.
[24] C Vita-Finzi and E. S. Higgs, 'Prehistoric Economy in the Mount Carmel Area of Palestine: Site Catchment Analysis', *Proceedings of the Prehistoric Society* 36 (1970): p. 5.
[25] Michael Chisholm, *Rural Settlement and Land Use: An Essay in Location*, 2nd ed. (London: Hutchinson University Library, 1968), pp. 20–21; J. H. von Thunen, *Der Isolierte Staat in Beziehung auf Landswirtschaft und Natinalökonomi* (Rostock: Perthes, 1826). English translation by C. M. Wartenberg: von Thünen's *Isolated State* (Oxford: Pergammon Press, 1966).

In archaeology, SCA has been used to predict site locations, analyse the environmental context of sites and to determine their function, size and distribution in order to suggest the finite distance a group would be willing to go to obtain resources; an inference about a site both economically and culturally as suggested by Tiffany and Abbott, Hodder and Barker.[26] SCA does not explicitly engage with cosmology but with economy. However, it can be argued that the reasoning that underpins SCA is in itself a way to view and approach the world and hence is a worldview or cosmology, albeit one that focuses on economy. Furthermore, the application of SCA may also allow some insight into a society's attitudes to its landscape, which in turn relates to its cosmology.

Malta: Von Thünen's general ideas of site catchment analysis may be applied to an island such as Malta in the Neolithic. Regarding the socio-economic capacity of a Maltese prehistoric village, Bonanno suggests that trade between one local village community and another would have had much less scope in the limited area of Malta and that each village community would mainly be self-sufficient.[27] Grima also argues in site catchment analytical terms, that the size of a Maltese prehistoric megalithic complex may be directly correlated to the opportunities offered by its local environment.[28] Another more complex element related to a wider perspective of a site catchment analysis, is the question why, in the first place, colonise the Maltese Archipelago when it according to Stoddart *et al.* had little to offer but isolation and poverty of resources compared to the bigger and more prosperous island of Sicily from where the first settlers emigrated.[29] As reported by Fenech, it is debatable how forested, if at all, Malta was during the Neolithic and there are indications of a steppe

---

[26] Joseph A. Tiffany and Larry R. Abbott, 'Site-Catchment Analysis: Applications to Iowa Archaeology', *Journal of Field Archaeology* 9, no. 3 (1982): p. 313; Ian Hodder, 'The Interpretation of Spatial Patterns in Archaeology: Two Examples', *The Royal Geographical Society (with the Institute of British Geographers)* 4, no. 4 (1972): p. 224; Graeme Barker, 'Cultural and Economic Change in the Prehistory of Central Italy', in C. Renfrew, ed., *The Explanation of Culture Change: Models in Prehistory* (Pittsburg: University of Pittsburg Press, 1973), pp. 359–60.
[27] Anthony Bonanno, 'A Socio-Economic Approache to Maltese Prehistory: The Temple Builders', *Malta Studies of its Heritage and History* (1986): pp. 20, 37.
[28] Reuben Grima, 'Landscape, Territories, and the Life-Histories of Monuments in Temple Period Malta', *Journal of Mediterranean Archaeology* 21, no. 1 (2008): p. 54.
[29] Simon Stoddart *et al.*, 'Cult in an Island Society: Prehistoric Malta in the Tarxien Period', *Cambridge Archaeological Journal* 3, no. 1 (1993): p. 5.

112 Different Approaches to Cosmology in Archaeology and Their Application to Maltese Prehistory

environment.[30] Further, the geological formation of the Archipelago lacks sufficient hard and precious cutting instruments such as flint and obsidian which all had to be imported.[31] Robb suggests that islands are more than physical land surrounded by water, but 'Islands are Ideas'.[32] Without debating this issue at any length, a plausible cause could be that the first settlers searched for a new island identity different from whence they came, more than exploring for an economic expansion. However, once established and settled, the concept of site catchment may have been an underlying factor for selecting dedicated land plots on the Archipelago.

4.2 Fragmentation

Fragmentation is a widely used concept in archaeology, anthropology and material culture to study how objects may be deliberately or accidentally broken and then re-used. Depending on the desired outcome of the user, Chapman suggests that the fragments affect theories of enchainment – the linking of person to person through object exchange.[33] Grinsell widens up the concept of fragmentation to ceremonial 'killing' of objects at funerals, but also introduces the killing of human beings and animals to accompany the dead and concludes, 'that there can be no doubt that in most of these instances the reason was to assist the deceased in his journey to the other world or when he reached the after-life', and 'to release the spirit in the object to accompany the dead to the after-life'. However, he admits it is difficult to get reliable information on the purposes behind a ceremonial 'killing'.[34] As such, the concept of 'fragmentation' as just outlined does

---

[30] Katrin Fenech, *Human-Induced Changes in the Environment and Landscape of the Maltese Islands from the Neolithic to the 15th Century AD* (Oxford: BAR, 2007), pp. 113–14.
[31] Anthony Bonanno, 'The Lure of the Islands: Malta's First Neolithic Solonisers', in Nellie Phoca-Cosmetatou, ed., *The First Mediterranean Islanders: Initial Occupation and Survival Strategies* (Oxford: University of Oxford School of Archaeology, 2011), p. 151.
[32] John Robb, 'Island Identities: Ritual, Travel and the Creation of Difference in Neolithic Malta', *European Journal of Archaeeology* 4, no. 2 (2001): p. 177.
[33] John Chapman, *Fragmentation in Archaeology: People, Places and Broken Objects in the Prehistory of South-Eastern Europe* (London: Routledge, 2000), pp. 4–5, 77–78.
[34] L. V. Grinsell, 'The Breaking of Objects as a Funerary Rite', *Folklore* 72, no. 3 (1961): pp. 475–76.

seem to relate to manifestations of a religious belief system or a cosmology.

Malta: As mentioned by Zammit, Bonanno and Malone *et al.*, the archaeological record of the Maltese Temple Period indicates a use of 'fragmentation' through possible intentional destructions of artefacts both from the temples above the ground and from burial practices.[35] Barrowclough mentions 'fragmentation' directly in the archaeological record, illustrating how imported greenstone axe pendants at the Tarxien Temples (Tarxien and Tarxien Cemetery Periods) were deliberately broken into two pieces as a token of an enchained social relationship through a cult membership, and that these precious objects could be further broken up and passed on to third parties.[36] That these fragmented objects were found in a cache in and around an area which Zammit originally named as the 'Oracle Room, emphasises the possibility that the fragmentation process may have been an integrated part of religious or cult ceremony.[37]

Malone and Stoddart refer to the destruction of about two metre standing and skirted figures from Tarxien as evidence of possible vandalism, however they point out that a finely carved stone figure retrieved at the hypogea of the Xagħra Circle (mainly Tarxien and Tarxien Cemetery Periods) on Gozo was smashed into fragments and crushed into burials in a pattern that could not have occurred naturally, with its location far away from the point of its breakage.[38] Xagħra Circle burials show the possible removal of selected body parts and the redistribution of long bones and skulls into secondary locations and also fragments of some 6841 animal bones, according to Stoddart et al.[39] The Ħal Saflieni Hypogeum (mainly Tarxien Period), though lacking original excavation reports from

---

[35] T. Zammit, 'The Hal-Tarxien Neolithic Temple, Malta. Archaeologia', *Archaeologia* 67 (1916): p. 133; Anthony Bonanno, 'In Search of an Identity: The Anthropomorphic Representations of Megalithic Malta', in D. Fenech, V. Fenech, and J.R. Grima, eds., *Lino: A Tribute (Festschrift in Honour of Lino Spitieri)* (Malta: PEG, 2008), p. 63; Caroline Malone *et al.*, *Mortuary Customs*, pp. 289–98.

[36] David A. Barrowclough, 'Putting Cult in Context: Ritual, Religion and Cult in Temple Period Malta', in D. A. Barrowclough and Caroline Malone, eds., *Cult in Context: Reconsidering Ritual in Archaeology* (Oxford: Oxbow, 2007), pp. 52–53.

[37] T. Zammit, 'Third Report on the Hal-Tarxien Excavations, Malta', *Archaeologia* 70 (1920): p. 182.

[38] Malone and Stoddart, 'Ritual Failure', pp. 77–79.

[39] Simon Stoddart *et al.*, 'The Human and Animal Remains', *Mortuary Customs*, pp. 329–30.

the lower levels, also indicates redistribution of human bones and skeletons.[40] As suggested by Malone and Stoddart, the mortuary practices at the Xagħra Circle embed a three level cosmology of the living, the dead and their ancestors of the earth and the sky.[41] Vella interlaces a possible intentional mutilation of a prehistoric statue at the Tas-Silg Temple (this temple had a continuous use and reuse from Tarxien Period all the way up to and including the Roman Period) as a process of iconoclasm and discontinuity, however concludes, 'the suggestion is attractive but inconclusive'.[42] During my visit to the Temple Period section of the Tas-Silg site with Prof. Bonanno in April 2016, a discussion about whether an area close to an about four metre long apparent altar formation showed indications of deliberate breakage. Metal tools had seemingly been applied to destroy or modify some boulders. Metal first arrived on Malta after the Temple Period in the Bronze Age, commencing at about 2,500 BCE. A section of the so-called 'Oracle hole' in the same area also seemed to have been smashed or reshaped on purpose. Some parts of this space showed visible signs of burn marks, difficult however to assess as to purpose or chronology (see Fig. 2). That these alterations or modifications are related to a cosmological change in belief or social system is debatable. Nevertheless, Malone and Stoddart suggest that there are indications of cultural changes and a possible breakdown at the end of the Temple Period at about 2,500 BCE and into the following Bronze Age Period.[43]

---

[40] Anthony Pace, ed., *The Ħal Saflieni Hypogeum: 4000 BC–2000 AD* (Malta: National Museum of Archaeology, Museums Department, Malta, 2000).
[41] Malone and Stoddart, 'Conclusion'.
[42] Nicholas Vella, 'Trunkless Legs of Stone: Debating Ritual Continuity at Tas-Silg, Malta', in Mifsud and Savona-Ventura, *Facets of Maltese Prehistory*, p. 230.
[43] Malone and Stoddart, 'Ritual Failure'.

Fig. 2 The alterations made to the so-called 'oracle hole'; the metal tool and the burned marks on some of the boulders. In the background the four-metre altar formation is also visible. Photo by T. Lomsdalen.

4.3 Access Analysis
The concept of Access Analysis is, according to Foster, based on morphological relations and considers the arrangement of different spaces as a pattern of accessibility in terms of the interconnections between spaces.[44] The theories and techniques for analysing spatial configurations conceived by Hillier and Hanson in their publication *The Social Logic of Space*, go one step further to being a manual for city, architectural or environmental planning, as they argue that space is the function of form of social solidarity, a product of the structural society, and that space has a certain social logic to it, raising the question; 'is there any sense in which space also determines society?'[45] Stöger proposes that Space Syntax in its theoretical form is partly rooted within structuralism and is based on two formal ideas reflecting both the objectivity of space and our intuitive

---

[44] Sally M. Foster, 'Analysis of Spatial Patterns in Buildings (Access Analysis) as an Insight into Social Structure: Examples from the Scottish Atlantic Iron Age', *Antiquity* 63, no. 238 (1989): p. 41.
[45] Bill Hillier and Julienne Hanson, *The Social Logic of Space* (Cambrige: Cambridge University Press, 1984), p. 22.

involvement with it; that space is an inherent facet of all human activity, and human space is not about properties of space; however the 'configuration of space' is the process through which space gains social significance with social consequences.[46]

The word 'cosmology' seems not to be explicitly mentioned, or even considered, in works that focus on Access Analysis, Space Syntax or in the syntactic analysis of structures. However, if patterns of space can be expected to reproduce social relations and transmission of ideas within social, cultural or religious and sacred boundaries, it can equally be argued that space and spatial order may be the fount of both creating and exchanging worldviews and belief systems. This is further underlined when applying Hillier's statement; 'the view of space is as true practically as it is philosophically'.[47]

Malta: In the study of spatial order and access patterns of the Maltese prehistoric temples, only Bonanno, Anderson and Stoddart have addressed the issue of access analysis.[48] Bonanno analysed the physical access analysis of areas within the temples. Anderson and Stoddart on the other hand went one step further by analysing the spatial temple areas with high and low visibility considering closed off areas for ritual performances; thus implying a concept of 'cosmology' based on a system of belief and an hierarchical social structure.

4.4 External Symbolic Storage
When Merlin Donald launched the concept of External Symbolic Storage (ESS), he did not relate it necessarily to archaeology as such, but to the

---

[46] Hanna Stöger, *Rethinking Ostia: A Spatial Enquiry into the Urban Society of Rome's Imperial Port-Town* (Amsterdam: Amsterdam University Press, 2011), pp. 41–42.

[47] Bill Hillier, *Space in the Machine: A Configuration Theory of Architecture* (Cambridge: Cambridge University Press, 1996), p. 345; Bill Hillier, ed. *Between Social Physics and Phenomenology*, Proceedings of the 5th International Space Syntax Symposium (Amsterdam: Techne Press, 2005).

[48] Bonanno *et al.*, 'Monuments', pp. 195–98; Anthony Bonanno, 'The Rise and Fall of Megalithism in Malta', in Karl W. Beinhauer, *et al.*, eds., *Studien Zur Megalithik. The Megalithic Phenomenon: Recent Research and Ethnoarchaeological Approaches* (Weissbach, Germany: Verlag Beier & Beran, 1999), pp. 105–6; Michael Anderson and Simon Stoddart, 'Mapping Cult Context: GIS Applications in Maltese Temples', in Barrowclough and Malone, *Cult in Context*.

cultural evolution of how the gradual process of human consciousness and cognition shifted from an internal to an external memory storage device. This started with the early Hominin species about two million years ago, and culminated about forty thousand years ago (Upper Palaeolithic) with a revolution in technology, symbols and material culture with a main cognitive driving force of externalisation of memory.[49] Donald, being a psychologist and neuroanthropologist, has nevertheless been accredited for his innovative work by Renfrew and also favourable referred to in the archaeological record by Zubrow and Daly, and d'Errico.[50] Donald further refers to the fact that about 100,000 years ago there appear, in the archaeological record, ritual artefacts of a quasi-symbolic nature, as well as adornments and costumes, which took a particular and vital symbolic function in ritual and religion.[51] Donald does not explicitly mention cosmology, but he relates mythology, ritual and symbolism – all of which are significant elements of a society's cosmology – to the evolution of cognition.

Rawson, an art historian, applying the concept of ESS, proposes that the idea of symbolic storage implies that ideas, beliefs and intentions exist prior to the artefacts that act as external storage.[52] However, the artefacts are not simply storage but are fully integrated into the process of constituting beliefs and bring them into being. Yet another example comes from Delano Smith, who interprets the Bulgarian Magoura prehistoric cave painting as a cosmological map, and therefore external storage of a symbolic nature.[53] She suggests that the study of maps from prehistoric

---

[49] Merlin Donald, *Origins of Modern Mind: Three Stages in the Evolution of Culture and Cognition*. (Cambridge, MA: Harvard University Press, 1991), p. 275.
[50] Colin Renfrew, 'Mind and Matter: Cognitive Archaeology and External Symbolic Storage', in Colin Renfrew and Chris Scarre, eds., *Cognition and Material Culture: The Archaeology of Symbolic Storage* (Cambridge: McDonald Institute Monographs, 1998), p. 1; Ezra B.W. Zubrow and Patrick T. Daly, 'Symbolic Behaviour: The Origin of a Spatial Perspective', in Renfrew and Scarre *Cognition and Material Culture*, pp. 157–59; Francesco d'Errico, 'Palaeolithic Origins of Artificial Memory Systems: An Evolutionary Perspective', in Renfrew and Scarre, *Cognition and Material Culture*, pp. 22–43.
[51] Donald, *Origins*, p. 277; Merlin Donald, *A Mind So Rare: The Evolution of Human Consciousness* (London: W.W. Norton, 2001), p. 262.
[52] Jessica Rawson, 'Chinese Burial Patterns: Sources of Information on Thought and Belief', in Renfrew and Scarre, *Cognition and Material Culture*, p. 107.
[53] Catherine Delano Smith, 'Imago Mundi's Logo the Babylonian Map of the World', *Imago Mundi* 48 (1996): pp. 45–49; Catherine Delano Smith, 'The Emergence of 'Maps' in European Rock Art: A Prehistroic Preoccuptaion with

periods reveals the breadth of human interest in all forms of space, not only the terrestrial, but also celestial and cosmological.[54]

Malta: In the Maltese context the only indications of tally-marks in prehistoric Malta is proposed by Ventura *et al.* on two vertical pillars at the temple at Mnajdra (Ġgantija and Tarxien Periods) which imply a sequence of heliacal risings which include, the Pleiades, Aldebaran, the Hyades (all in Taurus), Orion, Sirius and Murzim (in Canis Major), Arcturus, and Crux-Centaurus of the Pleiades.[55] Grima proposes the architectural and spatial layout of the Ħal Saflieni Hypogeum can be interpreted as a cosmological pilgrimage between the realm of the living and the dead, marked by its boundaries and transitional spaces.[56] The symbolic storage of 'The Tree of Life' which is related to the cosmic life force is a common motif on Mesopotamian and Egyptian pottery; and in Malta, where a similar design in red ochre covers part of the ceiling in the Neolithic hypogeum of Ħal Saflieni, it may also visualise another form of iconographic representation.[57] The archaeological record shows prehistoric plans and models of the temples as suggested by Trump, Ugolini and Pace.[58] However an open question is whether these plans were used as models before constructing the temples or afterwards (see Fig. 3). Examples of ESS that may be related to astronomy and cosmology are the

---

Place', *Imago Mundi* 34 (1982): p. 9; Catherine Delano Smith, 'Cartography in the Prehistoric Period in the Old World: Europe, the Middle East, and North America', in J.B. Harley David Woodward, ed., *The History of Cartography: Volume One, Cartography in Prehistoric, Ancient, and Medieval Europe and the Mediterranean* (Chicago: The University of Chicago Press, 1987), p. 92.
[54] Delano Smith, 'Emergence', p. 9.
[55] Frank Ventura, Georgio Fodera Serio and Michael Hoskin, 'Possible Tally Stones at Mnajdra, Malta', *Journal for the History of Astronomy* 24 (1993): pp. 171–83, p. 179.
[56] Reuben Grima, 'Journeys through the Underworld in Late Neolithic Malta', in George Nash and Andrew Townsend, eds., *Decoding Neolithic Atlantic & Mediterranean Island Ritual* (Oxford: Oxbow, 2016).
[57] Catherine Delano Smith, 'Prehistoric Maps and History of Cartography: An Introduction', in Woodward, *The History of Cartography*, p. 87.
[58] Luigi M. Ugolini, ed., *Origini Della Civilta Mediterranean Malta: Origins of Mediterranean Civilization* (Malta: Midesea Books, 2012), pp. 176–77; David Trump, 'Megalithic Architecture in Malta', in *Antiquity and Man: Essays in Honour of Glyn Daniel* (London: Thams and Hudson, 1981), p. 132; Anthony Pace, 'The Sites', in Cilia, *Malta before History*, p. 152.

so-called 'Solar Wheel' from the temple of Ħaġar Qim (clasified as Ġgantija Period) and the 'Star Stone' from the Tal-Qadi Temple (Tarxien Period) that may represent a lunar calendar or a star map as suggested by Ventura and Micallef (see Fig. 4).[59] Delano Smith rejects this interpretation without giving any specific explanation.[60]

Fig. 3 Possible models and graffiti of Maltese prehistoric temple design retrieved from the following sites; top left from Ta'Ħaġrat, top right from Ħaġar Qim, bottom left from Tarxien and bottom right from Skorba. All items displayed at the Archaeological Museum, Valletta, Malta. Photo: T. Lomsdalen.

---

[59] Frank Ventura, 'Temple Orientations', in Cilia, *Malta before History*, p. 312; Chris Micallef, 'The Tal-Qadi Stone: A Moon Calendar or Star Map', *The Oracle, The Journal of the Grupp Arkeoloġija Malti*, no. 2 (2001).
[60] Delano Smith, 'Cartography', p. 84.

120   Different Approaches to Cosmology in Archaeology and Their Application
      to Maltese Prehistory

Fig. 4 On the left side, the graffiti in the ceiling of the hypogea Ħal Saflieni. Top right, the so-called 'Star Stone' from Tal-Qadi and bottom right the 'Solar Wheel'. Centre bottom, a rim sherd of a bowl with a 5 pointed star decoration, provenance Tarxien. The last three artefacts mentioned are all displayed at the Archaeological Museum, Valletta, Malta. All photos by T. Lomsdalen.

4.5 Archaeology of Death
According to Boyer, death is the origin of religious concepts and all religions seem to have something to say about death; 'burying the dead in a ritual way is evidence for supernatural concepts – ancestors, spirits, gods – because we find a connection between these two phenomena in most human societies'.[61] In the study of death, Parker Pearson highlights the strange paradox that the physical remains of the dead are more likely to reveal more information about the life of individuals than about their death.[62] Indeed, funerary archaeology tends to focus on tangible quantities, such as the identification of age, gender, status and ethnicity from the contents of a burial, while a more innovative approach by Baker focuses on

---

[61] Pascal Boyer, *Religion Explained: The Human Instincts That Fashion Gods, Spirits and Ancestors* (London: Random House, 2001), p. 233.
[62] Michael Parker Pearson, *The Archaeology of Death and Burial* (Stroud, UK: The History Press, 2009), p. 3.

how the commodities of a culture, including worldview and religious belief system, can be recognised from its mortuary practices.⁶³
Tilley suggested that death, body symbolism and rituals accompanying inhumations generally play an important role in that they both imitate and mould social values where 'good death' and the proper healing and preservation of the ancestral remains has the same value as 'good marriage'; namely an act to secure the reproduction of society and social order.⁶⁴

Malta: The Maltese archaeological record on mortuary and burial practices contains a considerable amount of documentation relating to the entire Temple period which often linked cult, rituals, religion, belief and worldview (cosmology) to an ancestral connotation through shrines, depicted grave goods, various types of figurines and spatial order of skeletons and reorganised human bones.⁶⁵ Baldacchino and Evans systematically excavated five rock-cut tombs from the Żebbuġ Phase (4,100–3,700 BCE) in 1947, but did not infer symbolic interpretations from their findings.⁶⁶ The Xagħra Circle excavation in 1987–1994, provides material from the Pre-Temple Period Żebbuġ Phase, suggestive of it having been the origin of the long-term mortuary rituals that feature later in the Temple Period.⁶⁷ Another Maltese hypogea, Ħal Saflieni, was discovered in 1902 and was excavated by Magri who unexpectedly died and left no draft of his report, so Themistocles Zammit, the father of Maltese archaeology, concluded the excavation in 1910 and noted that intact deposits of human remains survived only on the Upper Level.⁶⁸ This extraordinary hypogea was once a burial temple of an estimated 6,000 to 7,000 people.⁶⁹

---

⁶³ Jill L. Baker, *The Funeral Kit: Mortuary Practices in the Archaeological Record* (Walnut Creek, CA: Left Coast Press, 2012), pp. 11–19.
⁶⁴ Christopher Tilley, *An Ethnography of the Neolithic: Early Prehistoric Societies in Southern Scandinavia* (Cambridge: Cambridge University Press, 1996), p. 246.
⁶⁵ Trump, *Skorba.*; Malone et al., eds., *Mortuary*; Stoddart et al., 'Cult'.
⁶⁶ J.G. Baldacchino and J.D. Evans, 'Prehistoric Tombs near Żebbug, Malta', *Papers of the British School at Rome* 22 (1954).
⁶⁷ Malone and Stoddart, 'Conclusion', pp. 362–63.
⁶⁸ Anthony Pace, *The Ħal Saflieni Hypogeum: Paola* (Malta: Heritage Malta, 2004), pp. 5–8.
⁶⁹ David H. Trump et al., 'New Light on Death in Prehistoric Malta: The Brochtorff Circle', *The Megalithic Builders of Western Europe* (Harper: San Francisco, 1993), p. 100.

122   Different Approaches to Cosmology in Archaeology and Their Application to Maltese Prehistory

The Maltese Temple Period, where the temples above the ground were for the living and the ones under the ground were for the dead, suggests a dichotomy between life and death, combined with a metaphoric passage on a cosmological scale of time and space, not only by the alteration of fragmented body parts over time within burial sites, but also by a possible processual procedure between temples within the wider geographical landscape including the Mediterranean connection to ancestral Sicily, as suggested by Stoddart and Malone, and Grima.[70]

The Maltese archaeological record on mortuary practice does indicate cult, rituals, shrines and 'oracle holes', implying a profound system of beliefs and worldviews all embedded into a cosmology of the values of life, death and ancestral afterlife. Besides, in Malta, where the approach to cosmology in the archaeological record became more and more emphasised from around 2000 onwards, it still seems that the wider archaeological literature does not easily interrelate mortuary practices with cosmology as a holistic belief system or worldview, and in many cases, excepting the Xagħra Circle, excludes a sky involvement.

4.6 Landscape archaeology
The geographers and part-time archaeological enthusiasts Aston and Rowley initially launched the concept of landscape archaeology in 1974.[71] However, it was not until the 1980s that landscape archaeology was widely cited in the archaeological literature, and its focus was mainly on human impact on the landscape.[72] From then on landscape archaeology emerged in its own right, parallel with the post-processual evolution in archaeology, embedded in the idea that human activity, societies and culture have a

---

[70] Simon Stoddart and Caroline Malone, 'Changing Beliefs in the Human Body in Prehistoric Malta 5000–1500 BC', in Dusan Boric and John Robb, eds., *Past Bodies: Body Centred Reserch in Archaeology* (Oxford: Oxbow, 2008); Reuben Grima, 'The Landscape Context of Megalithic Architecture', in Cilia, *Malta before History*.
[71] Michael Aston and Trevor Rowley, *Landscape Archaeology: An Introduction to Fieldwork Techniques on Post-Roman Landscapes* (Newton Abbot: David & Charles, 1974).
[72] Bruno David and Julian Thomas, 'Landscape Archaeology: Introduction', in Bruno David and Julian Thomas, eds., *Handbook of Landscape Archaeology* (Walnut Creek, CA: Left Coast Press, 2008), p. 27.

spatial dimension.[73] Landscape archaeology can have many facets. As suggested by David and Thomas, it is not only defining a physical place in the environment, but all its lived dimensions, how people visualise their world and how they engage with one another across space and time.[74] According to Layton and Ucko landscapes are particular ways of expressing conceptions of the world and they are also a means of referring to physical entities; the same physical landscape can be seen in many different ways by different people, often at the same time.[75] Casey brings in the notion 'placescape' relating to 'place-world', reflecting an historic or prehistoric world that is anchored in a given unique place.[76] Tilley's 1994 publication, *A Phenomenology of Landscape*, laid the foundation for a new approach to archaeological, anthropological and philosophical perception of landscape.[77] In a more recent publication Tilley advocates, 'The materiality of landscape always outruns us; the real turns into the surreal', implying that we understand landscapes through modes of embodied engagement.[78] Bradley, Richards and Thomas, and Harding further suggest that modern landscape archaeology encompasses areas like monumentality in landscape connected to political acts and concepts of unification, social transformation, and the landscape's spiritual, religious, cult, pilgrimage and holy significance influencing location and orientation of prehistoric monuments.[79] Crumley brings in the assumption of a cosmic frame in the

---

[73] Timothy Darvill, 'Pathways to Panoramic Past: A Brief History of Landscape Archaeology in Europe', in David and Thomas, *Handbook of Landscape Archaeology*, p. 60.
[74] David and Thomas, 'Philosophical', in David and Thomas, *Handbook of Landscape Archaeology*, p. 38.
[75] Robert Layton and Peter J. Ucko, 'Introduction: Gazing on the Landscape and Encountering the Environment', in Peter J. Ucko and Robert Layton, eds., *The Archaeology and Anthropology of Landscape: Shaping Your Landscape* (London: Routledge, 1999), p. 1.
[76] Edward S. Casey, 'Place in Landscape Archaeology: A Western Philosophical Prelude', in David and Thomas, *Handbook of Landscape Archaeology*.
[77] Christopher Tilley, *A Phenomenology of Landscape* (Oxford: Berg, 1994).
[78] Christopher Tilley and Kate Cameron-Daum, *An Anthropology of Landscape: The Extraordinary in the Ordinary* (London: UCL Press, 2017), p. 20.
[79] R. J. Bradley, 'Ritual, Time and History', *World Archaeology* 23 (1991); Colin Richards and Julian Thomas, 'The Stonehenge Landscape before Stonehenge', in Joshua Pollard, Andrew Meirion Jones, Michael J. Allan and Julie Gardiner, eds., *Image, Memory and Monumentality: Arachaeological Engagements with the Material World: A Celebration of the Academic Achievements for Professor Richard Bradley* (Oxford: Oxbow, 2012); Jan Harding, 'Henges, Rivers and

124   Different Approaches to Cosmology in Archaeology and Their Application to Maltese Prehistory

perception of a sacred landscape as it gives life a meaning.[80] The concept of sacred geography can also be applied in the context of spatiality of religion and a spiritual dimension in natural or constructed environments.[81] The thirty to forty temples in prehistoric Malta do bring in an association of sacredness to the landscape in which they were built. This description of the Maltese prehistoric sacred landscape is further validated by Snead and Preucel's more general statement that 'Landscape cannot be fully understood without reference to a world view (*cosmology*, added by this author) which integrates place and space in the production of meaning'.[82]

Malta: When it comes to landscape archaeology and prehistoric Malta, Grima seems to be the first to disclose not only a land- and seascape connotation to cosmology, but also the use of GIS and multivariate analysis on how priorities may have been chosen by the temple builders to select a specific site or location in the landscape.[83] Vassallo uses archaeoastronomy when investigating the Maltese prehistoric temples' positions in the landscape based on cardinal directions, alignments to rising and setting of the sun at the equinox and the solstices including demarcated points or features on the apparent horizon which may have been markers

---

Exchange in Neolithic Yorkshire' in Pollard *et al.*, *Image, Memory and Monumentality*.
[80] Carole L. Crumley, 'Sacred Landscapes: Constructed and Conceptualized', in Wendy Ashmore and A. Bernard Knapp, eds., *Archaeology of Landscape: Contemporary Perspectives* (Oxford: Blackwell, 1999), p. 270.
[81] Ellen Churchill Semple, 'The Templed Promontories of the Ancient Mediterranean', *Geographical Review, American Geographical Society* 17, no. 3 (1927); Nicholas C. Vella, 'The Lie of the Land: Ptolemy's Temple of Hercules in Malta', *ANES 39* (2002); Bob Trubshaw, *Sacred Places: Prehistory and Popular Imagination* (Loughborough, UK: Heart of Albion Press, 2005).
[82] James E. Snead and Robert W. Preucel, 'The Ideology of Settlement: Ancestral Keres Landscapes in the Northern Rio Grande', in Ashmore and Knapp, *Archaeology of Landscape*.
[83] Reuben Grima, 'Monuments in Search of a Landscape: The Landscape Context of Monumentality in Late Neolithic Malta' (PhD Thesis, University College London, 2005); Reuben Grima, 'Landscape and Ritual in Late Neolithic Malta', in Barrowclough and Malone, *Cult in Context*; Reuben Grima, 'The Prehistoric Islandscape', in Charles Cini and Jonathan Borg, eds., *The Maritime History of Malta: The First Millennia* (Malta: Salesians of Don Bosco and Heritage Malta, 2011).

influencing the site location in the landscape.[84] Bonanno *et al.* associate Maltese prehistoric monuments with the island's topology and seem not to include explicitly the position of the monuments in a human/landscape relation.[85]

5. Other 'Scape' concepts

This section will briefly describe some other 'scape' perceptions relevant to this paper as;'seascape', 'islandscape', 'taskscape' and 'skyscape' reflected in a holistic cosmological worldview.

5.1 Seascape

In 1978 Westerdahl was the first to launch the notion that 'the maritime cultural landscape' relates to the whole network of sailing routes, with ports and harbours along the coast relating to human activity.[86] Another Scandinavian, Wehlin, introduced the word 'seascape' breaking down the boundaries between landscape and the sea, as people living by water were very likely to create beliefs, myths and gods inspired by this element.[87] In prehistory, trade and exchange of goods and commodities was to fulfil needs, but also to create new or to maintain existing social relationships, constituting a maritime cultural landscape with exchanges of ideas, beliefs and culture, often with a sacred response, according to Farr.[88] The religious element to a maritime environment crosses wide gaps of time and space, and is a theme that is explored by Vella and Cassar who suggest that seafaring and seascape may have a religious element in a maritime environment.[89]

---

[84] Mario Vassallo, 'The Location of the Maltese Neolithic Temple Sites', *Sunday Times*, 26 August 2007.
[85] Bonanno *et al.*, 'Monuments'.
[86] Christer Westerdahl, 'The Maritime Cultural Landscape', *International Journal of Nautical Archaeology* 21, no. 1 (1992): p. 6.
[87] Joakim Wehlin, 'Approaching the Gotlandic Bronze Age from Sea. Future Possibilities from a Maritime Perspective', *Gotland University Press* 5 (2010): p. 89.
[88] Helen R. Farr, 'Seafaring as Social Action', *Journal of Maritime Archaeology* 1, no. 1 (2006): p. 86.
[89] Nicholas C. Vella, 'A Maritime Perspective: Looking for Hermes in an Ancient Seascape', *International Colloquium (1st: University of London)* (2001); Grace Cassar, 'Is the Seashore an Opening into the Sacred? Exploring Liminality of the Littoral', *Spica: Postgraduate Journal for Cosmology in Culture* 3, no. 1 (2015).

126 Different Approaches to Cosmology in Archaeology and Their Application to Maltese Prehistory

5.2 Islandscape
Vogiatzakis *et al.* suggest that islands around the world have matured by integrating land and sea, forming the concept of land- and seascape combined with cultural implications and ecological patterns created by humans, which merit the term 'islandscape', as it encompass all these aspects in a holistic manner.[90] In this context, the concept of islandscape is the particular integration of landscape and seascape that occurs only in an island environment. This does not mean it is an entirely separate entity, but that it is an example of an incremented combined level of two 'scapes'. In other words, land- and seascape can be present without islandscape, but islandscape cannot exist without land- and seascape. Due to the terrestrial limitations of an island and its relative isolation, a new and unique cultural integrity (islandscape) could develop in its totality as an integrated element of land- and seascape. The succeeding example from the Maltese Temple Period may be such a case.

The Maltese Archipelago is situated some 80 km south of Sicily. Subject to atmospheric conditions both islands are intervisible, however in prehistoric terms the archipelago could be considered remote from insular and mainland Italy.[91] During the Maltese Temple Period, the archipelago went through a deep unique cultural process unparalleled to any other areas in the Mediterranean.[92] These cultural and cosmological changes in context with 'islandscape' are outside the remit of this paper, however the question whether the archipelago's relative isolation had an impact on its cultural expression throughout its history has been subject to considerable scholarly debate, see for example Robb, Stoddart *et al.* and Trump.[93]

---

[90] Ioannis N.Vogiatzakis *et al.*, 'Characterizing Islandscapes: Conceptual and Methodological Challenges Exemplified in the Mediterranean', *Land* 6, no. 14 (2017).
[91] Grima, 'Prehistoric Islandscape'.
[92] Andrew Townsend, 'Searching Beyound the Artefact for Ritual Practices: Evidence for Riutal Surrounding the Unclothed Human Body on Prehistoric Malta During the Temple Period', in Nash and Townsend, *Decoding Neolithic Atlantic & Mediterranean Island Ritual*.
[93] Robb, 'Identities'; Stoddart *et al.*, 'Cult'; David H. Trump, 'The Insularity of Malta: A Matter of Geography and of Conscious Choice', *Treasures of Malta* 8, no. 2 (2002).

5.3 Taskscape
Ingold, who introduced the word 'taskscape' in 1993, defines it as an array of related features or activities with every task taking its position within an ensemble of tasks.[94] Activities and taskscape are to labour what landscape is to land, a concept of temporality and moving through a task or an environment, being land, sea or the sky.[95] Thomas suggests that the concept of taskscape may have its drawbacks as in some cases it may be extended or superseded and that; 'people are not always engaged in activities that have a clear objective'.[96] Nevertheless, both the term and the concept of taskscape could be adopted more frequently in both archaeology, archaeoastronomy and skyscape literature, as taskscape cannot be understood without a cognitive human contribution.

5.4 Skyscape
As Campion states, 'the sky is all around us. We would not be alive without it', and he continues by saying that it is widely assumed that the perception of the sky 'played an essential role' in ancient societies.[97] Harding *et al.* seem to be the first to use the term 'skyscape' within the ambit of archaeoastronomy relating the observation of the sky to life cycles, as an integral part of beliefs and local cult practices.[98] The concept of skyscape was further established by Campion and Silva for the Theoretical Archaeology Group conference (TAG) 2012.[99] For a contemporary observer of an ancient monument, skyscape may give a meaning and an indication of what the builders intended. As Silva put it, the skyscape plays an active role in structuring and being structured by

---

[94] Tim Ingold, 'The Temporality of the Landscape', *World Archaeology* 25, no. 2, Conceptions of Time and Ancient Society (1993); Tim Ingold, *The Perception of the Environment* (Oxon, Canada: Routledge, 2000).
[95] Tim Ingold, 'Taking Taskscape to Task', in Ulla Rajala and Philip Mills, eds., *Forms of Dwelling: 20 Years of Taskscapes in Archaeology* (Oxford: Oxbow Books, 2017).
[96] Julian Thomas, 'Concluding Remarks: Landscape, Taskscape, Life', in Rajala and Mills, *Forms of Dwelling*, p. 277.
[97] Nick Campion, 'Skyscapes: Locating Archaeoastronomy within Academia', in F. Silva and N. Campion, eds., *Skyscapes: The Role and Importance of the Sky in Archaeology* (Oxford: Oxbow, 2015), p. 8.
[98] Jan Harding *et al.*, 'Neolithic Cosmology and Monument Complex of Thornborough, North Yorkshire', *Archaeoastronomy* 20 (2006): p. 48.
[99] Fabio Silva, 'The Role and Importance of the Sky in Archaeology: An Introduction', in Silva and Campion *Skyscapes*, p. 4.

128   Different Approaches to Cosmology in Archaeology and Their Application to Maltese Prehistory

humans; myths may shape and be shaped by the sky and skyscape is an ideal depiction for metaphors, people's ideologies, beliefs and worldviews as they are open to control and thus tied to political strategy, as well as structuring time.[100]

5.5 Holistic Views
Brown suggests that the trinity of land, sea and sky cannot materialise through a scientific approach, but that it is a person's emotional experience materialised through observation and participation where place, time and space are all elements of a holistic awareness to the world one lives.[101] Silva adds the taskscape as a fourth element, bringing in notions of time, and argues that the union of this quadruple constitutes a holistic worldview or cosmology.[102] In affinity with the Maltese Archipelago, a fifth notion can be added, that of 'Islandscape for the ensemble of a holistic worldview, or in other words – a *cosmology* (see Fig. 5).

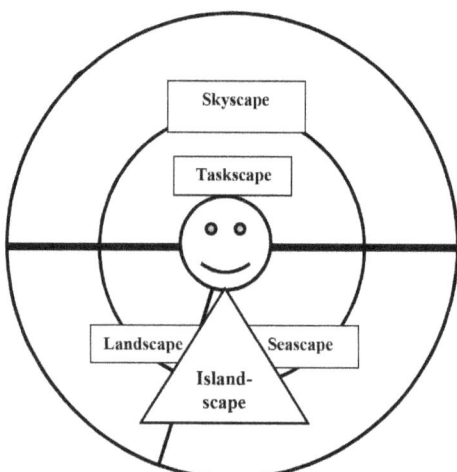

Fig.5 This figure, adapted from Brown and Silva, illustrates the human element's ensemble of the quintuple scape-concept consisting of land, sea, island, task and sky as proposed here.[103]

---

[100] Silva, 'Introduction', p. 3.
[101] Brown, 'Watching', pp. 22–23.
[102] Silva, 'Transition', p. 3.
[103] Daniel Brown, 'Skyscapes: Present and Past – From Sustainability to Interpreting Ancient Remains', in Silva and Campion, *Skyscapes*; Silva, 'Transition'.

Studies of alignments, orientations and the relationships between the sky, ancient monuments and landscape are carried out mainly within the purview of cultural astronomy, archaeoastronomy and ethnoastronomy where a considerable number of publications are, to a large extent, generated in the ambit of two academic organisations, the European Society of Cultural Astronomy (SEAC) and the International Society for Archaeoastronomy and Astronomy in Culture (ISAAC). According to Silva, both archaeoastronomers and landscape archaeologists have studied the orientation of European prehistoric structures, largely in isolation, and despite their similar interests, the two academic fields have failed to converge primarily due to differences of epistemology.[104] Belmonte on the other hand claims there is a conflict of interest as archaeoastronomy and cultural astronomy in its wider context abide in a 'no-man's land', and is neither endorsed by archaeology nor astronomy.[105] Both Henty and Silva have suggested ways to bridge this gap, mostly based on a holistic approach that combines archaeology and astronomy through a skyscape and cosmological concept.[106]

6. Conclusion

As referred to in Section 4.4, it is well documented in the archaeological record that the human mind, at least since the Upper Palaeolithic, possessed a cognitive ability to process complex spatial and temporal knowledge. Consequently, it is not inconceivable that some of that cognitive ability was directed not just at the landscapes below and around them but also to the skyscapes above them. This interconnection between the sky, humans and material culture – the very foundation of cosmology – is rarely explicitly explored in archaeology.

---

[104] Fabio Silva, 'A Tomb with a View: New Methods for Bridging the Gap between Land and Sky in Megalithic Archaeology', *Advances in Archaeologicial Practice: A Journal of the Society for American Archaeology* (2014): p. 24.

[105] Juan Antonio Belmonte, 'Is There a Conflict between Archaeology and Archaeoastronomy? An Astronomer's View', *Journal of Skyscape Archaeology* 2, no. 2 (2016): p. 259.

[106] Liz Henty, 'An Examination of the Divide between Archaeoastronomy and Archaeology', in Silva and Campion, *Skyscapes*, p. 30; Liz Henty, 'Skyscape Archaeology: An Emerging Interdiscipline for Archaeoastronomers and Archaeologists', *Journal of Physics: Conference Series* 685, no. 01200 (2016); Fabio Silva and Liz Henty, 'Editorial', *Journal of Skyscape Arcaheology* 1, no. 1 (2015): p. 1.

130   Different Approaches to Cosmology in Archaeology and Their Application to Maltese Prehistory

In Section 4, a number of theoretical concepts and analytical tools, developed and applied by archaeologists, have been presented. Each in its own way, engage with cosmology as a whole or, at least, with important elements of it, either explicitly or implicitly. Even processual concepts such as site catchment analysis, can be said to implicitly relate to cosmology in the sense that researchers try to uncover economic relationships between human societies and their landscape and therefore recover an important part of a society's cosmology. Such elements might find correlates and relationships among other elements such as the skyscape.

This work highlights the variety of ways in which archaeology engages with cosmological elements, some of which might prove useful to both archaeologists and archaeoastronomers in bridging the gap between their different but complementary epistemologies. Furthermore, the archaeological record of Malta's Temple Period and its general absence of cosmological considerations has been a primary concern of this research.

The preliminary results from this analysis seem to suggest a lack of involvement with the sky or cosmological considerations (with a few exceptions as already mentioned) within the general field of archaeological research and studies. However, further investigations ought to be conducted to reach a more consolidated conclusion.

Acknowledgements: I would like to thank Dr Reuben Grima, University of Malta and Dr Fabio Silva, University of Wales Trinity Saint David, UK, for always valuable input and directions. Finally, much gratitude to Heritage Malta for allowing me to conduct research on the prehistoric temple sites.

Petra Revisited: An Astronomical Approach to the Nabataean Cultic Calendar

Juan Antonio Belmonte and A. César González García

Abstract: Petra, the ancient Nabataean capital, has been one of our main research objectives since the first field campaign on site in 1996.[1] In December 2015 a new visit to the city was made to coincide with the winter solstice. Historical, ethnographic, epigraphic and archaeological records are compared in order to gain an insight on the Nabataean calendar. From this multi-source analysis two main points arise: the importance of both equinoxes and winter solstice within the lunisolar calendar and the relevance of some processions and pilgrimages. These combined with illumination effects observed and broadcasted at the principal monuments of Petra, and new important hierophanies, predicted in previous campaigns,[2] indicate the relevance of these dates at the time of the Nabataeans. Winter solstice was an important event in the Nabataean cultic calendar when a festival of the main deities of the city, the God Dushara and his partner the goddess Al-Uzza, was commemorated. This probably took the form of a pilgrimage, and related cultic activities, such as ascending from the temples at the centre of the city (presumably from Qsar el Bint and the Temple of the Winged Lions), to the Monastery (Ad-Deir) through an elaborated stone-carved processional way. The relevance of the spring and autumn equinox within the cultic calendar will also be emphasized in relationship to other sacred sites in Petra, such as the Zibb Atuff obelisks, and additional Nabataean sites.

In the two centuries before and after the birth of Christ, the ancient Nabataeans developed a singular culture in the harsh lands of Arabia Petraea at the frontier, then under the influence of the Hellenistic World. For generations, they carved out tombs and palaces for their kings and

---

[1] Juan A. Belmonte, 'Mediterranean archaeoastronomy & archaeotopography: Nabataean Petra', in A. Lebeuf & M. Ziolkowsky, eds., *Proc. V SEAC Meeting* (Gdansk: Institute of Archaeology, 1999): pp. 77–90.
[2] Juan A. Belmonte, A. César González García and Andrea Polcaro, 'Light and Shadows over Petra: Astronomy and Landscape in Nabataean Lands', *Nexus* 15 (2013): pp. 487–501.

Juan Antonio Belmonte and A. César González García, 'Petra Revisited: An Astronomical Approach to the Nabataean Cultic Calendar', *The Marriage of Astronomy and Culture,* a special issue of *Culture and Cosmos*, Vol. 21, nos. 1 and 2, 2017, pp. 131–50.
www.CultureAndCosmos.org

temples for their divinities in the sandstone slopes of the Shara Mountains and created one of the most fascinating places on Earth, the legendary city of Petra. There, they worshipped, above the rest of the gods, their powerful divinities Dushara and Al-Uzza.

The Nabataeans had a religion inspired by the forces of nature. This was a strange mixture of elements from pre-Islamic Arabs and Hellenistic, Egyptian and Middle Eastern influences. Divinities were often represented by stone blocks (betyls or *neshebet* in Nabataean) although in the late period (S. I-II CE), human or quasi-human forms were developed (Fig. 1).[3] The main male divinity was the god Dushara, or Dushares, very probably an astral god. His name means 'He of Shara', Shara being the mountain range bordering Petra to the east where the neighbourhood of Gaia, or Al-Ji (today Wadi Musa) was located. On certain occasions he seems to be a form of the god Al-Kutba ('the one who writes'). Dushara was identified by classical writers either with Zeus, Ares or Dionysos. Triclinium no. 17 in Bab es-Siq at Petra has an inscription dated 96/95 BCE devoted to Dushara: this is the oldest dated Nabatean inscription and probably the oldest dated betyl of the god.[4] According to Suidas [Theus Ares] 'the god Dushara is worshipped by them for him they honour above all others. The image is a black stone square and unshapen, four feet high by two feet broad – one foot in thickness – (see Fig. 1d). It is set on a base of wrought gold'.[5]

There has been much discussion regarding the head female divinity of the Nabataeans. In Bosra, the northern Nabataean capital during the reign of Rabel II (71–106 CE), the main goddess was Allat (or Al Lat), meaning simply 'The Goddess'. With a hypothetical solar character (As Sams, the Sun, was a female divinity in pre-Islamic Arabia), she has been identified with Athena and Atargatis (the Syrian Goddess). Her name has also been found in Palmyra (Syria), in Iriam (Wadi Rum, Jordan), and in Hatra (Iraq) and is indeed mentioned in the Quran. She is identified with the Alilat of Herodotus (Hist. III.8).

---

[3] John F. Healey, *The Religion of the Nabataeans: A Conspectus* (Leiden: Brill, 2001).
[4] Robert Wenning, 'The Betyls of Petra', *Bulletin of the American Schools of Oriental Research* 324 (2001): pp. 79–95.
[5] Robert G. Hoyland, *Arabia and the Arabs: from the Bronze Age to the coming of Islam* (Oxon: Routledge, 2001): p. 183.

Fig. 1 The principal Nabataean deities were a male god, Dushara, and a female goddess, Al-Uzza in Petra or Allat anywhere else. Other divinities were also worshipped, Manatu, goddess of fate among them. Panel (a) shows the carved betyls of these three deities at As-Siq. Panel (b) shows a relief of Manatu at Dejebel Madbah. Panel (c) a betyl of Al-Uzza discovered at the Winged Lion Temple. Panel (d) is the relief of Dushara at the Wadi Farasa trail. Panel (e) shows a Roman period female portrait of a deity found at the Qsar al-Bint temple, presumably the goddess Al-Uzza. Panel (f) shows a betyl of Isis-Al-Uzza found in excavations at Ez-Zantur Hill. Finally, panel (g) shows a Roman period young male sculpture, presumably Dushara, also found at the Qsar al-Bint temple. Diagram and images by the authors.

However, in Petra, this name is never found. Instead, inscriptions refer to the goddess Al-Uzza. Her name means 'The Most Powerful' and she was the personification of the planet Venus – often specifically assigned to the Evening Star, identified with Aphrodite and the Cananaean Astarte, and also with the Egyptian goddess Isis.[6] In inscriptions in Petra and Iriam, she is mentioned together with the 'Master of the House', Al-Kutba and, of course, Dushara of whom she could have been his partner or even his mother-cum-consort. It has been argued that Al-Uzza was possibly an alternative manifestation of Allat, e.g., the lion was the totem animal for both deities. Interestingly, she seldom had a double nature, being known as Al-Uzzatan, so that, according to many scholars, Allat and Uzza could go in tandem, being two faces of the same coin.[7]

What seems to be a completely different female personality is Manatu, a deity (or deities, since the word is a plural) who joins Dushara in several funereal inscriptions at Hegra (Median Saleh, Saudi Arabia).[8] She was the goddess of fortune and the numen of the city and is usually believed to be Dushara's daughter. Manatu, as Fate, was conceived as a trio of goddesses in antiquity.[9] It is possible that the moon was one of her manifestations and she could have been represented as three adjacent betyls in several niches across Petra and Hegra (see Fig. 1). Finally, stars were also very important in pre-Islamic Arabia and bright stars and asterisms such as Sirius (Sira') Canopus (Suhail) and the Pleiades (An-Nijm) were broadly used for guiding the caravans, establishing the dates of pilgrimages, regulating

---

[6] Gerald R. Hawting, *The Idea of Idolatry and the Emergence of Islam: From Polemic to History* (Cambridge: Cambridge University Press, 1999): pp. 141–44. Helmut Merklein and Yvonne Gerger, 'The Veneration Place of Isis at Wadi as-Siyyagh, Petra: New Research', *Studies in the History and Archaeology of Jordan* 7 (2001): pp. 421–32.

[7] See Hawting, *Idea of Idolatry*, p. 144. However, the Holy Quran reports in Sura 53: 'Have thou seen Al Lat, Al-Uzza and the other, Manat, the third one. These are only names that thou and thour ancestors had given to them. Allah has not put any power in them'. According to this, these three goddesses would be different personalities who have been identified as the daughters of Allah. However, this does not necessarily would apply for Nabataean times five centuries earlier. The controversy still continues. See, e.g., Michal Gawlikowski, 'Les dieux des Nabatéens', *Aufstieg und Niedergang der römischen Welt II* 18, no. 4 (1990): pp. 2659–77.

[8] John F. Healey, *The Nabataean Tomb Inscriptions of Mada'in Salih*, Journal of Semitic Studies Supplement 1 (Oxford University Press, 1993).

[9] Hawting, *Idea of Idolatry*, pp. 141–42.

calendars or making weather forecasts.[10] This could have been so in Nabataean times. Following this line of argument, it is worth mentioning that Nabataean Queens were often assimilated with Isis-Al-Uzza and that Sirius, in her name of Sopdet, was the main celestial aspect of Isis.

The Nabataean Kingdom had a lunisolar calendar of Babylonian type inherited of the Seleucid period.[11] We know very little of it apart from a few inscriptions and related historic information that will be discussed in subsequent sections, together with interesting ethnographic sources that will offer a new perspective. This information will give several clues for a coherent interpretation of the archaeological record within the context of cultural astronomy. The final outcome will be a proposal for a comprehensive Nabataean cultic calendar where pilgrimages would have played a most relevant role.

Ancient ethnohistoric sources
Stabro,[12] the great geographer of Amasiya, wrote that the Nabataeans 'worship the sun, building an altar on the rooftop of their houses, pouring libations on it every day and burning frankincense'. These sort of domestic cultic practices presumably had their official counterpart in the numerous high-places still standing in Petra and its surroundings. Some of them are of a size appropriated for a family group or clan, others have a much larger monumental character and, often, are located on mountain and hill tops, possibly related to important festivals.

Historical sources for cult practices in the Levant in the Hellenistic and Roman periods are scarce but most relevant. Lucianus [Dea Siria 49] reports that in the city of Hierapolis-Mambij, in northern Syria, 'the greatest festival they celebrate is that held in the opening of the spring'. Similarly, Procopius [2, 16] informed that the time for festivals 'was the season of the vernal equinox and at this season the Saracens always dedicated about two months to their god'. The Saracens was the name given by late Roman sources to the Arab speaking tribes who inhabited the ancient lands of the Nabataeans and still used, among others, the Nabataean Aramaic alphabet for their inscriptions. Hence, it is highly

---

[10] Miquel Forcada Nogués, T*ratado sobre los Anwa y los Tiempos de Ibn 'Asim* (Barcelona.: C.S.I.C. Fuentes Arábico-Hispanas 15, 1993).
[11] Mahdi Alzoubi, 'The Nabataean Timing System', *Acta orientalia* 69 (2016): pp. 301–9.
[12] Strabo [Geo 16, 4, 26].

probable that they were somehow related to the Nabataeans of the Kingdom period, if not their direct descendants.

This information can be further extended with the text of Protheus [3] who argued that the grammarian Nonnius reported that most of the Saracens gather at a certain sacred place – not explicitly mentioned – twice each year. The text continues: 'the first of these assemblies extends over a whole month and takes place about the middle of the spring, when the sun passes through the sign of Aries, while the other lasts two months; this they celebrate after the summer solstice'. The arrival of the new moon marked the time for these festivals.[13] To judge from the evidence of pre-Islamic northern Arabia, annual spring festivals are likely to have been held with aspects of pilgrimage attached.[14]

Another clue to the identity of the Saracens and their close ties with the Nabataeans is that the Christian apologist Jerome [Vita Hilarionis, 42-43] reported that they 'arrived at Elusa on the very day that the solemn festival [which] had brought all the people of the town to the temple of Venus; for the Saracens worship this goddess as the Morning Star and their race is dedicated to her cult'. It has been argued that the name of the city of Elusa is actually a Greek spelling for Al-Uzza and that the name of the city actually honoured the greatest Nabataean goddess.[15] Festivals dedicated to the greatest female divinity were amongst the most important in the Levant.

In this sense, the texts of the Christian apologist Epiphanius of Salamis (fourth century CE), who was born at Eleutheropolis (Bayt Jibrin) – a locality in ancient Edom and was hence a native of the region, are very important and elucidating. This is so even considering the time passed from the annexation of the Nabatean Kingdom and that his arguments were written to support his assertions and to show that the cult of the virgin had its pagan equivalences.[16] Epiphanius reported that 'in the idolatrous temple at Petra, ..., they praise the virgin with hymns in the Arabic language and call her khaabu ... in Arabic; and the child who is born of her they call Dusares. And this is also done that night in the city of Elusa, as it is there in Petra, ..., on the very night of the Epiphany [Contra Haeretici, Panarion, 51, 22]. Hence, according to this text, written c. 374–376 CE, the

---

[13] For an extended discussion of all these topics, see: Hoyland, *Arabia and the Arabs*, pp. 161–252 and note 6.1.
[14] Healey, *Religion*, p. 161.
[15] Hoyland, *Arabia and the Arabs*, note 6.1.
[16] Johannes H. Mordtmann, 'Dusares bei Epiphanius', *Zeitschrift der Deutschen Morgenländischen Gesellschaft* 29 (1875): pp. 99–106.

celebration of the birth of Dushara from the womb of the Virgin Mother (almost certainly his mother-cum-consort Al-Uzza) took place on the 6th January in Petra at dates closer to the winter solstice. This fact strongly suggested that Dushara was also probably considered a sort of solar deity.[17] In the same line of argument, it has been suggested that 'all these evidences may well point towards a solar character of the Nabataean chief god and incline us to consider the "very great festival on the very night of the Epiphany", mentioned in Epiphanius, as a survival of an earlier celebration of the winter solstice among the Nabataeans'.[18]

It rains very little in Petra and when it does, it is in the form of short showers concentrated in the winter months. The Nabataeans developed a fascinating system of water channels and cisterns to take the maximum advantage of this limited rainfall. Some are still functioning today. This system allowed them to develop a flourishing agriculture and to maintain a population of some 30,000 people throughout the year in Petra. It is hence highly possible that one of their most important festivals was celebrated at the time of the winter solstice, close to the period of maximum rainfall. Interestingly, Epiphanius also reports on the religious sect of Peraean Sampsaseans (or *Ash-shamsiun*, worshippers of the sun), 'which were also in Moabitis ad Nabatitis' and who continued the solar cult in the region after the fall of the Nabataean Kingdom [Panarion 53, 1 and 19].

Lastly and most relevantly, Epiphanius also informed that the fourth-century Arab tribes of southern Palestine and Transjordan performed pilgrimage during the month of Aggathalbaeith at a major sanctuary – 'al baeith', The House – in the region [Panarion 51, 24]. This could have been Petra, since Dushara is often called *mr' byt'*, Lord of the House in Nabataean inscriptions. The time of the pilgrimage corresponds to the month of Tishri, the first lunar month after the Autumn Equinox in the Nabataean calendar. The central day of the festival was on the 22nd day of Aggaathalbaeith, roughly corresponding to November 8th in the Julian calendar. This cult strongly resembles the later Muslim tradition of pilgrimage (*hajj*) at the month of Dhu al Hijja to the 'House of God' in Mecca, the Kaaba.[19]

Although it might be argued that all these sources are quite late for our interest, they clearly speak of a well-stablished tradition in the Nabataean

---

[17] Belmonte, *Nabataean Petra*, pp. 82–88.
[18] Moulay M. Janif, 'Sacred Time in Petra and Nabataea: Some Perspectives', *ARAM* 18–19 (2006/7): pp. 341–61.
[19] Janif, *Sacred Time*, p. 341.

lands that probably ought to be ascribed to earlier practices rooted to the apogee of the Nabatean Kingdom three centuries before. Consequently, we can be confident by assuming that the time of the equinoxes were important markers for the celebration of important pilgrimages in the subsequent lunar months and that a date close to the winter solstice was the moment for the celebration of one of the most important Nabataean festivals - the birth of their main deity Dushara from the 'Virgin' mother Al-Uzza. If this was the case, it certainly ought to be registered in epigraphy and in the archaeological record. We will return to these important issues later.

Petra 'today': the Ethnography
Jewish, Christian and Muslim traditions give a most relevant role to Aaron, Moses's brother. He died at the top of Mount Hor [Numbers 20: 22–29] which has been identified since time immemorial with the highest peak in the environment of Petra: Jabal Hārūn. This mountain has a close topographical relationship to the city as it was the most obvious high-place in the vicinity. The Jabal al-Nabī Hārūn, to give its complete Arabic name, is located c. 5 km south-west of Petra city center, easily attracting attention and stirring the imagination. The Finnish Jabal Haroun Project has carried out archaeological excavations of a Byzantine monastery located on the high plateau of the mountain since 1997. But the existence of the monastery represents only part of the whole spectrum of religious significance accorded to the mountain since Nabataean times, a significance that continued well into the Islamic period. The excavations revealed that the site was initially occupied by a major Nabataean shrine, dated to the apogee of the Kingdom. In the late fifth century CE, a Byzantine monastery was built at the site, but as early as the fourth century CE the mountain began to be associated with the biblical tradition of the Exodus, and attracted Christian pilgrimages.[20]

This importance persisted even after Muslims took control of the region in the mid-seventh century CE. This is demonstrated by the fact that in the beginning of the twentieth century CE some Bedouins still buried their dead with the bodies facing the Jabal, rather than Mecca, a custom clearly

---

[20] Zbigniew T. Fiema, 'Reinventing the Sacred: From Shrine to Monastery at Jabal Hārūn ', *Proceedings of the Seminar for Arabian Studies* 42 (2012): pp. 27–37.

confronting Islamic tradition.[21] This area was visited by Muslim pilgrims as early as the eighth century and the shrine at the very top of the mountain has a black obsidian stone (the 'Mirror of prophet Haroun') which is set in the north wall of the building and is still kissed by visitors.[22] This resembles the tradition in Mecca but also let us think of the black cult betyl of Dushara. Furthermore, nineteenth-century travelers to the site such as J. Ludwig Burckhardt or Gertrude Bell left clear references of the mountain as a most important landmark: its visibility was enough to sanctify a certain spot to slaughter a victim to Prophet Harun and built heaps of stones piled up. This tradition is perhaps a souvenir of Nabataean times since important high-places, such as the one of Madras, were established with a clear land- and skyscape connections to the mountain summit. Jabal Hārūn was indeed a pilgrimage goal of primordial importance.

In the early twentieth century, ethnographer Tawfik Canaan reported that twice in the year large numbers of Bedouin of many tribes flock to the sanctuary to make a pilgrimage, and to offer their prayers and vows. One of these *mawasim* (to use the local name) was the winter and the other the summer feast. The first falls in February (sic), and the second during the grape season and it was called *Darb an-Nabi Haroun*.[23] Interestingly, when the pilgrims ascend the mountain they sing:

> *O Aaron we are coming thirsty to you.*
> *In this summer heat driven by thirst.*
> *O Aaron! O great star!*
> *O father of high planets!*

This indeed is an astonishing set of verses since they clearly identify a relevant prophet of the monotheistic religions with a celestial body. This could easily be a reminiscence of much earlier Nabataean astral cults and related traditions.

More recent reports have slightly tingled Tawfik's arguments reporting that the two visits to Jabal Hārūn are made as a prayer asking for rain and

---

[21] Stewart Crawford, 'The Attitude of the Present Day Arab to the Shrine of "Mount Hor"', in George Livingstone Robinson, ed., *The Sarcophagus of an Ancient Civilization* (New York: MacMillan, 1930): pp. 285–300.

[22] Zeyad Al-Salameen and Hani Falahat, 'Religious Practices and Beliefs in Wadi Mousa Between the Late 19th and Early 20th Centuries', *Al Majaq al Urdunia* 3 (2009): pp. 170–204.

[23] Tawfik Canaan, 'Studies in the Topography and Folkrore of Petra', *Journal of the Palestine oriental Society* 9 (1929): pp. 136–218.

to offer Haroun the first grains and fruits. Hence the festivals, and the pilgrimages are made at the beginning of October (certainly the *Darb an-Nabi Haroun*) and in Spring rather than in February which apparently seems more logical.[24] Even today, ritual visits are made to the peak to ask for rain in times of draught.

Apart from the pilgrimage practices to Jabal Hārūn, the inhabitants, both Christians – now foregone – and Muslim Bedouins, of El-Ji (Wadi Musa) performed a procession lead by old and pious women and children at the beginning of the rainy season (i.e., close to the winter solstice).[25] This was called the *Amm al-Ghaith* procession as the goal was to ask the 'Mother of Rain' for showers during the drought season. During the procession, the participants sang:

Oh Mother of Rain!
Rain upon us: wet the mantle of our herdsman
Oh mother of Rain!
Rain upon us, with pouring rain allay our thirst
Oh Mother of rain!
Rain upon us.

Again we might be hearing echoes of a very ancient tradition, certainly pre-Islamic – due to God uniqueness – but not necessarily pre-Christian since Virgin Mary often plays that role in several ancient and modern Mediterranean cultures. It would be reasonable to assume that this is a reminiscence of a Nabataean cult related to fertility. Notwithstanding, a few researchers have claimed that *An-Nabi Haroun* and *Amm al-Gaith* must be Islamized aspects of some of the most important Nabataean deities.[26] From our point of view, if this were the case, they should have been Dushara and Al-Uzza, respectively.

This would mean that Jabal Hārūn might have sacred aspects related to Dushara and, possibly, to Al-Uzza as the main deities of the Nabataean pantheon at Petra. Hence, the festivals and pilgrimages connected to the mountain, even in the present day, would be clear souvenirs of the traditions which were analysed in the previous section. Is the *Darb an-*

---

[24] Al-Salameen and Falahat, *Religious Practices*, p. 183
[25] Al-Salameen and Falahat, *Religious Practices*, pp. 184–85.
[26] Anti Lahelma and Zbigniew T.Fiema, 'From Goddess to Prophet: 2000 Years of Continuity on the Mountain of Aaron near Petra, Jordan', *Temenos* 44 (2008): pp. 191–222.

Nabi Haroun a reminiscence of the pilgrimage certainly related to the month of Aggathalbaeith? Is the *Amm al-Gaith* procession a modern snapshot of a ritual celebrated since time immemorial in the area of Petra related to the birth of Dushara and the renewal of life and nature through the arrival of the vivifying rains? Is the Spring pilgrimage to the peak summit an echo of the Spring Festival celebrated by the Saracens? The answer to this related chain of questions is not easy. An answer in the affirmative would indeed be fantastic. Ethnographic and ethno-historic sources seem to back up each other but a doubt remains. Let us now turn into epigraphy and architecture for further insight.

The information of epigraphy: Nabataean calendar inscriptions
Nabataean inscriptions from Petra including calendar dates are surprisingly scanty. These include three examples only: an inscription in the shrine of Isis at Wadi Siyyagh dated in the 1st of Iyyar, in the 5th year of Obodas III, a dedication to Dushara found in the ruins of one of the churches, and hence presumably out of place, mentioning the month of Thebet, in the 11th year of Aretas IV, and finally another inscription found in the temple of the Winged Lions, at the city centre, dated to 4 (B)ab of the 37th year of Aretas IV.

However, this is not the case for the southern Nabataean capital of Hegra, today Mada'in Saleh, in the northwest of Saudi Arabia. There, the Nabataeans produced an incredible series of tombs of an outstanding beauty sculpted in the local sandstone for a period a little longer than a century or so with a maximum in the first century CE.[27] Of more than 80 monuments, as much as 22 of them have foundation inscriptions where the date of the tomb dedication was included. Seven months out of twelve possible are mentioned but only a few of them: Nisan, Iyyar, Abib or (B)ab and Thebet (*nysn, 'yr, b'b, Tbt*) are mentioned in several occasions.[28] An analysis of the data (Fig. 2) shows that this distribution cannot be random. For example, the probability that the month of Nisan is just by chance referred to in 8 of the 22 inscriptions is of less than 2% and of some 17% in the case of Iyyar and Thebet. This illustrates that these months should have had a sort of special character within the Nabatean calendar, at least as far as the cult of the dead was concerned.

---

[27] There is a need for serious archaeoastronomical research on the site, one of our group's dreams. A very preliminary approach can be sketched in: Ioannis Liritzis, F.M. Al Otaibi, B. Castro and A. Drivaliari, 'Nabatean Tombs Orientation by Remote Sensing: Provisional Results', *Mediterranean Archaeology and Archaeometry* 15 (2015): pp. 289–99.

[28] Healey, *Nabataean Tomb Inscriptions*, Appendix 1.

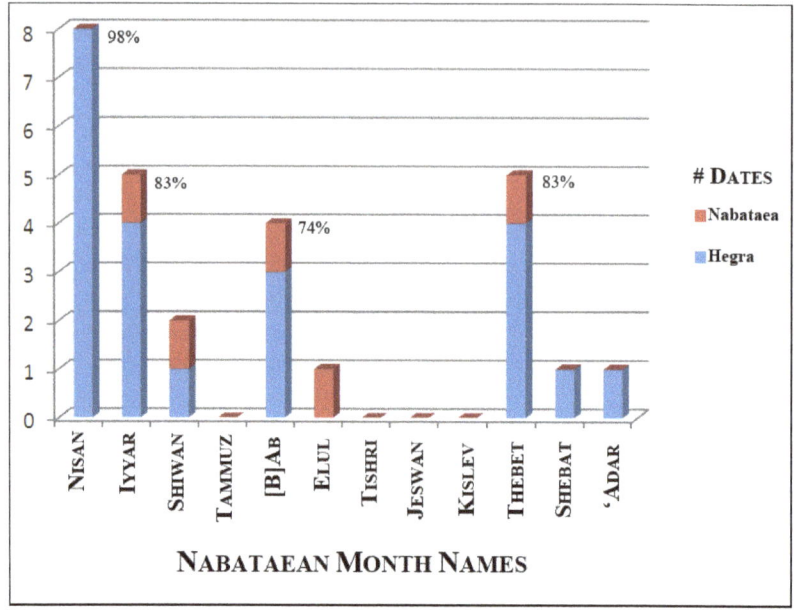

Fig. 2 Histogram of the months of the year mentioned in dated Nabataean inscriptions in the tombs of Hegra and elsewhere in Nabataea, including Petra, during the Kingdom period. Notice the relevance of the months of Nisan, Iyyar, (B)ab or Ab'ib and Thebet which are statistically significant (numbers indicate the probability of the month not being mentioned by chance). See the text for further discussions.

In the Babylonian calendar, the New Year was celebrated on the first moon after the vernal equinox, in Nisan.[29] The New Year at the city of Palmyra was celebrated the eleven first days of Nisan, the month that, in this particular case, included the spring equinox. Besides, the Hebrew New Year, presumably related to Passover on the first full-moon after the vernal equinox, was also celebrated in the first day of Nisan until it was moved to 1 Tishri in the fourth century CE. These historical facts, linked to the epigraphic proofs, strongly suggest that the months of Nisan (perhaps the first of the year including the vernal equinox), Iyyar (which on some occasions would be the first crescent of the year), and Thebet (the lunar

---

[29] Judith McKenzei, *The Nabataean Temple at Khirbet et-Tannur, Jordan: Final Report on Nelson Glueck's 1937 Excavation* AASOR 67 (Atlanta: American Schools of Oriental Research, 2013): p. 249.

month most likely including the winter solstice) must have been of importance within the context of the Nabataean cultic calendar. Shiwan and specially (B)ab, the first month after the summer solstice and the one including the heliacal rising of Sirius could also be relevant (see Fig. 2).

Interestingly, the few epigraphic sources do agree with what had been stressed in the two previous sections related to ethno-historic and ethnographic sources, emphasizing the importance of the vernal equinox and the winter solstice together with the months related to these two important time markers. Only the absence in the inscriptions of the month of Tishri is indeed challenging. It is now the time to complete the puzzle, moving to the terrain to look for the last clues offered by skyscape archaeology.

Land and skyscape in Petra: pilgrimages and hierophanies

As has been extensively discussed in previous sections, Nabataeans certainly celebrated religious festivals and performed pilgrimages to sacred shrines at the time of these celebrations. However, direct evidence on the terrain has so far been very scanty if it were not for the imposing ascending routes to several sacred sites in various spots of the city of Petra (Fig. 3).[30] The question is how, why and when these pilgrimage routes could have been used.

Petra has an impressive access when coming from the east: the ravine of As-Siq. This was not only the main entrance to Petra but also a Via Sacra in its own sanctity because of its high cliffs and narrow gorge.[31] The large amount of niches – a good number of them including betyls or composite divine images (see Fig. 1) –, and high-reliefs carved on its wall for more than a mile are explicit enough in this sense. Siq has been followed by pilgrims during the *Darb an-Nabi Haroun* for centuries and certainly earlier, and it is still admired today by hundreds of astonished visitors when entering the city.

---

[30] Lamia El Khoury, 'Nabatean Pilgrimages as Seen Through Their Archaeological Remains', *ARAM Periodical* 19 (2007): pp. 325–40
[31] Wenning, *Betyls of Petra*, pp. 79–95.

144 Petra Revisited: An Astronomical Approach to the Nabataean Cultic Calendar

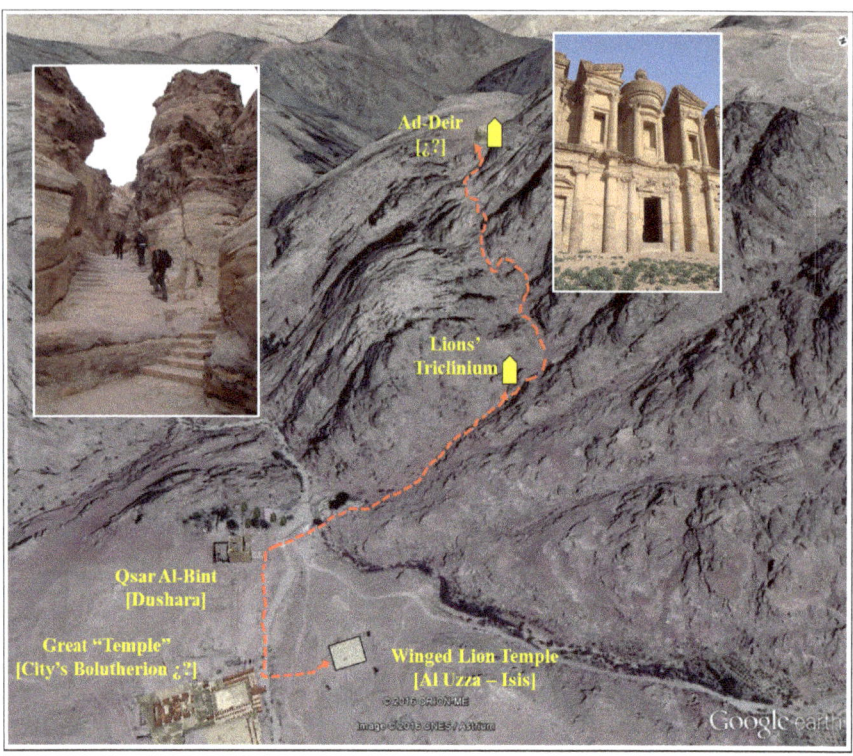

Fig. 3 The proposed pilgrimage route – mostly carved on the rock (upper left) – from the city centre (temples of Al-Uzza and Dushara) up to the mountain plateau where the Monastery (Ad-Deir) was sculpted on the sandstone cliffs (upper right) - the route passed by the Lions' Triclinium. This could be the most important Nabataean pilgrimage route in Petra related to Dushara's birth festival from his mother cum-consort Al-Uzza. Diagram is by the authors based upon an image courtesy of Google Earth.

Going out from the Outer Siq, and enclosing the nucleus of Petra to the east and southeast, there are a couple of flattened mountains with summits of slightly different heights. They are called Jabal Madbah and Jabal Khubza; both were reached by impressive mostly rock-carved monumental stairways and their summits were plagued by open-air shrines with conspicuous sky and landscape implications.[32] The latter was a sacred

---

[32] Belmonte, González García and Polcaro, *Light and Shadows*, pp. 497–502, and Table 1.

space for Dushara and Al-Uzza as demonstrated by local inscriptions and had several high-places scattered on the site which mostly were overlooking the city. They were orientated to the sunsets in early autumn (mid-October), corresponding to the month of Tishri.

Over the highest summit of the Jabal Madbah, there is one of the World's most impressive and best preserved Mountain High Places (a 'madbah', in Arabic) which was erected c. 7 CE. It consists of a large flat excavated court and a couple of rock carved altars. The larger altar is free on its four sides and it is facing west in the direction where the Jabal Hārūn can be seen in the foreground. Sunset over Jabal Hārūn summit occurs in 7/8 October in the Gregorian Proleptic calendar, which corresponds again to a period most likely coinciding with the first lunar month after the autumn equinox, i.e., Tishri, or even the month of Aggathalbaeith of later historic sources. Besides, the possible temporal link with the later tradition of *Darb an-Nabi Haroun* is indeed appealing and fascinating.[33] Near the 'madbah', and facing west, there is a beautiful rock-carved shrine, showing a triple-betyl framed by two pillars surmounted by crescent moons (see Fig. 1). Once again, early autumn sunsets, or the corresponding new moon visibilities, were observable from this spot. Consequently, this sacred area could have acted as a perfect site for ritual celestial observation at the time of autumn or spring pilgrimages.

Noteworthy, the lower and southern summit of this fascinating mountain was completely sculpted by the Nabataeans producing a nearly flat surface over which two large obelisks of more than six metres in height were left. These 'Zibb Attuf' look like memorial *nefeshes* for the dead without a pedestal, similar to those crowning other Nabataean monuments, but probably were neither *nepheses* nor *neshebets* (betyls). They may have been left standing in the quarry in respect and reverence for Dushara or, perhaps Al-Uzza.[34] The carving of these behemoths represented a huge challenge for the abilities of the Nabataean sculptors and there must have been a very important reason for their creation. In this sense, our team discovered two decades ago, and later confirmed, that the two obelisks are orientated precisely E-W ($\delta \approx \frac{1}{4}°$), so that only at dawn or dusk on the days

---

[33] A huge ceremonial platform with a similar orientation pattern (to the east in this case) has recently been identified in the valley below close to the route to Jabal Hārūn. See, Sara Parcak and Christofer A. Tuttle, 'Hiding in Plain Sight: The Discovery of a New Monumental Structure at Petra, Jordan, Using WorldView-1 and WorldView-2 Satellite Imagery', *Bulletin of the American School of Oriental Research* 375 (2016): pp. 35–51.

[34] Belmonte, *Nabataean Petra*, p. 77; Wenning, *Betyls of Petra*, p. 92.

closer to the equinoxes, the shadow of one of the obelisks would touch the other.[35] This was, and still would be, a perfect astronomical marker.

The centre of the city of Petra spread over a vast area *circa* one km wide between the slopes of Jabal al Khubza and the cliffs of Umm al Biyara. Here, three large free-standing structures (see Fig. 3) were erected by the Nabataeans in the first century CE. Among them are the Qsar al Bint, dated to c. 40 CE, and the Temple of the Winged Lions, dated to 27/28 CE, presumably the temples of Dushara and Al-Uzza, respectively.[36] As shown in Fig. 3, one of the most important pilgrimage ascending routes was the one departing from the area of these temples, at the city centre, and marching up the mountains to a high plateau open to the western horizon where the most imposing monument of Petra was built: Ad-Deir. Here, a possible astronomical orientation related to the time of the winter solstice was discovered two decades ago (Fig. 4).[37]

As the processional way goes up to Ad-Deir there are several attention foci. One of them is the Lions' Triclinium which is located in a chasm beside the path (see Fig. 3). Nicely carved in the sandstone, it has a reworked niche for a cult statue in its back wall (see Fig. 4). Data obtained in 2011 clearly showed that a light and shadow effect could be produced at the niche at the moment of the winter solstice thanks to the light entering across a now worn oculus in the façade.[38] In December 2015 the effect was observed in all its splendour at that precise moment. This hierophany is only produced at sunrise on dates close to the winter solstice since the triclinium would be in darkness for the rest of sunrises throughout the year.[39] Still more, considering the variation of the ecliptic obliquity, the effect must have been still more spectacular in Nabataean times (see Fig. 4). So, either on the way up in the early morning of the day of the pilgrimage, or on the way down at dawn, after a night of celebration at the

---

[35] Belmonte, *Nabataean Petra*, Table 1; Belmonte, González García and Polcaro, *Light and Shadows*, Table 1.

[36] See, Ian Browning, *Petra* (London: Chatto and Windus, 1989); and Peter Alpass, 'The Basileion of Isis and the Religious Art of Nabataean Petra', *Syria* 87 (2010), pp. 93–113.

[37] Belmonte, *Nabataean Petra*, Fig. 6. Qsar al Bint and the Winged lion temples were also astronomically orientated to conspicuous star and asterisms, Table 1.

[38] Belmonte, González García and Polcaro, *Light and Shadows*, Table 1 and pp. 489–95.

[39] This could be appreciated by McKenzie's team in late January when the effect is not so impressive. McKenzie, *Khirbet et-Tannur*, p. 250.

Deir, the procession could stop at the Lions' Triclinium to glimpse such a spectacle, probably related to the rituals associated with the winter solstice and the birth of Dushara.

Fig. 4 The winter solstice light and shadow effect at the Lions' Triclinium. (a) Image of the façade with the 'restored' oculus. (b) Relief of one of the lions framing the main gate. (c) The light of the rising sun disk enters through the oculus above the gate. (d) This light illuminates the niche for a divine representation carved of the back of the sanctuary. The effect should have been much more impressive at Nabataean times, when the declination of the sun was 12' lower, with the image of the oculus exactly framing the niche. See the text for further details.

But the final objective of the pilgrimage was certainly Ad-Deir. Is this the temple of one of the Nabataean divinities or the unfinished burial place or cenotaph of one of their last kings? Its use as a church in the Byzantine period and its internal distribution strongly support Browning's assessment: Ad-Deir was 'a prominent festival venue, with an elaborated staged ascent to it and a vast court in front of it'.[40] Its astronomical orientation and the light and shadow effects produced at the moment of the winter solstice both at the môtab for the cult betyls (see Fig. 5) and at the nearby western horizon seem to ratify this line of argument.[41] The Nabataeans worshipped the môtab (*mwtb'*), the podium on which the stele/betyl was erected, the equivalent of the seat or throne of the deity.[42]

---

[40] Browning, *Petra*, pp. 190–95.
[41] Belmonte, González García and Polcaro, *Light and Shadows*, pp. 495–96.
[42] Healey, *Religion*, p. 158.

148   Petra Revisited: An Astronomical Approach to the Nabataean Cultic
      Calendar

Actually, traces of a rock-carved cult betyl, with Dushara's block proportions – possibly destroyed when the site was converted into a church, could still be appreciated at the centre of the môtab. The light and shadow effect of the double sunset phenomenon produced at Ad-Deir reaches its maximum precision and beauty at this position precisely, as we could verify in new observations taken at the moment of the winter solstice in December 2015 (Fig. 5). Interestingly, a very important fact is that the vaulted niche with the môtab was carefully positioned slightly off-centre relative to the doorway so that this impressive effect can occur.[43] It was hence an architectural design.

Fig. 5 Winter solstice sunset light and shadow effect in the interior of the Monastery. The first instance of the double sunset effect upon the môtab can be appreciated in the second row. Diagram by the authors.

The hierophany is spectacular and would have been observable for nearly a week before and after the winter solstice. Winter solstice sunset, as observed from the môtab itself, is produced in a peculiar way on a modified rock with the aspect of the head of a lion – the sacred animal of Al-Uzza. At present, the sun sets at least twice, first in the axis of the monument and then re-appears in the northernmost corner of the rock before its final disappearing. It is interesting to note that the anthropic modification of this rock at the area where this phenomenon occurs, as

---

[43] McKenzie, *Khirbet et-Tannur*, p. 250.

seen from the môtab of Ad-Deir, allows for such observation. The phenomenon would have been still more impressive two thousand years ago when the northern limb of the disk of the sun had a declination close to −23°½.[44] This ensemble of solar hierophanies confirms the idea of the Monastery as one of the most important sacred enclosures of the Nabataean realm and certainly of Petra as the goal of one of the most important pilgrimage routes. Ad-Deir possibly was the ideal place to celebrate 'on the very night of the Epiphany' – in dates close to the winter solstice – the birth of Dushara from his own mother-cum-consort Al-Uzza, the goddess of fertility. Echoes of this ancestral ritual could still be alive in the area of Petra as shown by the tradition of the *Amm al-Gaith* procession.

Conclusion: The Nabataean cultic calendar
The combination of classical historiography, ethnography, epigraphy and the archaeological record, interpreted at the light of cultural astronomy, and the authors would like to insist on that, strongly supports the idea of a Nabataean calendar centered in the cult of their deities and their ancestors framed within the Babylonian lunisolar calendar of the Seleucid Empire, but not fully restricted by it since the Nabataean calendar seems to have started in Nisan.[45] According to our proposal, the principal astronomical milestones of this calendar would have been:

- New Year's Eve on 1 Nisan;
- Full Moon after Spring Equinox (14 Nisan), in a clear parallelism to Jewish Passover;
- 1st Crescent after Spring Equinox (1 Nisan – New Year's Eve – or 1 Iyyar);
- with certain doubts, since evidence is not strong, the 1st Crescent including Summer Solstice (1 Tammuz);
- 1st Crescent after Summer Solstice (1 (B)ab);
- Heliacal Rising of Sirius (Isis) in c. 20/7 Julian. Hence in (B)ab.
- 1st Crescent after Autumn Equinox (1 Tishri);
- Aggathalbaeith in the 1st Lunar Month after Autumn Equinox (most likely Tishri);
- Winter Solstice and the Birth of Dushara, as celebrated during the Epiphany in late Roman times;
- 1st Crescent or Full Moon of Thebet, the lunar month after, or including, the Winter Solstice.

---

[44] Belmonte, González García and Polcaro, *Light and Shadows*, Fig. 6 & p. 496.
[45] Almost all the calendars of Asia Minor and the Near East in the Hellenistic and Roman Periods began the year in autumn. The Nabatean seems to have been an exception. Sacha Stern, *Calendars in Antiquity* (Oxford: OUP, 2012): p. 236.

150   Petra Revisited: An Astronomical Approach to the Nabataean Cultic
       Calendar

These dates would have been the moments for main festivals and celebrations which, in at least three occasions – the two equinoxes and the winter solstice – would have been performed in the form of ritual pilgrimages at attractive sacred spots in the vicinity of the city such as Ad-Deir (and the Lions' Triclinium), the high-places at Jabal Madbah and Jabal Khubza, or the summit of Jabal Hārūn proper. It would indeed be useful to analyse possible agricultural festivals, or any other kind of cultic festivities, but the evidence is so scarce and variegated that a detailed analysis would conflict with the objective of this paper centered on astronomical phenomena as calendar time-markers.

The Nabataean Kingdom had retained a lunisolar calendar of Babylonian type, but in 106 CE the Kingdom was annexed by Emperor Trajan. As implied by Babatha's Archive,[46] a new era was soon initiated 'according to Provincia Arabia' with New Year's Eve at 1 Nisan / Xandikos, corresponding to March 22 in the Julian Calendar and with a fully solar structure (length of 365 days with a bissextile after each four years).[47] Nabataean Lunar dates needed to adapt to these new circumstances, and they did it very well as proved by the Khirbet et-Tannur *parapegma*. However, this is another story![48]

Acknowledgments. The authors would like to acknowledge the useful suggestions of two anonymous referees and the excellent editing work of Frank Prendergast. This work has been financed under the framework of the projects P/310793 'Arqueoastronomía' of the IAC, and AYA2015-66787 'Orientatio ad Sidera IV' of the Spanish MINECO. JAB would like to thank the 'Sacred Sites' filming team in Petra in December 2015, and especially David Ryan, for such an unforgettable experience.

---

[46] Stern, *Calendars*, p. 112.
[47] Alan Edouard Samuel, *Greek and Roman Chronology: Calendar and Years in Classical Antiquity*, Handbuch der Altertumswissenschaft 1, part 7 (Munich: Beck, 1972), p. 177.
[48] Juan A. Belmonte, A. César González-García and Andrea Rodríguez Antón, 'Arabia adquisita: The Romanization of the Nabataean Cultic Calendar and the Tannur "Zodiac" Paradigm', invited paper delivered at the *First International Workshop on Archaeoastronomy of the Roman World*, Milan, November 2016, Springer-Verlag, in press.

Investigation of Solstice Horizon Interactions at Chacoan Monumental Architecture

Andrew M. Munro, Tony Hull, J. McKim Malville,
F. Joan Mathien and Cherilynn Morrow

Abstract: Multiple monumental structures built during the ninth through the twelfth centuries CE at Chaco Canyon, NM, are in locations where solstice sunrise or sunset visually interacts with horizon foresights. We report on the results of field survey and photo-confirmation of seven solstice foresight interactions at Chacoan Great Houses. These include two 'Early Bonito phase' (850–1040 CE) sites. A June solstice sunset (JSSS) horizon foresight is visible from Pueblo Bonito, including early (ninth century) portions of the structure. December solstice sunrise (DSSR) interacts with a foresight visible from Hungo Pavi kiva A. We also confirm solstice horizon foresights at five additional 'Late Bonito phase' (1100–1140 CE) Great Houses. These include DSSR at Kin Sabe, DSSR at the Peñasco Blanco McElmo unit, DSSR at the proposed Chetro Ketl McElmo unit (west room block), December solstice sunset (DSSS) at Hillside Ruin, and June solstice sunrise (JSSR) at Rabbit Ruin. Hillside Ruin, the Chetro Ketl McElmo unit, and Rabbit Ruin also participate in inter-site alignments to the cardinal directions.

Integration of this data with previous temporal analyses further highlights contrasting cultural intent between periods. A minority of Early and Classic Bonito phase Great Houses (850–1100 CE) are known to have been built at solstice foresight observing locations. During the Late Bonito phase, ten of thirteen (77%) newly built Great Houses are at such locations. Six of thirteen (46%) participate in inter-site alignments to the cardinal directions. Three of thirteen (23%) participate in both the inter-site cardinal alignment and solstitial foresight traditions. This provides direct evidence of common social intent and the growing importance of solar events for Late Bonito phase people at Chaco. These structures may indicate an interest in theophanies, i.e., conjoining the sacred sun at solstice with the cardinal directions and prominent features of the sacred landscape. They bolster the argument for either centralized leadership by an astronomically adept Late Bonito elite, or a religious revival among the Chacoan people after the severe drought of the 1090s CE.

Under the terms of a U.S. National Park Service field research permit, some location-specific site data has been deliberately withheld in this document, as required by the U.S. Archaeological Resources Protection Act of 1979.

Andrew M. Munro, Tony Hull, J. McKim Malville, F. Joan Mathien and Cherilynn Morrow, 'Investigation of Solstice Horizon Interactions at Chacoan Monumental Architecture', *The Marriage of Astronomy and Culture,* a special issue of *Culture and Cosmos,* Vol. 21, nos. 1 and 2, 2017, pp. 151–71.
www.CultureAndCosmos.org

152 Investigation of Solstice Horizon Interactions at Chacoan Monumental Architecture

Introduction
John Fritz[1] developed the first model for Chaco's ideological landscape. He suggested that fundamental relationships between individuals and the cosmos are encoded in architecture and the landscape, utilizing 'symbolic resonance' wherein experiences of a particular pattern at one scale can invoke experiences and meaning at other scales. Thus, the alignment of interior features of a kiva (i.e., round room, frequently identified as ritual space) to the cardinal directions of north and south is repeated both within Great Houses and on an inter-site scale. Within and beyond the canyon, further examples of intra-site and inter-site alignments to the cardinal directions have been proposed.[2]

Fieldwork conducted under National Park Service and BLM permits between 2008 and 2010 provided additional evidence that Chacoan Great Houses are not randomly oriented or placed. The majority of Great Houses conform to one or more of four traditions that are derivative of or dependant on observational astronomy. They are either: 1) front facing to the south-southeast (most to 151°–161°, a subset to 170°–172°), 2) front facing to the east-southeast (most to 113°–116°), 3) individually aligned and/or inter-site aligned to the cardinal directions, and/or 4) built at a location where solstice sunrise and/or sunset is observed to interact with a horizon foresight. Temporal analysis of how Great House building starts are associated with these four traditions provides new insights into shifts of cultural focus with respect to cosmology over time at Chaco.[3]

---

[1] John, M. Fritz, 'Paleopsychology Today: Ideational Systems and Human Adaptation in Prehistory', in Charles L. Redman, ed., *Social Archaeology: Beyond Subsistence and Dating* (New York: Academic Press, 1978), pp. 37–59; John M. Fritz, 'Chaco Canyon and Vijayanagara: Proposing Spatial Meaning in Two Societies', in Donald W. Ingersoll, and Gordon Bronitsky, eds., *Mirror and Metaphor* (Lanham: University Press of America, 1987), pp. 314–49.
[2] Stephen H. Lekson, *The Chaco Meridian: Centers of Political Power in the Ancient Southwest* (Walnut Creek, CA: Alta Mira Press, 1999); Anna Sofaer, *Chaco Astronomy, An Ancient American Cosmology* (Santa Fe: Ocean Tree Books, 2008); Ray A. Williamson, *Living in the Sky: The Cosmos of the American Indian* (Norman: University of Oklahoma Press, 1984), pp. 132–40.
[3] Andrew M. Munro and J. McKim Malville, 'Ancestors and the Sun: Astronomy, Architecture and Culture at Chaco Canyon', in Clive Ruggles, ed., *Archaeoastronomy and Ethnoastronomy: Building Bridges between Cultures*, (Cambridge: Cambridge University Press, 2011), pp. 255–65; Andrew M. Munro, 'The Astronomical Context of the Archaeology and Architecture of the Chacoan Culture' (PhD Thesis, James Cook University, 2012); J. McKim Malville and

Andrew M. Munro, Tony Hull, J. McKim Malville, F. Joan Mathien and
Cherilynn Morrow

To further test the identified pattern of astronomical associations with monumental architecture at Chaco, one previously published solstitial association required validation, and untested Late Bonito phase Great Houses documented at Chaco needed to be assessed.[4] This work was conducted under National Park Service permits during 2014 and 2015. During this study, we serendipitously identified an additional solstitial horizon foresight visible from Pueblo Bonito, validated William Calvin's published solstice horizon foresight at Hungo Pavi,[5] and assessed five additional Late Bonito or 'McElmo' phase sites. Each of the five was evaluated using published and archival material to assess their likely association to the Late Bonito period and tested for solstitial horizons. They include Kin Sabe, the Peñasco Blanco McElmo unit, a potential Chetro Ketl McElmo unit, Hillside Ruin, and Rabbit Ruin. We believe that the cumulative set of surveys conducted since 2008 now includes all Late Bonito Great Houses identified or proposed within the canyon.

The Late Bonito or 'McElmo' phase was defined by Vivian and Mathews and initially dated to approximately 1050 to 1124 CE. This determination was based on the presence of significant amounts of McElmo black-on-white pottery and architecture described as a compact, multi-story plan with kivas enclosed in house blocks. The masonry is cored, sometimes banded, but typically composed of large blocks of pecked and 'dimpled' sandstone with chinking between stones, which stands in contrast to the smaller tabular sandstone wall veneers in earlier Chacoan Great Houses. Kin Kletso was considered a typical example of a Late Bonito site with this type of construction and pottery.[6]

Later re-evaluation of this definition by Lekson refined the dating of these structures to the period 1100–1140 CE, and offered a suggested function of the buildings. His analysis of five sites from this period included

---

Andrew M. Munro, 'Houses of the Sun and the Collapse of the Chacoan Culture', in Fabio Silva, Kim Malville, Tore Lomsdalen and Frank Ventura, eds., *The Materiality of the Sky, Proceedings of the 22nd Annual SEAC Conference* (Ceredigion: Sophia Centre Press, 2016), pp. 245–53.
[4] Munro, 'The Astronomical Context of the Archaeology and Architecture of the Chacoan Culture', pp. 227–29.
[5] William H. Calvin, *How the Shaman Stole the Moon* (New York: Bantam, 1991), pp. 125–38.
[6] Gordon Vivian and Tom W. Mathews, *Kin Kletso: A Pueblo III Community in Chaco Canyon, New Mexico* (Globe, AZ: Southwest Parks and Monuments Association, Technical Series 6(1), 1965), pp. 109–10.

Tsin Kletsin, Wijiji, Kin Kletso, Casa Chiquita, and New Alto. Lekson suggested that these sites are often associated with roads and probably functioned as storage facilities.[7] Van Dyke has contested this idea.[8]

It is notable that the majority of the Late Bonito phase Great Houses previously tested are located at workable observation points for horizon foresights that mark solstice sunrises or sunsets. These include Wijiji and Kin Kletso,[9] as well as Casa Chiquita, Headquarters Site A, and Bis sa'ani. In addition, while Robert's Small Pueblo does not mark an observation point for such a horizon marker, it is 125 m from the workable DSSR observation site at 29SJ 2538/2539.[10] The only two previously tested Late Bonito phase Great Houses that have no demonstrated solstitial associations (New Alto and Tsin Kletsin) are both involved in inter-site alignments to the cardinal directions.[11]

Field Survey Methods
The field survey methods used for this study were consistent with those applied in Munro and Malville's previously reported work at Chaco. A Suunto Tandem magnetic compass and clinometer were used for preliminary measurements to identify features as potentially worthy of accurate survey. Measurement of inter-site azimuths and horizon features was performed using a Wild T-2 theodolite and a GPS receiver. All surveys were conducted in relation to observable surface features, and (for the unexcavated Chetro Ketl and reduced Kin Sabe sites) unmarked but well-documented surface locations.

Survey of inter-site alignment azimuths and horizon features was a primary objective. When measuring inter-site alignments or points to determine azimuths to potential horizon foresights, both the horizon altitude

---

[7] Stephen H. Lekson, *Great Pueblo Architecture of Chaco Canyon, New Mexico* (Albuquerque: University of New Mexico Press, 1984), pp. 224–56, 267–69.
[8] Ruth M. Van Dyke, 'Memory, Meaning, and Masonry: The Late Bonito Chacoan Landscape', *American Antiquity* 69 (2004), pp. 413–31; Ruth M. Van Dyke, *The Chaco Experience, Landscape and Ideology at the Center Place* (Santa Fe: School for Advanced Research Press, 2008).
[9] J. McKim Malville, *A Guide to Prehistoric Astronomy in the Southwest*, revised ed., (Boulder: Johnson Books, 2008), pp. 70–74.
[10] Munro and Malville, 'Ancestors and the Sun', pp. 255–65.
[11] Fritz, 'Paleopsychology Today'; Sofaer, *Chaco Astronomy, An Ancient American Cosmology*, pp. 95–98.

and azimuth were recorded for each point. In addition, each point was measured and recorded four times to enable calculation of standard error.

To establish the orientation of the reference line relative to true north, we obtained sequences of at least four measurements of the azimuth and altitude of the sun ('sun sights'), timed by the GPS receiver and/or reference to NIST time standard station WWV. The altitude measurements provide an additional check on the field measurements. Standard surveying procedures are followed by setting up back sights to be used as reference points at the start and completion of the sun sights.

Confirmatory photography was conducted for any solar horizon foresights identified for solstices. Photographic confirmation was routinely performed using the method suggested by professional photographer Patrick René; a standard #11 Welder's Shade and exposure bracketing to obtain clear definition of the solar disk against the horizon.[12]

Pueblo Bonito (860–925 CE)

Pueblo Bonito is the largest and best studied monumental structure at Chaco Canyon. It was one of the earliest Great Houses at Chaco, and it expanded and gradually reoriented over centuries.[13] Pueblo Bonito has a well-documented JSSR horizon foresight identified by Zeilik in the 1980s.[14] Based on the survey results from Kin Sabe (see below), a prominent mesa top feature on the western horizon had been identified by our field team as a possible foresight for JSSS from some location in the vicinity. Just prior to sunset on 20 June 2015 three team members observed the sunset from a public trail and parking area adjacent to Pueblo del Arroyo. The team wished to determine whether the horizon feature might operate as a JSSS foresight as observed from one of the Great Houses in the 'downtown Chaco' area. Based on visual observations of shadow

---

[12] Andrew M. Munro and J. McKim Malville, 'Archaeoastronomy in the Field: Methodologies Applied in Chaco Canyon', *Journal of Cosmology* 9 (2010): pp. 2147–59, at http://journalofcosmology.com/AncientAstronomy115.html [accessed 1 Aug. 2010]; Munro, 'The Astronomical Context of the Archaeology and Architecture of the Chacoan Culture', pp. 96–114.

[13] John R. Stein, Dabney Ford, and Richard Friedman, 'Reconstructing Pueblo Bonito', in Jill E. Neitzel, ed., *Pueblo Bonito: Center of the Chacoan World* (Washington, DC: Smithsonian Institution, 2003), pp. 33–60.

[14] Michael Zeilik, 'Keeping a Seasonal Calendar at Pueblo Bonito', *Archaeoastronomy; Supplement to the Journal for the History of Astronomy* 9 (1986) p. 79.

156   Investigation of Solstice Horizon Interactions at Chacoan Monumental
       Architecture

casting, the team determined that JSSS should be observable using this foresight from Pueblo Bonito.

On 21 June 2015 the team visually observed and photographically confirmed that a workable JSSS visual alignment exists to this foresight as viewed from the west wall of Pueblo Bonito. The foresight alignment is fully functional as viewed from the northwest section of that wall, which corresponds to the first phase of construction for the structure, dating to the late ninth or early tenth century CE.

Hungo Pavi (990–1010 CE)
William Calvin described a series of compass surveys he conducted from post-Chacoan Pueblo 'Anasazi' cliff structures inset into cliff alcoves or caves. Based on compass surveys, he proposed that kivas within ancestral Pueblo cliff dwellings at sites well to the north of Chaco, including Split-Level ruin, Perfect Kiva, and Betatakin, may have been sited to take advantage of alcove corners as sunrise or sunset foresights for December solstice. While he recognized that no convincing statistical case could be made for single foresight alignments in individual kivas, Calvin did suggest that a more convincing case could be made by identifying a pattern among a set of kivas. With these ideas in mind, his compass surveys among the structures at Chaco led him to propose a pair of solstitial horizon foresights visible from one of the kivas in the Great House of Hungo Pavi. He first describes the use of a foresight to the southeast; a, 'distant cliff rose up like a headland, forming a distinct step from the distant canyon floor'. Calvin noted that his proposed Hungo Pavi foresight works to 'corner' the sun. As it rose, it was 'cornered' in the frame established by the headland with the sun's disk intersecting both the horizontal horizon and the left side of the stepped foresight. Based upon a topographic analysis, Calvin went on to propose that the multi-story tower kiva of Tsin Kletsin Great House on South Mesa could have operated as a DSSS foresight as observed from Hungo Pavi.[15]

On 27 October 2014, our team conducted horizon theodolite surveys from the kiva depression in the west end of the plaza at Hungo Pavi, as well as from 'kiva A' of the structure. These locations were chosen to correspond to an unspecified observing location within a Hungo Pavi kiva provided by Calvin for his reported DSSR observation. Based on analysis of the survey

---

[15] Calvin, *How the Shaman Stole the Moon*.

results, we predicted that a prominent mesa edge would operate as a DSSR foresight from kiva A, in keeping with Calvin's reported alignment.

On 21 December 2014, team members observed and photographically confirmed the sunrise from Hungo Pavi kiva A. The DSSR alignment is not visible from the entire building due to parallax and the distance to the foresight; kiva A may have been specifically sited to be associated with this visual solstitial alignment.

Kin Sabe (est. 1100–1130 CE)
Because of its location on the south side of the Chaco Wash across from Pueblo del Arroyo, Kin Sabe has been subject to considerable erosion since it was first mentioned in 1877 by W. H. Jackson. Today, Kin Sabe is almost entirely reduced by a combination of erosion and materials salvage during the 1970s.

Because of its precarious position at the edge of the Chaco Wash, Kin Sabe did receive attention from archaeologists participating in major projects in the canyon for more than a century, including Pepper, Postlethwaite, and Hayes.[16] Nonetheless, the site is seldom mentioned in recently published literature. A hand drawn sketch map by Nels Nelson and photographs from the 1920 excavations under the direction of Edgar L. Hewett of the School of American Research/Museum of New Mexico provide our best evidence for this structure. Nelson's 1916 site plan shows three rows that are six rooms long and a kiva along the east side. The Chaco Research Archive has web-published some forty nine photographs associated with the site from multiple late nineteenth and early twentieth century expeditions. Judd visited the site in August 1920. During the evening of 12 August, two front walls collapsed. As documented by photographs on file at the Palace of the Governors Photo Archive, sometime after 12 August 1920 Hewett's excavation team uncovered at least six rooms. These figures and other photographs in this collection show before, during, and after views, as well as two masonry walls, each exhibiting late masonry types. These photographs from the 1920 excavation indicate that placement in the Late Bonito phase is reasonable based on both architectural form and masonry style. The photos depict a structure entirely consistent with Late Bonito 'McElmo unit' construction.

---

[16] Alden Hayes and Charles Harding III, *Excavation of Bc 51 [Kin Sabe or Bc 263]* (Chaco Culture NHP Museum Collection, Archive 86089/VA2103B, 1937).

158   Investigation of Solstice Horizon Interactions at Chacoan Monumental
       Architecture

The exact alignment of the site is difficult to assess, but its location is well documented in the archival material.[17]

On 31 October 2014, our team conducted a horizon theodolite survey at the now-reduced Kin Sabe Great House. The survey was conducted with the instrument placed at a documented Park Service site survey pin 3 m west of the wash's rim. It is notable that a small volume of residual large-block masonry is now exposed at the edge of the wash in this location, directly above a deposit of black-on-white potsherds that have apparently been deposited through ongoing erosion. Based on archived site data, including Judd's photographs and aerial imagery, we infer that this location was 3 m to 5 m directly west of the northwest corner of Kin Sabe prior to its reduction.

Fig. 1. Horizon foresight visible from the location of Kin Sabe, with confirmatory filtered DSSR photograph inset.

---

[17] Neil M. Judd, 'The Passing of a Small PIII Ruin', *Plateau* 39, no. 3 (1967): pp. 131–33;   The Chaco Research Archive, 'Kin Sabe' at http://www.chacoarchive.org/cra/chaco-sites/kin-sabe/ [accessed 20 Oct. 2016].

Based on analysis of the survey results, we predicted that a prominent mesa edge would operate as a DSSR foresight. On 20, 21, and 22 December 2014 team members attempted to collect photographic data to confirm this prediction, but obscuring clouds were present. On 23 December 2014, team members returned and successfully captured the sunrise photographically. Fig. 1 presents a photograph of the horizon as seen from the theodolite's position. The inset, filtered image provides confirmation of the observable DSSR event.

Peñasco Blanco McElmo Unit (est. 1100–1140 CE)
The McElmo room block at Peñasco Blanco is dated to the Late Bonito phase based on its architectural form and masonry style by Lekson. He noted that this building is placed at the top of a manmade terrace, and that, 'If, as seems likely, the terrace was prepared for the McElmo Ruin, this constitutes a great deal of site preparation for a rather small building'.[18]

On 29 October 2014, we conducted a horizon theodolite survey from the west end of this McElmo Unit. The theodolite was positioned at the centre of the standing west wall section. Based on analysis of the survey results we determined that a pair of small but prominent points on an otherwise flat horizon section should interact with the rising December Solstice sun. On 22 December 2014, the team observed the sunrise at the Peñasco Blanco McElmo Unit. However, unfiltered photography yielded overexposed images, and the filtered images did not clearly demonstrate the sun's position with respect to the foresights. On 24 December 2014, we returned and successfully captured the sunrise photographically. Fig. 2 presents an unfiltered photograph of DSSR as the Sun first broke the horizon centred between the two foresights on the horizon.

Each of the two foresights is approximately one solar disk from the Sun as it first breaks the horizon; visually the Sun bisects the space between them. To our knowledge this dual-foresight bracketing approach is unique in the literature. Absent the previously identified pattern of solstice sunrise and sunset associations with Late Bonito McElmo Units, we might hesitate to propose this as an operating DSSR foresight. Notwithstanding, this bracketing marker is less subtle than some of the horizon markers documented by McCluskey among Hopi Sun watchers at Walpi and

---

[18] Lekson, *Great Pueblo Architecture of Chaco Canyon, New Mexico*, pp. 94–109.

Shungopavi.[19] It is certainly functional and visually repeatable. It will be useful to determine if this type of foresight is present at additional Great Houses as they are surveyed in the future. If such foresights are identified at additional ancestral Puebloan sites, that will increase the degree to which this proposed foresight is convincing.

Fig. 2. Paired horizon foresights visible from the Peñasco Blanco McElmo unit bracket the unfiltered image of DSSR first light. Photo by Cherilynn Morrow, used with permission.

Proposed west room block at Chetro Ketl (est. 1100–1140 CE)
Excavations at Chetro Ketl were carried out under the direction of Edgar L. Hewett, initially in 1920–21 and again from 1929–37, the latter as part of the University of New Mexico field school. Paul Reiter, one of Hewett's field school students, prepared a plan view of this Great House that

---

[19] McCluskey, Stephen C., 'The Astronomy of the Hopi Indians', *The Journal for the History of Astronomy* 8 (1977): pp. 174–95.

includes the west building. Another student, Robert Coffin, created an interpretation of this site that Hewett would use in later publications.[20]

Because the west building/McElmo component at Chetro Ketl has not been excavated, dating for this component is by inference. Based on the original definition of ceramics found at Kin Kletso and comparison with those recovered from a 1920 sample from Chetro Ketl, Vivian and Mathews concluded a McElmo structural component was present. In his examination of the architecture of Great Houses in the canyon, Lekson states the ground plan for the construction phase dating to 1115–1140 CE includes the room block west of the site. A review of information that led to these conclusions follows.

The pottery assemblage for the McElmo period was defined as consisting of Escavada black-on-white, Gallup black-on-white, Chaco black-on-white (all mineral based painted ceramics), and McElmo black-on-white (a carbon based paint). Vivian and Mathews thought these types co-existed from 1025 to 1175 CE.[21] They proposed that:

> McElmo black-on-white as such, or as proto-Mesa Verde, has not previously been reported from this site [Chetro Ketl] (Hawley 1934). This appears to have been an oversight. Through the kindness of the late Stanley Stubbs we were able to examine 1,099 still unwashed sherds from Hewett's 1920 excavations in two locations in Chetro Ketl. This is an admittedly small sample from the tens of thousands of sherds that must have been taken from the work there.[22]

The sherds were from Kiva II (Layer 4 and to 3 feet from floor) and Room 2 of the main structure. Kiva II and Room 2 are part of later additions to Chetro Ketl. In all three samples, McElmo black-on-white was the dominant type, with Chaco black-on-white, Escavada black-on-white, and Gallup black-on-white present in slightly lesser proportions. Vivian and Mathews thought these data established the presence of McElmo black-on-white in at least some proveniences of Chetro Ketl in

---

[20] Edgar L. Hewett, *The Chaco Canyon and Its Monuments* (Albuquerque: University of New Mexico and School of American Research, 1936); Paul Reiter, 'The Ancient Pueblo of Chetro Ketl' (MA Thesis, University of New Mexico, 1933).
[21] Vivian and Mathews, *Kin Kletso*, p. 75; Lekson, *Great Pueblo Architecture of Chaco Canyon, New Mexico*, pp. 152–92.
[22] Vivian and Mathews, *Kin Kletso*, p. 77.

approximately the same proportions as in other contemporaneous Chaco Great Houses.[23]

Research by the National Park Service Chaco Project updates these statements. In his discussion of the architecture of the Great Houses, Lekson summarized the different forms of building construction through time. Between 1075 and 1115 CE the most massive construction phase included two sites, Pueblo del Arroyo and Wijiji, which predate the McElmo style ground plan that includes smaller rooms surrounding a circular structure/kiva. Lekson includes one of these structures outside the west wall of Chetro Ketl which he dates to the latter period, 1115–1140 CE.[24] Lekson indicates:

> The first major twelfth century building in the central canyon was the Kiva G complex at Chetro Ketl (Chetro Ketl XIIA, 1110–1115). This unit is constructed almost entirely of pecked massive sandstone, in a 'McElmo' style that would easily be lost at Kin Kletso, the 'McElmo' phase type site.[25]

In his presentation of the individual stages of construction at Chetro Ketl, Lekson does not elaborate on the subdivisions for Stage XII (dated 1090–1095 CE). However, he does discuss stage XIIIA (1095–1105 CE) and indicates Kivas G-1 and G-2 fell in this period. The Late Bonito or McElmo phase, thus, falls later than Vivian and Mathews suggested.

The question remains whether or not we can date the structure west of Chetro Ketl to the Late Bonito period. On his map of Chetro Ketl (his Figure 4.39) Lekson includes the unexcavated block located west of the main pueblo. It shows a circular structure within a rectangle, somewhat similar in form to the Kiva G complex (his Figure 4.41 a). In his discussion of the McElmo period, Lekson includes, 'the small room block appended to the west wing', as consistent in ground plan with, 'construction of a number of separate buildings, characterized by many small interior rooms and comparatively few round rooms', that are dated between 1115 and 1140 CE.[26]

---

[23] Vivian and Mathews, *Kin Kletso*, p. 79.
[24] Lekson, *Great Pueblo Architecture of Chaco Canyon, New Mexico*, pp. 66–73.
[25] Lekson, *Great Pueblo Architecture of Chaco Canyon, New Mexico*, p. 72.
[26] Stephen H. Lekson, *The Architecture and Dendrochronology of Chetro Ketl, Chaco Canyon, New Mexico. Reports of the Chaco Center, No. 6* (Albuquerque: Division of Cultural Research, National Park Service, 1983); Lekson, *Great Pueblo Architecture of Chaco Canyon, New Mexico*, pp. 153, 156, 263, 268.

In summary, this unexcavated building is plausibly assigned to the Late Bonito phase based on its rectilinear form, single kiva depression, and pottery types. With regard to the orientation of the building, there is some room for discussion. Because the room block has not been excavated, archaeologists can easily interpret piles of rock coming from walls at slightly different orientations. Considering that maps by Paul Reiter, Reginald Fisher, and Stephen Lekson were made at different times and under different field circumstances, this is not surprising. A photogrammetric map of the Chetro Ketl community was published by Drager and Lyons. Although the McElmo rectangular feature just west of the west wall of Chetro Ketl near the current visitor trail is small, the illustration was enlarged for analysis. Because the map has been rectified, it shows the most accurate placement and alignment of this component, and thus, we used it and Lekson's derived map to locate the structure.[27]

On 28 October 2014, our team conducted horizon theodolite survey from the proposed McElmo room block adjacent to the southwest corner of Chetro Ketl. The theodolite survey was conducted with the instrument placed at the westernmost end of the crest of a berm (likely a reduced wall), just south of the public trail that traverses the unexcavated structure. The specific location was identified based on surface features to be the east end of the southern wall of the proposed structure.

Based on analysis of the survey results, we predicted that a prominence on a stepped horizon section would operate as a DSSR foresight.

On 20 December 2014, our team collected photographic data in an effort to confirm this prediction. Fig. 3 presents a photograph of the horizon as seen from the theodolite's position. The inset filtered image provides confirmation of the observable DSSR event. Of note, the foresight is the same topographic feature that operates as the right side foresight for the 'bracketing' DSSR alignment at the Peñasco Blanco McElmo unit. While the inset image is adequate to confirm the alignment, the image quality is poor due to cloudy conditions. The solstice event is visible from the entire footprint of the structure.

---

[27] Dwight L. Drager and Thomas R. Lyons, *Remote Sensing. Photogrammetry in Archeology: The Chaco Mapping Project. Supplement No. 10 to Remote Sensing: A Handbook for Archaeologists and Cultural Resources Managers* (Albuquerque: Branch of Remote Sensing, National Park Service, 1985).

164   Investigation of Solstice Horizon Interactions at Chacoan Monumental
        Architecture

Fig. 3. Horizon foresight visible from the Chetro Ketl McElmo unit with confirmatory filtered DSSR photograph inset. Photos by Julia Munro, used with permission.

Hillside Ruin, 29SJ1175 (est. 1100–1140 CE)
What was known about Hillside Ruin prior to the NPS Chaco Project is reported by Judd, who provided a plan view that shows the southern and western walls and the kiva that was excavated. Masonry exposed in three trenches at the west end of the site was not the typical Bonito style; it rested on an adobe foundation that covered some units of the northeast foundation complex. Surface sherds, though few in number, were considered proto-Mesa Verde. Judd considered Hillside Ruin to be later than Pueblo Bonito and its unfinished northeast complex. Judd's photograph, Plate 47, clearly illustrates typical McElmo style masonry; Plate 45 indicates the location of three tests at the west side of the ruin.[28]

During his survey of this ruin for the NPS Chaco Project, Windes suggested Hillside was probably a two-story pueblo, mostly located at the

---
[28] Neil M. Judd, *The Architecture of Pueblo Bonito* (Washington, DC: Smithsonian Institution, 1964), pp. 146–53.

base of talus on bottom land, with the room block oriented east-west. It was probably two rooms deep. Windes suggests Hillside Ruin included twenty to thirty rooms, kivas, courtyards, and perhaps platforms and ramps.[29]

Windes' interpretation has not been unchallenged. A study by Stein and his colleagues consider this complex as part of a set of platforms that are part of a final component at Pueblo Bonito. They presented a plan view and perspective indicating how this late period (1115–1215 CE) may have been integrated with earlier construction at Pueblo Bonito. Roads and ramps are an important factor in their interpretative view.[30]

No matter which interpretation of Hillside Ruin is accepted, all investigators assign this site to the Late Bonito phase based on the masonry style pictured in Judd's photographs.

The potential for a DSSS foresight interaction from the vicinity of Hillside was identified by two of our team members while observing DSSS in 2014 on public trails in the area. On 20 December 2015 our team conducted horizon theodolite survey from Hillside. The theodolite survey was conducted with the instrument placed at the southeast extreme of the large kiva depression at the east end of the site.

Based on analysis of the survey results we subsequently confirmed that a prominent mesa edge on the horizon would operate as a DSSS foresight. Using GIS analysis, we also confirmed that Hillside Ruin lies directly on the north-south inter-site alignment to the cardinal directions between Casa Rinconada and New Alto identified by Sofaer.[31] Furthermore, it is possible to establish a cardinal east-west line from Pueblo Bonito, thorough Hillside and the Chetro Ketl McElmo Unit, to the front wall of Chetro Ketl itself. Thus, both Hillside and the Chetro Ketl McElmo unit lie on the cardinal east-west line between Pueblo Bonito and Chetro Ketl identified by Fritz.[32]

On 20 December 2015, our team also collected photographic data from both the east and west ends of Hillside Ruin. Fig. 4 presents a photograph of the horizon profile as seen from the theodolite's position. The inset,

---

[29] Thomas C. Windes, 'This Old House: Construction and Abandonment at Pueblo Bonito', in Jill E. Neitzel, ed., *Pueblo Bonito: Center of the Chacoan World* (Washington, DC: Smithsonian Institution, 2003), pp. 14–32.

[30] Stein, Ford, and Friedman, 'Reconstructing Pueblo Bonito', pp. 33–60; John R. Stein, Richard Friedman, Taft Blackhorse, and Richard Loose 'Revisiting Downtown Chaco', in Stephen H. Lekson, ed., *The Architecture of Chaco Canyon, New Mexico* (Salt Lake City: University of Utah Press, 2007), pp. 199–223.

[31] Sofaer, *Chaco Astronomy, An Ancient American Cosmology*, pp. 95–97.

[32] Fritz, 'Paleopsychology Today'.

filtered image provides confirmation of the DSSS event. The interaction is visible throughout the building's entire footprint.

Fig. 4. Horizon foresight visible from Hillside Ruin with confirmatory filtered DSSS photograph inset.

Rabbit Ruin, 29SJ390 (est. 1100–1140 CE)
Located north of Pueblo Alto, Rabbit Ruin is one of several larger houses in the Pueblo Alto Community. No excavations have been carried out at the site, but wall clearing was included as part of the NPS Chaco Project investigation at the Pueblo Alto Community. Dated to 1100–1140 CE, the site has approximately forty-one rooms and five kivas. It is located near road segment 43, which runs northwest from Pueblo Alto, passing Rabbit Ruin, and on to seeps on the east side of Clys Canyon. This road probably continues from the opposite side of the wash across North Mesa to Peñasco Blanco. Windes dated use of this road segment to 1050–1100 CE.

The configuration of the three sections of Rabbit Ruin, with two separate room blocks and part of a third, is typical of the McElmo pattern of rooms enclosing a kiva(s). Masonry was composed of rectangular blocks of dimpled sandstone in the typical McElmo style. In the easternmost mound, one tree-ring date is available and it indicates that construction of Kiva 3 occurred sometime after 1088 CE. The few ceramics found suggest an early 1100s CE construction. The small rooms were suggestive of a habitation site rather than a storage facility.[33]

Lekson includes Rabbit Ruin in the Late Bonito phase that he dates to 1115–1140 CE. He bases his decision on the masonry style exposed during the wall clearing and stabilization project at this site.[34]

On 30 October 2014, our team conducted horizon theodolite survey from the west kiva of the east room block ('kiva 3') at Rabbit Ruin. The theodolite survey was conducted with the instrument placed at the north side of kiva 3. The survey included the eastern and southern horizon, which incorporate views of both Pueblo Alto and New Alto.

Based on analysis of the survey results, we predicted that a visible mesa edge on a mostly flat horizon section would operate as a JSSR foresight. By combining the survey results with GIS analysis, we also confirmed that the east wall of Rabbit Ruin lies due north of the west wall of Pueblo Alto and is therefore integrated into the emblematic north-south inter-site alignment to the cardinal directions between Tsin Kletsin and Pueblo Alto, first identified by Fritz.[35]

On 19 June 2015, our team collected photographic data in an effort to confirm the JSSR prediction. Smoky haze due to forest fires west of our location rendered filtered images useless. On 20 June 2015, we collected additional photographic data in an effort to confirm the prediction. On this date, the haze provided enough natural attenuation to enable collection of unfiltered images that clearly show a defined solar disk against the horizon feature. Confirmatory photographic images were obtained; the interaction is visible throughout the building's entire footprint.

Summary and Interpretation: Patterns in Time
The following positive fieldwork results were obtained at Chaco Canyon during this study. We determined that a JSSS horizon marker is observable from the west wall of Pueblo Bonito. The previously published DSSR alignment to a horizon foresight from Hungo Pavi is confirmed and

---

[33] Thomas C. Windes, *Investigations at the Pueblo Alto Complex, Chaco Canyon, New Mexico, 1975–1979. Volume I: Summary of Tests and Excavations at the Pueblo Alto Community* (Santa Fe: National Park Service, Publications in Archeology 18F, Chaco Canyon Studies, 1987), pp. 77–79, 94, 133.
[34] Lekson, *Great Pueblo Architecture of Chaco Canyon, New Mexico*, p. 72.
[35] Fritz, 'Paleopsychology Today'.

operates from kiva A. A DSSR horizon foresight interaction was observable from Kin Sabe. A pair of horizon foresights bracket DSSR as observed from the McElmo unit northeast of Peñasco Blanco. A DSSR horizon foresight interaction is observable from the proposed unexcavated McElmo unit immediately west of Chetro Ketl. A DSSR horizon foresight interaction is observable from Hillside Ruin, and a JSSR horizon foresight interaction is observable on the nearly-flat horizon from Rabbit Ruin. Three of the studied structures also participate in inter-site alignments to the cardinal directions including Rabbit Ruin, Hillside Ruin, and the Chetro Ketl McElmo room block.

Previous work has demonstrated that Chacoan Great Houses and Great Kivas are not randomly oriented or placed. The overwhelming majority of studied monumental structures at Chaco conform to one or more of four traditions that are derivative of observational astronomy. They are either: 1) front facing to the south-southeast ('SSE', most to 151°–161°, a subset to 170°–172°), 2) front facing to the east-southeast ('ESE', most to 113°–116°), 3) individually aligned and/or inter-site aligned to the cardinal directions, and/or 4) built at a location that enables visual observation of an interaction between the Sun's disk and a horizon foresight during a solstice sunrise or sunset.[36]

Fig. 5 presents a histogram of Chacoan architectural associations with these four traditions. If we are correct in inferring Late Bonito phase construction dates for Kin Sabe, Rabbit Ruin, the McElmo room block at Peñasco Blanco, and the potential McElmo Unit at Chetro Ketl, we now see that ten new monumental structures were built at Chaco during this period that have horizon foresights for solstice sunrises or sunsets. As shown, the previously discussed four-tradition pattern is significantly strengthened by the newly acquired data, including reinforcement of the remarkable association of Late Bonito monumental architecture with solstices and inter-site alignments to the cardinal directions.

---

[36] J. McKim Malville and Andrew M. Munro, 'Cultural Identity, Continuity, and Astronomy in Chaco Canyon', *Archaeoastronomy: The Journal of Astronomy in Culture* 23 (2010): pp. 62–81; Munro, 'The Astronomical Context of the Archaeology and Architecture of the Chacoan Culture'; Andrew M. Munro and J. McKim Malville, 'Change and Continuity in the Architecture and Cosmology of Chaco Canyon', paper presented at the 2014 Theoretical Archaeology Group (TAG) conference in Manchester, UK; Malville and Munro, 'Houses of the Sun and the Collapse of the Chacoan Culture'.

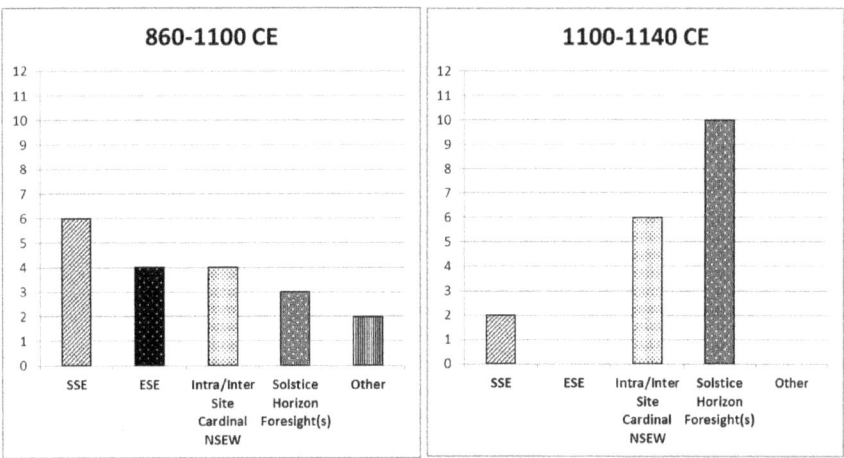

Fig. 5. Cumulative associations of Chacoan monumental building starts and major reorientations with four traditions.

The figure includes a cumulative count of associations with the traditions; for example, Pueblo Bonito is counted once for its SSE orientation, once for its solstice horizon foresights, and once for its reorientation to the cardinal directions. It should be noted that for the period 860–1100 CE, a total of four likely monumental Chacoan structures with solstice horizon foresights have been identified. They include Pueblo Bonito and Hungo Pavi (discussed in this study), as well Casa del Rio[37] and the Great Kiva at Marcia's Rincon.[38] Because it has not yet been possible to photographically confirm the proposed solstitial alignment at Casa del Rio, it is not included in the histogram.

The consistent associations of Late Bonito phase architecture with solstice observation locations and inter-site alignments to the cardinal directions are very provocative. These findings provide direct support for the idea that central planning occurred at Chaco during the Late Bonito phase, and Late Bonito Great Houses were constructed as monumental

---

[37] Andrew M. Munro and J. McKim Malville, 'Calendrical Stations in Chaco Canyon', *Archaeoastronomy: The Journal of Astronomy in Culture* 23 (2010): pp. 91–106.

[38] J. McKim Malville, 'The Enigmas of Fajada Butte', in Gregory E. Munson, Todd W. Bostwick, and Tony Hull, eds., *Astronomy and Ceremony in the Prehistoric Southwest: Revisited* (Albuquerque: Maxwell Museum of Anthropology, 2014), pp. 29–42.

architecture designed to incorporate cosmological references and ritual power. Lekson suggested these Late Bonito Great Houses served as storage facilities.[39] We believe that their purpose and function were much more complex, and their astronomical associations provide circumstantial support for Van Dyke's suggestion that Late Bonito Great House construction was undertaken to 'restore confidence in the rituals' that occurred in Chaco.[40]

Two of us (Munro and Malville) have previously identified this type of observable solstitial horizon alignment simply as 'calendrical'.[41] With the benefit of a much larger data set, we have increased confidence in a deeper inferential interpretation. Certainly, there was no functional benefit to generating a massive building program to create new Great Houses as solstice and cardinal direction markers. Chacoan people clearly had the capacity to find the cardinal directions and identify solar horizon calendrical markers. The overwhelming majority of the horizon profiles associated with the solstitial Great House sites are almost entirely flat – not terribly useful for finding dates other than the solstices. Two exceptions of note are Headquarters Site A, where foresights provide a granular solar calendar for almost the entire year including a mid-May date useful to help identify planting time, and Casa Chiquita, where June Solstice sunrise and sunset markers are also accompanied by a mid-May foresight.[42]

At least two Great Houses and one Great Kiva were constructed at such solstitial observation locations at Chaco in the late ninth or early tenth centuries. However, most monumental structures built at Chaco between 860 and 1100 CE do not manifest this characteristic. Rather, their front facing directional alignments appear to have had greatest importance. During the Late Bonito phase after 1100 CE, ten new, more compact monumental structures were erected in single construction phases that are now known to manifest this solstitial characteristic. Six are involved in inter-site cardinal alignments, and three of the six manifest both characteristics. This is evidence for a profound shift in cultural focus after 1100 CE. It provides unequivocal evidence for common social intent, and

---

[39] Lekson, *Great Pueblo Architecture of Chaco Canyon, New Mexico*, p. 269.
[40] Van Dyke, 'Memory, Meaning, and Masonry: The Late Bonito Chacoan Landscape', p. 423.
[41] Munro and Malville, 'Calendrical Stations in Chaco Canyon'.
[42] Munro, 'The Astronomical Context of the Archaeology and Architecture of the Chacoan Culture', pp. 125–30, 158–61.

may be indicative of astronomically-adept central leadership or of a spontaneous religious revival.

We consider the meaning and intent of the Late Bonito Great Houses to be of major importance in understanding the nature of Chacoan culture and how it changed in the early twelfth century. In particular, what did the Late Bonito Great Houses actually mean with their solstitial associations and cardinal direction alignments? What may they tell us about the religious and political processes that occurred during the decline in power and influence of the inhabitants of Chaco Canyon? It may be that these Great Houses represent an effort to construct theophanies to bring Chaco and its people into closer alignment with their own cosmovision as a 'centre place'. The construction of the sites may have been performative, and if they hosted ritual activities, those rituals would likely have been performative as well. These sites indicate a level of cultural continuity with modern Pueblo peoples for whom the solstices are calendrical anchors. December solstice is a date of special ritual importance. While the details of historic period December solstice ceremonials differ among Pueblo peoples, most relate this time to veneration of the Sun and performative rituals in support of its return for another agricultural season.[43]

Acknowledgements
This work includes data collected under National Park Service permits 14-CHCU-01, and 16-CHCU-02. It could not have been completed without the support of volunteers Gene McCracken, Donald D. Munro IV, and Julia Munro.

---

[43] McCluskey 'The Astronomy of the Hopi Indians', pp. 176–81; Alfonso Ortiz, *The Tewa World* (Chicago: University of Chicago Press, 1972), pp. 104–8, 555–56; Elsie C. Parsons, *Pueblo Indian Religion* (Lincoln: University of Nebraska Press, Bison Books edition, 1996), pp. 102, 122–23, 212, 496, 555–56.

Archaeoastronomy and Cultural Astronomy as Scientific Disciplines: Falsifiability and Photo Documentation

J. McKim Malville and John L. Ninnemann

Abstract: Archaeoastronomy is a discipline born at the intersection of cultural anthropology and the science of astronomy. As such, practitioners apply a variety of approaches. It is agreed, however, that casual naked-eye observation is not enough to convincingly assert the significance of prehistoric structures, alignments, and symbols. Although they can be equally creative, science differs from literary fiction in its strong preference for hypotheses that are testable and falsifiable by reproducible evidence. Digital photography is one of several tools in the field that bridges the gap between observation, essential documentation, and a search for meaning. A digital file yields both an archival image and unalterable EXIF time/date metadata. We present examples of the utility of digital photography in our studies of archaeoastronomy in the southwestern United States featuring Chimney Rock Pueblo, Yucca House Pueblo, Yellow Jacket Pueblo, Cliff Palace at Mesa Verde National Park, and Fajada Butte in Chaco Canyon.

Introduction

This paper addresses the necessity of including falsifiability in developing theories in archaeoastronomy and considers the usefulness of photography as one of several possible approaches for testing hypotheses. In our studies of the archaeoastronomy of the American Southwest, we have found photographic documentation to be a useful technique for testing predictions based upon field measurements. This paper discusses the importance of testable hypotheses and falsifiability in science, in general, and in archaeoastronomy, in particular. We consider the Russian proverb, 'trust but verify' (*Doveryai, no proveryai*) to be valuable advice.[1]

---

[1] S. Massie, *Trust but Verify: Regan, Russia, and Me: A Personal Memoir*. (Maine: Authors Publishing, 2013).

Reproducibility of results is a fundamental feature of science and should be an essential element of the methodology of archaeoastronomy.[2]

In addition to the imperative of falsifiability of hypotheses in archaeoastronomy, cultural context and holistic perspective have become increasingly important in understanding the meaning of the sky for the ancient peoples. Whenever dealing with possible astronomical alignments, photography can assist in documenting the archaeological context as well as providing a broad view of the geographic setting. Unlike laboratory research, which may primarily involve the acquisition of data points, the phenomenon in archaeoastronomy is often so multivalent and visually complex that its essence needs to be documented through photography. A research programme that yields only a number such as the declination of the Sun or Moon is probably an anachronism in today's archaeoastronomy with its emphasis on emic approaches, involving context, indigenous meaning, and the implications of human agency.

We note that photography is but one (albeit a very powerful one) of a number of approaches that can lead to the falsification of a hypothesis. A further benefit of photographic documentation of the actual predicted event is to experience the phenomenon *in situ*, giving the researcher additional interpretative power by sharing the experience. Photography of the back sight also provides important documentation as well as possibly revealing light and shadow effects. Examples of this methodology as applied in the American Southwest are given in detail.

Falsifiability of a Scientific Hypothesis

> *No amount of experimentation can ever prove me right; a single experiment can prove me wrong.* —Albert Einstein[3]

> *The criterion of the scientific status of a theory is its falsifiability, or refutability, or testability.* —Karl Popper[4]

---

[2] I. Lakatos, 'Falsification and the Methodology of Scientific Research', in I. Lakatos and A. Musgrave, eds., *Criticism and the Growth of Knowledge* (London: Cambridge Univ. Press, 1970), pp. 91–196.

[3] Charles M. Arthur Wynn and Sidney Harris, *The Five Biggest Ideas in Science* (New York: John Wiley and Sons, 1997), p. 107.

[4] Karl Popper, *Conjectures and Refutations: The Growth of Scientific Knowledge*, (New York: Routledge and Kegan Paul, 1963), pp.33–39.

The law of relativity is supposed to be true at all energies, but someday somebody may come along and say how stupid we were. We do not know where we are 'stupid' until we 'stick our neck out', and so the whole idea is to put our neck out. —Richard Feynman[5]

Scientific ideas can never be proven true; they can only be falsified. A basic feature of the scientific enterprise is the willingness to test ones hypothesis, to 'stick out ones neck' and risk having a favorite proposal rendered false. Science moves forward by falsification. A number of years ago, the physicist John Platt published a paper in *Science* titled, 'Strong Inference: Certain Methods of Scientific Thinking May Produce Much More Rapid Progress than Others'.[6] By strong inference Platt meant the systematic application of the old-fashioned method of inductive inference that was proposed by Francis Bacon (1561–1626).[7] The necessity of testing of scientific hypotheses was emphasised by Karl Popper.[8] Strong inference can be summarised as follows: devising multiple hypotheses to explain an observation, performing an experiment or observation that can falsify one or more of those hypotheses, recycling the procedure to make sequential hypotheses to gather more data about the original phenomenon. There was nothing really new about this proposal except for Platt's insistence on the systematic application of this procedure and, especially, robust falsification. Because science advances by disproof, there is no point in advancing hypotheses that are not falsifiable. Popper stresses the issue of demarcation, distinguishing the scientific from the unscientific – and makes *falsifiability* the demarcation criterion. What is unfalsifiable is classified as unscientific, and the practice of declaring an unfalsifiable theory to be scientifically true is pseudoscience. Every 'good' scientific theory contains a number of prohibitions. The more a theory forbids, the better it is. A theory, hypothesis, or proposal that is not refutable by any conceivable method is non-scientific.

---

[5] R. P. Feynman, R. R. B. Leighton and M. Sands, *The Feynman Lectures on Physics* (Reading: Addison-/weslkey, 1963), pp. 38–39.
[6] John R. Platt, 'Strong Inference: Certain Methods of Scientific Thinking May Produce Much More Rapid Progress than Others', *Science* 146 (1964): pp. 347–52.
[7] F. Bacon, *The New Organon and Related Writings* (New York: Liberal Arts Press, 1960).
[8] K. R. Popper, *The Logic of Scientific Discovery* (New York: Routledge, 2002).

There are those who object to Popper's ideas about falsification.[9] Some argue that Popper was unduly influenced by the revolutionary changes in physics that were occurring in the 1920s and 1930s and that hypotheses in science today are less dominated by fundamental premises such as that of General and Special Relativity. At the Oxford VII archaeoastronomy conference in Flagstaff, Arizona, David Whitley, a philosopher of science interested in rock art suggested that the proper methodology in science today is 'Post Positivism'.[10] He commented,

> Due to an unfortunate historical quirk, however, the earlier methods advocated before World War II by well-known philosophers like Popper and Hempel became enshrined in most physical and social science departments and they continued to be taught, at least into the 1990s despite the fact that philosophy of science *per se* had moved on.[11]

A concern shared by Whitley is the possibility that laboratory measurement is value-driven and therefore experiments that appear to falsify a hypothesis are themselves flawed by the motivations of the experimenter. Whitley suggests that hypotheses should be accepted or rejected not by falsification but by 'inference of the best hypothesis' from multiple lines of evidence. He contends, 'The use of a single data set or collection technique potentially yields a result that is a predetermined outcome of the assumptions of that approach'.[12]

Philosophers of science appear to separate themselves, in general, from practising scientists in their apparent distrust of empirical data and their preference for correction rather than elimination.[13] Falsifiability seems to be too disputative for philosophers. But, scientists are often disputative in their search for flaws in accepted theories. Abandoning the criterion of falsifiability could mislead the public about the nature of science and open

---

[9] See for example, David S. Whitley, 'Issues in Archaeoastronomy and Rock Art', in Todd Bostwick and Bryan Bates, eds., *Viewing the Sky Through Past and Present Cultures* (Phoenix: Pueblo Grande Museum Anthropological Papers No. 15, 2006); James Blachowicz, 'There is No Scientific Method', *New York Times*, (4 July 2016); James Blachowitz, 'Elimination, Correction, and Popper's Evolutionary Epistemology', *International Studies in the Philosophy of Science* 9 (1995): pp. 5–17.
[10] Whitley, 'Issues in Archaeoastronomy'.
[11] Whitley, 'Issues in Archaeoastronomy', pp. 86–87.
[12] Whitley, 'Issues in Archaeoastronomy', p. 87.
[13] Blackowicz, 'Elimination, Correction, and Popper's Evolutionary Epistemology'.

the door for pseudoscientists to claim that their ideas are equally scientifically legitimate.

Photo Documentation Methodology in Archaeoastronomy

When dealing with horizon phenomena our methodology should consist of calibrated photographs of both the horizon event as well as the back sight, so that other investigators could fully reproduce the data. In our studies of the American Southwest, once we had predicted a horizon phenomenon by means of Sun sights using a second-of-arc theodolite, we considered that our project was not completed until we had verified the prediction by means of photographing the event at the pre-cited location and date. Because of the effect of the changing obliquity of the ecliptic, the horizon photograph should have a well-calibrated azimuth scale. In the case of the America Southwest at the latitude of Chaco Canyon, Chimney Rock, and other sites, the effect of changes of obliquity over a period of 10 centuries is close to 10 minutes of arc, one-third the diameter of the sun, and hence in most cases what one sees today is close to that seen by the Ancestral Pueblos.[14] For earlier epochs at other sites the differences in azimuth can be calculated and applied to the photograph. Thus, the photograph of the moon rising above a recumbent stone or a trilithon of Stonehenge can be similarly documented.

Almost as important as photographic documentation of the foresight, is that of the back sight. Such a record can identify its precise location and its archaeological features. Photographic documentation is reproducible, and with digital cameras it is highly trustworthy. EXIF (exchangeable image file) metadata is produced by many cameras and smart phones. Most importantly the date, time, and often the location are recorded. Any attempt at photo editing or image modification would not escape detection. Metadata provides transparency and reproducibility of the results. Below we give an example of metadata (Table 1).

---

[14] Per the Jet Propulsion Laboratory, the obliquity is given by $e = 23°26'21.406" - 46".8368T - .000183T^2 + .002003T^3 +.....$ where T is the number of Julian centuries since J2000. Astronomical Almanac, (USA: US Governmental Printing Office, B52, 2010).

178 Archaeoastronomy and Cultural Astronomy as Scientific Disciplines: Falsifiability and Photo Documentation

DEVICE	Nikon D5300
ISO	1250
SHUTTER SPEED	1/30 sec f/8
FOCAL LENGTH	105 mm
DATE	9/27/16
TIME	7:19:09 PM
LOCATION	Chimney Rock
SIZE	22.0 MB

Table 1. EXIF METADATA (Peterson Ridge, Chimney Rock)

Archaeoastronomy of the American Southwest
This paper describes almost three decades of collaboration between Malville and Ninnemann in exploring the archaeoastronomy of the Ancestral Pueblos. Using Sun sights with a Wild T-2 second of arc theodolite, Malville made predictions, and Ninneman the expert photographer, checked up on them in the field, often in very challenging conditions. We have chosen five places to illustrate our work because of the limitations placed on figures in this publication. There are other Chacoan Great Houses that mark the solstices, most of which have been reported at SEAC conferences.[15] In our work we have also emphasized the importance of documenting back sights. We are interested in the processes by which astronomical observations were made and exactly where, when, and why an observer-designer stood. Unless archaeologically noteworthy back sights can be found, there can be no confidence that intentionality was involved. This uncertainty is particularly vexing in the case of claims that certain walls and lines of Great Houses in Chaco Canyon were

---

[15] J. M Malville and A. Munro, 'Houses of the Sun and the Collapse of Chacoan Culture', in Fabio Silva, Kim Malville, Tore Lomsdalen and Frank Ventura, eds., *The Materiality of the Sky, Proceedings of the 22nd Annual SEAC Conference.* (Ceredigion, Wales: Sophia Centre Press, 2016); Andrew Munro, Tony Hull, J. M. Malville, F. Joan Mathien, and Cherilynn Morrow, 'Investigation of Solstice Horizon Interactions at Chacoan Monumental Architecture', in this volume.

intentionally aligned to the solstices and lunar standstills.[16] In contrast, the north wall of the Great House at Chimney Rock is aligned with June Solstice Sunrise as viewed from a very identifiable bedrock basin. The alignment of walls to the cardinal directions, such as north-south wall of Pueblo Bonito and the alignment of its Great Kiva, is a different matter because such orientations could have been established by means of shadow casting using a *gnomon*.[17]

1. Winter solstice over Fajada Butte

Fajada Butte (Fig. 1) is the single most dramatic topographic feature of Chaco Canyon, and it may have been a spiritual magnet for early migrants or seasonal visitors. The butte was the sacred centre of the Fajada Gap community, a group of 54 small house sites which flourished in the 900s and early 1000s CE.[18] The community contains one of the earliest Great Kivas to be constructed in Chaco Canyon. The presence of a Great Kiva and evidence of careful observations of the sky at a small house community is noteworthy because it has been generally believed that Chaco's 'high' culture had developed during the period of Great House construction. Previous investigations of astronomy in Chaco Canyon had primarily focused either upon the three-slab site of Fajada Butte or upon Great Houses and Great Kivas.[19] Very little has been written about the astronomy of the small house communities that preceded the Classic Bonito phase. In 2011 it appeared to Malville that the December solstice sun might rise over Fajada Butte as viewed from the isolated Great Kiva in Marcia's Rincon, so a permit was obtained from the National Park Service to test this prediction, which was photographically confirmed by Ninnemann and G.B. Cornucopia on very cold December mornings of 2011.[20]

---

[16] J. M. Malville and A. Munro, 'Cultural Identity, Continuity, and Astronomy in Chaco Canyon', *Archaeoastronomy: The Journal of Astronomy in Culture* 23 (2010): pp. 62–68.
[17] J. M. Malville, *Guide to Prehistoric Astronomy in the Southwest* (Boulder: Johnson Books, 2008), p. 56.
[18] Thomas C. Windes, *The Spadefood Toad Site: Investigations at 29SJ 629 in Marcia's Rincon and the Fajada Gap Pueblo II Community, Chaco Canyon, New Mexico*. Reports of the Chaco Center No. 12, Branch of Cultural Research, Division of Anthropology, National Park Service (Santa Fe, 1993).
[19] Malville, *Prehistoric Astronomy*.
[20] J. M. Malville, 'The Enigmas of Fajada Butte', in Gregory E. Munson, Todd W. Bostwick, and Tony Hull, eds., *Astronomy and Ceremony in the Prehistoric*

180   Archaeoastronomy and Cultural Astronomy as Scientific Disciplines:
       Falsifiability and Photo Documentation

Fig. 1. Theodolite at the edge of the Great Kiva of Marcia's Rincon. The theodolite is located at the back sight above the likely antechamber of the unexcavated structure. Inset: Filtered images of December Solstice Sunrise over Fajada Butte. Both photographs by Ninnemann.

Construction of the Great Kiva in Marcia's Rincon occurred before the start of rapid growth of construction of the Great Houses sometime between 1020 and 1040 CE. During the Classic Bonito Phase, 1020–1100 CE, Chaco Canyon became the centre for periodic festivals, pilgrimages, and trade fairs, which drew in participants from outlying communities in the San Juan Basin. The regional festivals probably took place near winter solstice when the agricultural fields were fallow and the San Juan River could be easily crossed.[21]

Modest celebrations may have taken place at some of the early Great Houses and small house sites. In particular, Fajada Butte may have been an important element in early ceremonialism in Chaco Canyon. It dominates the view as one approaches the canyon from the north or south and has the

---

Southwest: Revisited (Albuquerque: Maxwell Museum of Anthropology, 2014), pp. 29–42.
[21] Malville, *Prehistoric Astronomy*, pp. 49–79.

quality of verticality that mythologies around the world associate with sacred mountains and the dwelling places of the gods.[22]

In Chaco Canyon, there are some 140 stairways that are associated with Pueblo occupation. Many do not seem to serve practical purposes: they are too steep and precarious to descend carrying a load. In the words of Hayes, 'Stairways are commonly associated with roads, but in many cases, rather elaborate arrangements for getting up or down a cliff were found where no other evidence of a road was found either above or below'.[23] It is possible that some of these stairways are examples of ritual ascent or descent, intended for ritual movement rather than for movement of people or trade goods into or out of the canyon.

Ritual ascents on its southwestern ramp could have taken place starting in the 900s CE. The ramp begins at the toe of the lowest talus rock slope close to a fire box with fire-reddened vertical slabs. The three sections of the ramp rise a total of 280 vertical feet and involve a mixture of masonry retaining walls, masonry stairs, carved hand and toe holds, carved stairs, and wooden platforms.[24] At the top of the ramp there is another fire reddened slab box, of which, each side is one metre.

The unexcavated Great Kiva of Marcia's Rincon (29SJ 1253) is the second largest of the 21 Great Kivas to have been built in Chaco Canyon.[25] Its diameter of 20 m is more than a standard deviation from the average of all the Great Kivas in the canyon, which is 15.8 m. The surface ceramics of the site were dominated by Red Mesa Black-on-white, which date from the 900s to early 1000s CE. The location of the Great Kiva places it within the shadow cast by Fajada Butte at sunrise on December solstice. Celebrants emerging from the Great Kiva at dawn, an experience miming their ancestors when they emerged from the worlds beneath ours, would have observed the Sun rising over the summit of Fajada Butte, and perhaps fires burning in the two fireboxes.

---

[22] Ruth M. Van Dyke, *The Chaco Experience, Landscape and Ideology at the Center Place* (Santa Fe: School for Advanced Research Press, 2008); Mircea Eliade, *Patterns in Comparative Religion* (New York: World Publishing, 1963), p. 99.

[23] Alden C. Hayes, 'A Survey of Chaco Canyon Archaeology', in A.C. Hayes, David M. Brugge and W. J. Judge, eds., *Archaeological Surveys of Chaco Canyon, New Mexico* (Albuquerque: University of New Mexico Press 1981), pp. 1–68.

[24] Dabney Ford, 'Architecture on Fajada Butte', in Windes, *The Spadefoot Toad Site*.

[25] Ruth Van Dyke, 'Great Kivas in Time, Space, and Society', in Stephen Lekson, ed., *The Architecture of Chaco Canyon, New Mexico* (Salt Lake City: The University of Utah Press, 2007), pp. 93–126.

2. The Sun and Moon at Chimney Rock

In the spring of 1988, the archaeologist Frank Eddy, who had excavated Chimney Rock in the 1970s, suggested that the high mesa with its excellent horizon had the feeling of an astronomical observatory.[26] Inspection of the topographic map of the Chimney Rock mesa suggested that the sun at June solstice might rise between the twin rock towers as viewed from the Great House. However, a visit on solstice dawn falsified that hypothesis. The Sun rose some distance to the south. Finally, near the end of summer, calculations provided encouragement that the major standstill Moon might rise between the towers, and an expedition to Chimney Rock on the night of August 8, 1988, confirmed that prediction with the photograph contained in Figure 2.[27]

Over a period of several years, we located two spots for observing the June solstice sunrise. At one location, marked by the only tower on the high mesa, the Sun rises at a well-defined depression on the horizon. At the second location, marked by a basin carved into the bedrock, the June solstice Sun rises along the northern wall of the Great House.

---

[26] Frank Eddy, 'Archaeological Investigations at Chimney Rock Mesa 1970–1972', *Memoirs of the Colorado Archaeological Society* 1 (Boulder: US Forest Service and the University of Colorado, 1977).

[27] J. M. Malville, 'Chimney Rock and the Ontology of Skyscapes: How Astronomy, Trade, and Pilgrimage Transformed Chimney Rock, Southwestern Colorado', *Journal of Skyscape Archaeology* 1, no. 1 (2016): pp. 39–64.

J. McKim Malville and John L. Ninnemann 183

Fig. 2. Major lunar standstills as viewed from the Chimney Rock Great House. Inserts: left, discovery photograph, 8 September 1988 by Malville; right, 7 January 2007 by Ninnemann.

Across the Piedra River and upward to the top of Peterson Ridge one finds the unexcavated C-shaped unit-type pueblo, 5AA8.[28] Twice a year, near the dates of equinoxes, the Sun can be seen rising between the two chimneys. This is one of nine pueblos built along the rim overlooking the Piedra River. In addition to being larger it is the only one that is rotated away from north-south to the east to face the Chimneys. Its two-storey room block overlooks an enclosed plaza containing a kiva depression and beyond to the double chimneys. There are two larger kiva depressions located to the north and south of the structure, indicating the special ceremonial role of the site.

We speculated that a similar structure might have been built on Peterson Ridge to observe June solstice sunrise. Our team searched the ridge and located the spot from which one could observe the June solstice sun beyond the Chimneys. At that location, the chimneys are in line and one only sees a single spire. Although it is a very dramatic sight, a very thorough search revealed that there are no archaeological features in this area. Without an identifiable back sight, that spot cannot be identified as a prehistoric Sun watching station. We conclude that the Chimney Rock people could not cross the Piedra River around June solstice when it was in

---

[28] Malville, 'Chimney Rock and the Ontology of Skyscapes', pp. 46–48.

flood and that the residents of Peterson Ridge had different priorities and were not interested in constructing a monumental building dedicated to the Sun.

3. Yellow Jacket
The Yellow Jacket Pueblo, 5MT5, lies on a 100-acre peninsula at the head of Yellow Jacket Canyon in southwestern Colorado in the Great Sage Plain. During the Pueblo III period it may have contained as many as 1200 rooms and 195 kivas.[29] Its population has been estimated at between 850 CE and 1360 CE. For a time, it was the largest settlement of the ancient Pueblo world. Its fertile neighborhood may have been the breadbasket for the ancient Mesa Verde world and beyond. Today the region is known as the 'Pinto Bean Capitol of the World'. The red soil at Yellow Jacket is fertile loess blown in from the Kayenta region around Monument Valley.[30]

Fig. 3. Sunrise over El Diente and Lizard Head. Inset. The shaped monolith of Yellow Jacket, 5MT5 is the back sight. Photography by Ninnemann

During that summer of 1986 we confirmed that the top of a standing monolith at the south end of the pueblo had been shaped into a wedge, which aligned with the first gleam of the Sun on June solstice. Also, we

---

[29] Malville, *Prehistoric Astronomy*, pp. 102–21.
[30] Malville, *Prehistoric Astronomy*, pp. 102–21.

discovered in the deep sagebrush along the solstice line three additional fallen monoliths, a wall, and a small shrine at the edge of the mesa. A line, established by 20 survey points was within 1/10 of a degree of our estimate of the first gleam of the sun on summer solstice of 1200 CE. We also replicated the fallen wall adjacent to the monolith and photographed the shadow of the sharp top of the monolith at solstice sunrise. That wall could have contained marks that identified the days near solstice. To the south of El Diente is a prominent spire, Lizard Head Peak (Fig. 3). It is such a dramatic spire on the horizon, it seemed possible that it may also have been incorporated in the solar ceremonialism of Yellow Jacket. The Sun rises above Lizard Head on June 5, a little more than two weeks before June solstice. As was perhaps the case in Chaco Canyon, sunrise above Lizard Head may have marked the 30-day period when the next full Moon was the full Moon closest to solstice; the occasion for celebrations and trade fairs.

4. December Solstice Sunset over Towaoc at Yucca House

Yucca House is a unexcavated Pueblo III village with as many as 600 rooms and 102 kivas. It lies in the significant area between Mesa Verde and the Sleeping Ute Mountain, the southern gateway to and from the Great Sage Plain in Southwestern Colorado. Early migrants to Chaco Canyon from the Dolores River would have had to pass through it. Some 20 miles to the north the Yellow Jacket community was probably one of the major exporters of food. Heavely laden porters could cover that distance in two days, making Yucca House a likely place to be a port of trade between north and south. For several centuries a community at this location probably played a role in facilitating and/or controlling north-south trade that passed through the corridor.

The Yucca House complex consists of a lower house with a great kiva and an L-shaped room block and the much larger western complex which consists of a D-shaped pueblo which surrounds the upper house, a massive two story structure (Fig. 4).[31] The rectangular upper house contains two kivas, which are larger than average, and there is no evidence of residential rooms. The height, massiveness and presence of two over-size kivas suggest a special function, similar perhaps to the Sun Temple of Mesa Verde. The Yucca House size, 440 sq metres, is smaller than the Sun Temple, but both structures contain two interior circular rooms. This structure may have been the first to have been built in the area, and it is

---

[31] D. Glowacki, Yucca House (5MT5006) Mapping Project Report. Crow Canyon Archaeological Centre, June 21, 2001.

possible that its location may have been established by the setting Sun over the spire of Towaoc. Immediately to the south of this building is the major spring of the area. The structures that are now visible at Yucca House may have been built up around this original structure.

A tree ring date of 1263 places the western structure in the late Pueblo III period when life was getting very difficult throughout the Ancestral Pueblo world. The climate was getting dryer and colder, and crops were failing. Famine and competition for scarce resources led to social instability and violence. The residents of the nearby pueblos of Sand Canyon (built in 1250 and abandoned sometime after 1277) and Goodman Point (built in 1260 and abandoned less than 20 years later) built their villages in defensible locations, surrounded by protective walls that included springs. Isolated homesteads, once the norm in the Northern San Juan, were abandoned as people aggregated into these more secure villages.[32]

Fig. 4. Yucca House viewed from the west. Inset: December solstice sunset over Towaoc. The star marks the place in the upper house where the photograph of sunset was taken. The dashed line leads to Towaoc. (Reconstruction of Yucca House, courtesy of Dennis Holloway). Photographs by Ninnemann.

---

[32] J. McKim Malville, 'Astronomy and Abandonment in the Pueblo III World', in *Goodman Point Paleohydrology*, Appendix D Wright Paleohydrological Institute, (Denver: Wright Water Engineers, 2011), pp.1–17.

5. The Sun and Moon at Cliff Palace and the Sun Temple

The previous four sites involved prominent natural horizon features. Our last case study involves the Sun Temple, a prominent artificial horizon feature visible from Cliff Palace, Mesa Verde. Located on the promontory between Cliff and Fewkes Canyons of Chapin Mesa, the Sun Temple has been one of the challenging enigmas of the Mesa Verde. There are two towers, which may have been the initial construction at the site. They may have been intended as mimicking the two natural rock towers of Chimney Rock since the major standstill moon was connected with both. These round towers were eventually surrounded by a wall with a height of 4.2 metres, according to Munson's reconstruction.[33] Most extraordinarily, the high wall apparently had no external doorways. Excavated by Fewkes during three months in the summer of 1915, the Sun Temple yielded no evidence of occupation or domestic activity.[34] Fewkes named it Sun Temple because of the eroded rock on its south-east corner that looks like rays of the Sun. With a square area of 660 sq m, the Sun Temple is the largest exclusively ceremonial structure on the Mesa Verde and, quite possibly, the largest of the Ancestral Pueblos. It is nearly twice the size of the largest of the Great Kivas.

Cliff Palace faces to the southwest and receives little sunlight during summer solstice. But for its residents, eager to receive heat from the sun on a winter afternoon, the greatest amount of sunlight and solar heating would occur during the coldest months. It is an auspicious coincidence that the Sun rock is located approximately at the place on the horizon where the sun would set at winter solstice.

At the extreme southern end of the Cliff Palace enclosure, just where the modern trail heads upward to the canyon rim, there is a smooth trapezoidal platform, at the centre of which is a pecked basin with a diameter of 8 cm and depth of 3 cm. A person standing on the platform

---

[33] Gregory Munson, 'Legacy Documentation: Using Historical Resources in a Cultural Astronomy Project', in Clive Ruggles, ed., *Archaeoastronomy and Ethnoastronomy:Building Bridges between Cultures* (Cambridge: Cambridge University Press, 2011), pp. 265–74.

[34] J. W Fewkes, 'A Sun Temple in the Mesa Verde National Park', *Art and Archaeology* 3 (1916): pp. 341–46.

over the basin would have seen the winter solstice Sun setting over the centre of the Sun Temple (Fig. 5).[35]

Fig. 5 Cliff Palace, Mesa Verde. Inset on left: pecked basin and December solstice sunset; inset on right: Major lunar standstill moonset. Both back sights are shown. Photographs by Ninnemann.

Close to summer solstice another astronomical ceremony may have been held in Cliff Palace to watch the setting of the full Moon at its most southern position. The full Moon is always opposite the Sun, so that when the sun is furthest north at summer solstice, the full Moon is furthest south.

During the years approaching the major standstill of the Moon, residents of Cliff Palace would have seen the full moonset at summer solstice gradually move southward toward the Sun Temple on the horizon. During two to three years around the major standstill, the full Moon at summer solstice would set over the Sun Temple as seen from Cliff Palace. At dawn near summer solstice, when life was just stirring in the cliff dwellings, the setting full Moon would have provided a spectacular sight on the opposite horizon. One of the best places for staging a public viewing of the moonset is the open area near the four-storey square tower of Cliff Palace.

---

[35] J. M. Malville, 'Astronomy and Social Integration Among the Anasazi', in Jack E. Smith and A. Hutchinson, eds., *Proceedings of the Anasazi Symposium*, 1991 (Mesa Verde: Mesa Verde Museum Association, 1993), pp. 155–66.

A dramatic visual reciprocity could have occurred for several years around the time of the major standstill on the mornings of the full Moon near summer solstice. A sun priest may have greeted solstice Sun rising over the dark cavern of Cliff Palace from the centre of the Sun Temple. Observers in Cliff Palace could have had a view of that celebrant on the Sun Temple silhouetted against the setting full Moon. If that person was holding a reflecting device such as a mica or pyrite mirror, the flash of sunrise would signal sunrise just as the Moon was setting. A feature of the Sun Temple, which suggests intentionality of design, is that the line tangent to its two interior circular rooms aligns approximately with the major standstill moonset and with the four-storey square tower of Cliff Palace.

Other Non-Photographic Tests
Research questions were tested in ways other than photography; such as the dating of the construction of the Great House of Chimney Rock. The preponderance of tree ring dates of the Great House and vicinity of the Great Kiva coincide with the years of major lunar standstill. Three other instances of one date each, 1011, 1018, and 1070 CE may be related to the dates of scavenged dead-fall trees and may not document intentional astronomical events. It is important to note that dates of 1011 and 1018 CE are not consistent with archaeological evidence of the first occupation of the upper mesa sometime after 1050 CE.

Year CE	Number	Location	Lunar event
1011	1	Great House	Minor lunar standstill
1018	1	Great House	Major lunar standstill
1070	1	Great House	
1076	1	Great House	Major lunar standstill
1077	10	Building 16, close to the Great Kiva	Major lunar standstill
1093	20	Great House	Major lunar standstill

Table 2. Cutting Dates at Chimney Rock[36]

The northeast face of Piedra del Sol in Chaco Canyon contains a large spiral petroglyph, which marks June solstice and June 5, establishing a period of one month centred on June solstice sunrise. The south face of Piedra del Sol contains a petroglyph that may record a coronal mass

---

[36] Malville, 'Chimney Rock and the Ontology of Skyscapes', p. 58.

ejection during the total solar eclipse of July 11, 1097 CE.[37] That prediction could have been falsified if the Sun were in its quiet phase at that time. Analysis of various indictors of solar activity such as aurora, naked eye sunspots, and ^{14}C abundance shows that the eclipse of 1097 CE occurred during a period of high solar activity. Two other total eclipses have occurred that also appear to have occurred during coronal mass ejections.[38]

Results

Four of the five cases we have discussed involve the conjunction of the Sun or Moon with natural rock spires. The fifth, the Sun Temple, visible as an imposing structure on the south-western horizon of the Cliff Palace, with its two towers surrounded by an impenetrable high wall may have been considered to have been uniquely powerful and sacred. When these terrestrial objects (or beings) were conjoined with the Sun and Moon, the spectacle may have been a memorable meeting of gods, a visual theophany.[39] In Table 3 we summarise our photo-documentation of predictions in the American Southwest. In a parallel and comprehensive series of investigations Munro and colleagues have demonstrated with photo-documentation that ten of the thirteen Great Houses built in the Late Bonito period (after 1100 CE) were located at places where the solstice sun rose or set at prominent features on the horizon.[40] Whatever the explicit meaning, the reoccurrence of these well documented solar and lunar events suggests that they were not due to chance or idiosyncratic individual actions. Rather, they appear to illuminate deeply ingrained patterns of behaviour and cultural norms among the Ancestral Puebloans.

---

[37] Malville, *Prehistoric Astronomy*, pp. 64–70.
[38] J. M. Malville, and J. Vaquero, 'Piedra del Sol: The Solar Eclipse Petroglyph in Chaco Canyon', *Mediterranean Archaeology and Archaeometry* 14, no. 3 (2014): pp. 189–96.
[39] Eliade, *Patterns in Comparative Religion*, p. 126.
[40] Munro *et al.*, 'Investigation of Solstice Horizon Interactions at Chacoan Monumental Architecture'.

PREDICTIONS	PHOTO-DOCUMENTATION
Sunrise over Fajada Butte from the Great Kiva in Marsha's Rincon	Confirm
Moonrise between Chimney Rock spires from the Great House	Confirm
Sunrise between Chimney Rock Spires from Peterson Ridge	Confirm
June solstice sunrise between Chimney Rock Spires from Great House	False
June solstice sunrise above merged Chimney Rock spires from Peterson Ridge	False
December solstice sunrise marked on horizon from Wijiji	Confirm
December solstice sunrise from Kin Kletso	Confirm
December solstice sunset over the Sun Temple	Confirm
Major standstill moonset over Sun the Temple	Confirm
June solstice sunrise over Lizard Head from Yellow Jacket	Confirm
December solstice sunset over Towaoc from Yucca House	Confirm
June solstice sunrise at Piedra del Sol	Confirm

Table 3. Predictions and Testing of Sites by Photo-Documentation

Concluding Remarks

An anonymous reviewer of this paper commented that photographic documentation is 'icing on the cake and is used, where possible and when time permits'. The reviewer continued: 'The claims made by the author that digital photos are essential and 'gap bridging' is overstating the case'. We regret that this reviewer does not agree with our recommendation that photographic documentation should be part of standard methodology in cultural astronomy. An understanding of the cultural meaning of astronomical phenomena in any culture should be an imperative in most cases, and thick descriptions accompanied by careful photography can assist in that search for meaning.[41]

---

[41] J. M. Malville, 'Reading Alien Landscapes: Thick Versus Thin Descriptions in Archaeoastronomy', in F. Pimenta, N. Ribeiro, F. Silva, N. Campion, A.

192 Archaeoastronomy and Cultural Astronomy as Scientific Disciplines:
Falsifiability and Photo Documentation

We have emphasized the importance of photographic documentation of both fore sight and back sight for the purposes of reproducibility of the measurement, recordation of archaeological features, and identification of possible light and shadow phenomena. Claims of alignments of walls to astronomical events must remain problematic until convincing back sights can be identified in the archaeological record.

Notwithstanding the arguments of professional philosophers of science, we believe that most scientists agree that falsifiability is a *sina qua non* of science or, stated differently, that non-falsifiable science is an oxymoron. An important issue for scholars in archaeoastronomy and cultural astronomy who are interested in maintaining these fields as scientific, is that of falsifiability, and it can be approached very simply in the form of a question: what potential evidence (perhaps photographic) would persuade you that your hypothesis is wrong?

Acknowledgements
We thank G.B. Cornucopia, ranger *extraordinaire* of Chaco Canyon, for his continuing support of our research, Frank Eddy for his guidance and encouragement at Chimney Rock, Jim Judge for his wise council, and Frank Occhipinti for his important work at Mesa Verde, such as discovering the back sight of December solstice sunset at Cliff Palace.

Joaquinito and L Tirapicos, eds., *Stars and Stones: Voyages in Archaeoastronomy and Cultural Astronomy* (Oxford: BAR International Series 2720, 2015).

Caves, Liminality, and the Sun in the Inca World

Steven R. Gullberg and J. McKim Malville

Abstract: Caves were liminal features of the Inca sacred landscape, connecting this world with the underworld. They were places for making contact with ancestors and the powers of creation. In this paper we examine caves in southeastern Peru for solar orientations and cosmological context, with recourse to the concept of liminality that appears central to cave use. The cave within Kenko Grande has ceremonial steps adjacent to an altar upon which sunlight climbs at midday in June. A rear entrance and altar are illuminated at the time of the solar equinox sunrises. Lacco has three caves which have one solsticial orientation and two light-tubes. A primary opening in the cave at Lanlakuyok faces sunrise at the time of the equinoxes. Tambomachay contains a major fountain and a cave with a platform oriented to December solstice sunrise. Rumiwasi Bajo contains a number of niches and a nine-meter-long passageway oriented close to the June solstice sunset, while the other door opens to December solstice sunrise. Choquequilla is a complex cave opening to December solstice sunrise. The Royal Mausoleum is one of the major shrines of Machu Picchu and opens to June solstice sunrise. Intimachay is a cave with a constructed opening for the December solstice sunrise. The Temple of the Condor contains a cave approximately open to the anti-zenith sunrise. The Gran Caverna includes both an upper and a lower cave oriented for June solstice sunset. There are two caves at the River Intihuatana that, while part of an astronomically oriented complex, don't have solstitial nor equinoctial orientations, nor do they have interior carvings. We end the paper by considering the role of caves and liminality in Inca cosmology.

1. Introduction

Caves connected the Inca world of the present to the ancestors and the cosmological forces of the underworld. Some contain carved steps, which are further symbolic of movement between three cosmological worlds, *hanan*, *hurin*, and *ucu pacha*.[1] In the major Inca creation story, the creator god made human beings from rock in the vicinity of Lake Titicaca and sent them through underground tunnels to emerge from caves, springs, and

---

[1] Gary Urton, *At the Crossroads of Earth and Sky: An Andean Cosmology* (Austin: University of Texas Press, 1981).

rivers in different places in the Andes. The ancestors of the Incas emerged from the center-most of three caves at a place called Pacariqtambo, 'The Inn of Dawn'.

The Inca considered the remains of deceased to be like seeds (called *mallki*, 'seedlings') that were planted in caves, sometimes intended as offerings for the Earth Mother, Pachamama, as well, apparently, to affirm the non-duality of the living and dead. In their search for riches, the Spanish destroyed the tombs and burials that they could locate and burned royal mummies. Fortunately, they missed about a hundred caves discovered by Bingham in Machu Picchu, with burials in them. The caves were modest, sometimes only rock alcoves, and the burials were not of royalty.[2]

In this study we have investigated caves in the Cusco basin associated with known huacas (shrines) as well as the major caves associated with huacas in the Sacred Valley and Machu Picchu. The number of such natural structures associated with huacas that have orientations to either the solstices or equinoxes is somewhat surprising. It appears that in the selection of caves to be huacas some preference was given to those with solar orientations. The significance of the passage of the Sun into the dark interior of these caves is not explicated in Inca Origin Myths. The phenomenon may be understood in terms of empowerment and animation of interior spaces, such as is suggested in double-jamb doorways and windows at Machu Picchu and Llactapata.

2. Inca Cosmology

The Sun, Moon, planets, and stars appeared to rise from the ground and subsequently return to it when viewed from the ancient perspective of a flat Earth. The underworld, therefore, becomes a natural extension of the path traveled by these celestial bodies.[3] In effect, the cosmos of the Incas existed in three distinct worlds – that of Ucu Pacha, the underworld; Kay

---

[2] George R. Miller, 'Food for the Dead. Tools for the Afterlife: Zooarchaeology at Machu Picchu', in Richard L. Burger and Lucy C. Salazar, eds., *The 1912 Yale Peruvian Scientific Expedition Collections from Machu Picchu* (New Haven: Yale University Publications in Anthropology 85, 2005), pp. 1–63.

[3] Michael A. Rappenglück, 'Copying the Cosmos: The Archaic Concepts of the Sacred Cave Across Cultures', in Herman Jung and Michael A. Rappenglück, eds., *Symbolon – Jahrbuch der Gesellschaft für wissenschaftliche Symbolforschung: Neue Folge, Band 16 – Signaturen des Lebens: Bilder und Zeichen von Kosmos und Bios und Symbole des Alltags – Alltag der Symbole* (Frankfurt am Main: Peter Lang, 2007), pp. 63–84.

Pacha, the here and now; and Hanan Pacha, the world above.[4] There are many extant examples of symbolic sets of three stairs representing transition between the three worlds of the Incas' being. Caves figure prominently in Inca origin myths and were thought also to be chthonic connectors to the underworld.[5]

Water empowered the shrines known as huacas through a life-energizing force that could be used to provide sentience to the inanimate or renew power to the living.[6] The world's water cycled through the heavens and Earth in its journey down the Vilcanota with return via the Milky Way.[7]

Inca cosmology viewed the Milky Way as a river flowing across the night sky in a very literal sense. They saw earthly waters as being drawn into the heavens and then later returned to Earth following a celestial rejuvenation. The Earth was thought to float in a cosmic ocean.[8] When the celestial river's orientation was such that it dipped into that ocean, waters were drawn into the sky. 'The Milky Way is therefore an integral part of the continuing recycling of water throughout the Quechua universe'.[9]

2.1. Sacred Landscape

The Incas venerated natural features such as mountains, outcroppings, caves, springs, and rivers, all believed to be endowed with sacred powers. Most of all the Incas revered mountains and the great entities within them. Sacred mountains are prominent on the horizons of Cusco and Machu Picchu and the Inca's great reverence for the Earth was no better displayed than in their worship of these majestic snow peaks. Quechua populations today view mountains either as powerful deities themselves or the residences of deities. They are worshipped as ancestors, sources of water and weather, and in the case of Ausangate, the father of alpacas and lamas.

---

[4] Urton, *At the Crossroads of Earth and Sky*.
[5] Maartin J. D. Van de Guchte, 'Carving the World: Inca Monumental Sculpture and Landscape' (PhD Dissertation, University of Illinois, 1990).
[6] J. McKim Malville, 'Animating the Inanimate: Camay and Astronomical Huacas of Peru', in Jose Alberto Rubiño-Martín, Juan Antonio Belmonte, Francisco Prada, and A. Anexton Alberdi, eds., *Cosmology Across Cultures, ASP Conference Series, 409* (San Francisco: Astronomical Society of the Pacific, 2009), pp. 261–66.
[7] Frank Salomon and George L. Urioste, *Introductory Essay in The Huarochiri Manuscript: A Testament of Ancient and Colonial Andean Religion* (Austin: University of Texas Press, 1991).
[8] Urton, *At the Crossroads of Earth and Sky*.
[9] Urton, *At the Crossroads of Earth and Sky*, p. 60.

Mountains were often venerated as the most important of deities throughout the empire.[10]

Similar to sacred mountains, many rock outcrops were also understood to be hierophanies, or manifestations of the sacred. Eliade's definition of hierophanies seems particularly apt: the breaking through of the sacred into the mundane world. Rock outcrops were indeed places where the underworld is breaking through into this world.[11] Pachacuti may have believed that he could improve upon these stones and, as the son of the Sun and co-creator of the land, he could modify and enhance the work of the creator. The carved huacas are bedrock features with their roots in the earth, an important aspect of the symbolism involving three worlds. They also seem to be laid out across the landscape in meaningful patterns. A limited number of motifs were used in the shaping of huacas suggesting that the carvings were not a form of mindless or inventive graffiti, but elements in a symbolic language with cosmological significance.

2.2. Carved and Uncarved Rocks

Inca emperors felt it their right to improve upon nature by sculpting in situ outcrops that often became huacas.[12] Improvements to rocks appear to have been state-controlled and likely guided by a certain class of artisans as evidence does not suggest innovation. These methods also were not sudden inventions, but instead had developed over time with knowledge acquired from other societies.[13]

The Incas used carved rocks as a vehicle for promoting state ideology and the solar religion. They were symbols of commemoration, mediation with the cosmos, and state identity, all the while remaining part of the Incas' perception of their sacred relationship with nature and the land.[14]

A rock, once carved, became a hierophany and was worshipped by the Incas. Uncarved rocks could also become sacred and animated such as a

---

[10] Johan Reinhard, 'Sacred Mountains: An Ethno-Archaeological Study of High Andean Ruins', *Mountain Research and Development* 5, no. 4, (1985): pp. 299–317.
[11] Mircea Eliade, *Patterns in Comparative Religion* (New York: New American Library, 1974).
[12] César Paternosto, *The Stone and the Thread: Andean Roots of Abstract Art* (Austin: University of Texas Press, 1996).
[13] Susan Niles, *Callachaca: Style and Status in an Inca Community* (Iowa City: University of Iowa Press, 1987).
[14] Van de Guchte, 'Carving the World: Inca Monumental Sculpture and Landscape'.

white boulder in the center of the Palace of Huayna Capac.[15] Embedded in the Earth, these manifestations of the sacred were connected with the powers of the underworld.

Sculpted steps in huacas, were an important part of cosmological symbolism regarding the three worlds of the Incas.[16] Ritual stairs are a dominant motif, perhaps expressing movement from the underworld to the Earth to the heavens and are often associated with these three realms. Carved stairs are quite common and frequently non-functional, such as those on inaccessible cliff sides at Ollantaytambo and within a cave at Machu Picchu. They often include three steps, corresponding with the three worlds, and likely were symbolic representations of this cosmology. Carvings of condors, pumas, and serpents proliferate as representatives of these respective spiritual domains. Carved rocks also had geopolitical function as they were used to mark territory during expansion of the Inca empire. They were elements in a language used to communicate political concepts and to transmit state ideology.[17]

2.3. The Meaning and Function of Caves

As with numerous other ancient cultures, caves represented a connection with ancestors and primeval forces and this played centrally in Inca origin myths and ritual emergence.[18] Caves were residences of ancestors with niches for mummies and served as locations for food and offerings to the dead. They were huacas empowered and animated by sunlight.

Caves are lineal places serving as openings to other worlds, passages that can carry one from one realm to another. The word comes from the Latin word *limen*, meaning a threshold. These passageways may involve the frightening (and transformative) ambiguity and disorientation that can occur when one crosses into a new and unfamiliar space and time. Moving through a cave, such as that near the summit of Huayna Picchu, could mean a rebirth, cleansing, or entry into the realm of the gods. The idea of

---

[15] J. McKim Malville, 'Astronomy of Inca Royal Estates I: The Sacred Valley', in C.L.N. Ruggles, ed., *Handbook of Archaeoastronomy and Ethnoastronomy* (Heidelberg: Springer, 2014), pp. 865–77.
[16] Paternosto, *The Stone and the Thread: Andean Roots of Abstract Art*.
[17] Jessica Joyce Christie, *Memory Landscapes of the Inca Carves Outcrops* (Lanham: Rowan and Littlefield, 2016).
[18] Michael A. Rappenglück, 'Cave and Cosmos, a Geotopic Model of the World in Ancient Cultures', in Mauro Peppino Zedda and Juan Antonio Belmonte, eds., *Lights and Shadows in Cultural Astronomy: Proceedings of the SEAC 2005* (Isili: Associazione Archeofila Sarda, 2007), pp. 241–49.

liminality was extensively developed by Victor Turner in his analysis of pilgrimage and the unsettling experience of traveling into unfamiliar landscapes.[19] While in the liminal state, human beings have a heightened awareness of their surroundings and are open to transformative suggestions from the environment or their companions. Liminality can involve places as well as experiences. Liminal places can range from springs, caves, shores, rivers, crossroads, bridges, and sacred spaces such as temples. Caves and springs, especially, were viewed as openings or doorways to the place of origin of the world and to the realm of ancestors and deities. In the Andean world caves and windows are considered symbolically equivalent. The frame of a window or doorway (especially a double jamb doorway) is like the entry to a cave.[20] The Temple of Three Windows at Machu Picchu may thus have been intended as a representation of the three caves at Pacariqtambo.

3. Methodology

As is clear from the foregoing, caves were multivalent features in Inca and Andean cultures. Our approach has been to frame these various meanings in term of multiple hypotheses to be tested. We have studied the caves associated with known huacas in the Cusco valley, the Sacred Valley, and Machu Picchu and asked which of the following hypotheses for the meaning of Inca caves stands up to scrutiny:

1. Analogues to the three caves in the origin myth of the Inca;
2. Entry into the underworld; places for offerings and communication with Pachamama;
3. Transformative ritual passageways;
4. Metaphorical windows; penetration of light into darkness;
5. Geopolitical markers of Inca hegemony;
6. Sites for ancestral mummies placed like seeds in the Earth.

Magnetic bearings and inclinations were initially recorded with a Suunto Tandem Compass Clinometer Survey Tool, a liquid-filled precision compass and clinometer. Measurements were validated with a Wild

---

[19] Victor Turner, 'Liminality and Communitas', in *The Ritual Process: Structure and Anti-Structure* (New Brunswick: Aldine Transaction Press, 2008); Victor Turner, *Process, Performance, and Pilgrimage: A Study in Comparative Symbology* (New Delhi: Concept, 1979).
[20] Carolyn Dean, *A Culture of Stone: Inka Perspectives on Rock*. (Durham: Duke University Press, 2010).

Heerbrugg T2 Theodolite. Solar horizon positions and inclinations were verified trigonometrically for position and time of sunrises on the actual horizon. Photo-documentation was accomplished with a Canon 8-magapixel digital camera and tripod. Global positioning was made with a Garmin GPS. GPS Azimuths were validated trigonometrically. Magnetic declinations were taken from the National Oceanic and Atmospheric Administration: National Environmental Satellite, Data and Information Service – National Geophysical Data Center.

4. The Caves of Cusco
4.1. Kenko Grande (Patallacta, Ch. 1:2)
The first cave examined is within Kenko Grande, also known as Patallacta and classified in the Cusco ceque and huaca system of Bauer as Chinchaysuyu 1:2, or Ch. 1:2.[21] It is located at S 13° 30.53' W 071° 58.24' and 3614 metres above sea level. Visible from the central Coricancha sun temple of Cusco, Kenko Grande is a sculpted limestone outcrop north of the city that incorporates carvings and crevices as well as a stone monolith surrounded by a series of niches along an arced plaza.

Kenko Grande exhibits a visually dramatic phenomenon at the time of the June solstice which is known locally as 'the awakening of the puma'. The Inca venerated the condor, puma and snake as representing cosmological correlations with the sky, earth, and underworld.[22] Located atop the huaca and carved into the stone are two carved cylinders perhaps designed at gnomons for effects of light and shadow. The cylinders are about twenty-five centimeters high and are spaced thirty-five centimeters apart. In close proximity is a small wall with a fissure aligned for the sunrise at the June solstice. Light from the morning Sun passes through the fissure and first touches the left side of the left cylinder. As the Sun continues to rise, its rays move across the cylinder and then illuminate the opposite one as well. The cylinders are situated in such a way that the glowing pair and the relative shadows now resemble a puma – 'the puma's awakening'.

The top of Kenko Grande exhibits many once-fine carvings including an offering/divination channel flowing to a cave. A common motif of carved huacas is a straight or zigzag channel through which liquids, most probably the revered maize corn beer chicha, could flow. The current of

---

[21] Brian S. Bauer, *The Sacred Landscape of the Inca: The Cusco Ceque System* (Austin: University of Texas Press, 1998).
[22] Van de Guchte, 'Carving the World: Inca Monumental Sculpture and Landscape'.

energy necessary to establish harmony and maintain equilibrium in the world was stimulated by the pouring of liquid offerings into these channels. The channel begins with a cup-like basin, extending first for 130 centimeters and then 145 centimeters before splitting into two eighty centimeter branches. *Kenko* in Quechua means a zigzag, which was also known as a *paqcha*. It is a remarkable ritual device in which liquids would be poured to flow downward to feed and honor the earth[23].

Located within Kenko Grande is a sculpturally-enhanced cave with two entrances, two altars, niches, and a set of three ritual stairs (see Fig. 1). Locals maintain that light entering the cave within a niche at the northwest end of the cave was reflected with gold or silver plates in order to illuminate the entire chamber. The cave includes an additional niche to the right of the altar large enough to hold a mummy.[24]

Fig. 1 Interior cave of Kenko Grande.

---

[23] Paternosto, *The Stone and the Thread: Andean Roots of Abstract Art*.
[24] Steven R. Gullberg and J. McKim Malville, 'The Astronomy of Peruvian Huacas', in Wayne Orchiston, Tsuko Nakamura, and Richard Strom, eds., *Highlighting the History of Astronomy in the Asia-Pacific Region* (New York: Springer, 2011), pp. 85–118.

The primary altar was carved and polished with three ritual stairs at its northwestern end. Approaching local noon nearing the June solstice sunlight enters the cave, approaches the altar and climbs the three stairs. Ritual stairways are common features, symbolizing shamanic movement between the three worlds. Such a cave could serve as a portal for communication with Ucu Pacha, the world below. Light enters through an opening approximately 341.5° in azimuth and an inclination of +40°.

At times of the solar equinoxes the cave's secondary altar is illuminated at sunrise. The opposite entrance of the cave near the main altar is oriented to the equinox sunset. The meaning of this remains uncertain as the ethnohistorical record is ambiguous with regard to Inca interest in equinox horizon observations, as discussed elsewhere.[25]

4.2 Lacco (Chuquimarca, An. 3:4)
Lacco, Antisuyu 3:4, is a carved limestone outcropping incorporating several astronomical orientations. It was barely visible from the Coricancha. The elaborate huaca, also known as Salonpuncu, is located above Cusco to the northeast of Kenko Grande. Zuidema and Aveni suggest Lacco as being at or near the site of Chuquimarca (An 3:4).[26]

> [An-3:4] The fourth was called Chuquimarca; it was a temple of the Sun on the hill of Manto calla, in which they said that the Sun descended many times to sleep. For this reason, in addition to everything else, they offered it children.[27]

Lacco is the largest limestone outcrop in the vicinity, covering an area of about 1670 square meters, and contains elaborately carved stairways, caves and niches. Three caves were modified with altars and exhibit astronomical orientations.

Lacco, on its northeast face, has a cave opening which is oriented for the June solstice sunrise. The greatest eastward angle as viewed through the cave's entrance is 78°, therefore at least some sunlight is admitted to a

---

[25] Gullberg, 'The Astronomy of Peruvian Huacas'.
[26] Tom R. Zuidema, 'The Inca Calendar', in Anthony Aveni, ed., *Native American Astronomy* (Austin: University of Texas Press, 1977), pp. 219–59; Anthony Aveni, 'Horizon Astronomy in Incaic Cusco', in Ray A. Williamson, ed., *Archaeoastronomy in the Americas* (Los Altos: Ballena Press, 1981), pp. 305–18.
[27] Bernabe Cobo, *Inca Religion and Customs*. 1653. Translated by Roland Hamilton, (Austin: University of Texas Press, 1990), p. 65.

portion of the interior for several days before and after the solstice. The Sun centers on the cave opening at the time of the June solstice. During this period sunlight enters the portal, illuminating the altar and cave interior. When observed, the process began very quickly at 06:25 and persisted until the last vestiges of light disappeared from the altar's stone surface nearly two hours later at 08:24. The northeast cave is located at S 13° 30.34' W 071° 57.86' and 3650 metres above sea level.

Immediately to the east of the northeast cave is the north opening of the crevasse running across Lacco. It also is oriented approximately to the June solstice sunrise. Two ceremonial thrones are situated in front of and below two large steps leading to the northeast cave's opening. The thrones might have been occupied during the rising of the solstice Sun.

Lacco's southwest cave contains a small altar below a light-tube oriented to the ecliptic at times when the Sun's path is between 70° and 75° above the horizon. When properly positioned, either the Sun or a full moon can illuminate the altar within. A crescent moon was viewed through the light tube on 30 October 2006, near the date of the zenith Sun. The cave's door is oriented towards 235° of azimuth. The southwest cave is located at S 13° 30.35' W 071° 57.89' and 3662 metres above sea level.

A small altar was carved within the cave and is oriented so that it is illuminated by light passing through the tube. The elevation from the altar to the lower edge of the light tube is approximately 70° at an azimuth of 211°. The cave includes two recesses cut into its western wall.

Lacco's second cave with light-tube illumination is also known as Temple de la Luna, the Temple of the Moon. This chamber is the most elaborate found within the huaca and the remains of carvings of both a puma and a snake adorn its entrance, the serpent symbolizing passage into the realm of Ucu Pacha.[28] The shaft aligns vertically to admit the light of the zenith Sun at local noon. The light-tube is directed at a finely carved altar, which also can be illuminated by the Moon. The southeast cave is located at S 13° 30.36' W 071° 57.86' and 3655 metres above sea level.

4.3 Lanlakuyok (Amaromarcaguaci, An. 1:7)
Van de Guchte has suggested that Lanlakuyok is a candidate for Amaromarcaguaci (An 1:7).[29]

---

[28] Paternosto, *The Stone and the Thread: Andean Roots of Abstract Art.*
[29] Van de Guchte, 'Carving the World: Inca Monumental Sculpture and Landscape'.

[An-1:7] The seventh *guaca* was called Amaromarcaguaci; this was a house of Amaro Tupa Inca, which was on the road of the Andes.[30]

Lanlakuyok is a large carved outcrop along the road to Pisac that has an extensive system of passageways within it illuminated by occasional tubes or openings to the sky. A primary cave opening faces sunrise at the time of an equinox on an azimuth of 91.5°.

4.4 Tambomachay (Cirocaya, Ch. 1:4)

The cave of Tambomachay is associated with an Inca platform and staircase and Bauer suggests it to be the best candidate for Cirocaya (Ch 1:4).[31]

[Ch-1:4] The fourth *guaca* was called Cirocaya. It is a cave of stone from which they believed the hail issued. Hence, at the season when they were afraid of it, all went to sacrifice in the cave so that hail should not come out and destroy their crops.[32]

The cave opening looks out on a bearing of 135° while the 28°/208° oriented platform in front of it more directly faces the December solstice sunrise. The site is positioned with a clear view of the rest of Tambomachay, Puca Pucara, and parts of the Cusco valley. A stairway leads to the south end of the platform. A boulder rests centrally on the terrace of the platform and a wall was constructed on the side of the terrace closest to the outcrop. The cave is shallow and its interior is extensively eroded.

Bauer proposes that the highly structured fountain at Tambomachay may have been Quinoapuquiu (An 1:10).[33]

[An-1:10] The tenth *guaca* was called Quinoapuquiu; it was a fountain near Tambo Machay which consists of two springs. Universal sacrifice was made to it, except children.[34]

The fountain is situated on the main level of the complex, well below the cave and platform.

---

[30] Cobo, *Inca Religion and Customs*, p. 63.
[31] Bauer, *The Sacred Landscape of the Inca: The Cusco Ceque System*.
[32] Cobo, *Inca Religion and Customs*, p. 54.
[33] Bauer, *The Sacred Landscape of the Inca: The Cusco Ceque System*.
[34] Cobo, *Inca Religion and Customs*, p. 64.

4.5 Rumiwasi Bajo (Comovilca, An. 6:2)

To the east of Cusco above San Sebastián lies Rumiwasi Bajo. The rock of Rumiwasi Bajo contains a number of niches and a nine-meter-long cave passageway. One doorway to the passageway through the huaca looks out close to the June solstice sunset, but is 12° off. The other doorway opens to the December solstice sunrise. Niles proposes this as Comovilca, An. 6:2.[35]

5. Other Caves
5.1 Chinchero

Beyond the ceque system of Cusco is Chinchero. One of the first tasks before each new Inca was the establishment of his royal residence.[36] Topa Inca, the son of Pachacuti, claimed the Chinchero valley as the site for his estate and soon set about construction of its palace, courtyard, support buildings and agricultural terraces. Also at the site are several intricately carved rock huacas. The style of architecture and design suggests a view of nature similar to that of Pachacuti, in which natural rock and landscape features were included in structural forms.

There are two major carved rocks; the first to the south of the plaza, Titikaka, has two carved stairways, one of which leads upward to the top of the rock through a cave with an axis approximately north and south. On the top there are a series of cut rectangular trays similar to those of Kenko and Lacco. To the southwest is a second large carved stone, Chinkana, containing an elaborate stairway, enclosures, altars, and carved trays. At its lower end is a flowing stream beneath carved teeth. A triangular basin opens approximately toward sunset on the December solstice. Above and to the south are carved stones known as Mesakaka and Kondorkaka. In contrast to the solstitial orientations of the Cusco valley, our field research showed the majority of the features of Chinchero to emphasise cardinal directions. Still, solstitial orientations are not totally absent. The two primary carved rock huacas, Titikaka and Chinkana lie approximately on the axis of the June solstice sunrise and December solstice sunset. Both huacas are very large and were carved in-situ. The existence of this orientation was not lost upon the Incas while developing this site.

Chinkana lies low on the eastern end of Chinchero's central valley and has been extensively carved on all sides with such as seats, stairs, shelves

---

[35] Bauer, *The Sacred Landscape of the Inca: The Cusco Ceque System.*
[36] Susan Niles, *The Shape of Inca History: Narrative and Architecture in an Andean Empire* (Iowa City: University of Iowa Press, 1999).

and niches. Chinkana is located at S 13° 23.27' W 072° 02.58' and 3724 metres above sea level.

Approximately 360 meters west of Chinkana and north of the great plaza of Capallanpampa is Chinchero's largest carved outcrop, Titikaka.[37] Titikaka displays many carvings including two prominent stairways, one external, and the other within a break in the center of the rock leading to its top. This cave-like central stairway exhibits figurative carving. The upper surface of the stone is extensively carved and displays several examples of seats or trays and animals such as a condor and a snake. Titikaka is located at S 13° 23.35' W 072° 02.80' and 3753 metres above sea level. At the base of the rock, lower on the western side, are more carvings, a large crevasse and an opening to a shallow cave. The crevasse looks out on a 278.0° bearing and the cave opens to 254.5°. A large niche was carved near the mouth of the cave and looks out at 293.5°.

5.2 Choquequilla
Above the Rio Huarocondo, five kilometers southeast of Ollantaytambo and fourteen kilometers west of Urubamba, are the ruins of Choquequilla. Choquequilla is located at S 13° 17.53' W 79° 13.93' and 3627 metres above sea level. This remote huaca lies within the mouth of a cave opening to the approximate direction of the December solstice sunrise. The intricately carved shrine faces inward toward the cave, away from the horizon, and is flanked to the south by a wall constructed with two rows of four double-jambed niches, emphasizing the site's significance. The roof of the cave is formed by two relatively flat stone faces that form an inverted 'V'. Light from the December solstice Sun as it rises above the opposing horizon brightly illuminates the cave and huaca (see Fig. 2).

The cave is situated on the mountainside above agricultural terraces that have fallen into disuse. A central staircase ascends the terraces and at the top a trail proceeds north to the cave. A small masonry structure with a door and windows is situated immediately to the cave's north. The carved rock of black granite is said to be among the finest examples in existence and exhibits great symmetry and exquisite carving.[38] The sculpting closely resembles that of the Baño de la Ñusta at Ollantaytambo, but the Choquequilla rock has been damaged by looters.

---

[37] John Hemming and Edward Ranney, *Monuments of the Incas* (Albuquerque: University of New Mexico Press, 1982).
[38] Paternosto, *The Stone and the Thread: Andean Roots of Abstract Art.*

Fig. 2 Cave opening at Choquequilla.

Paternosto calls this 'the cave of *Choqequilla*, the Golden Moon', and Van de Guchte calls it the 'Moon Temple' of Choquequilla.[39] The cave opens to the December solstice sunrise and the carved stone is slightly offset at 130°. The horizon is inclined +32.0°. The rise of the Sun illuminates the cave brightly. The light of a rising moon could create a dramatic effect.

6. The Caves of Machu Picchu
6.1 The Royal Mausoleum
The Torreon/Royal Mausoleum complex was one of the major huacas of Machu Picchu. The huaca consists of a very fine masonry wall that crowns the top of a large rock. The wall is beautifully fitted into the rock and contains two windows, one of which opens to June solstice sunrise and the rising of the Pleiades. A stone surrounded by the walled enclosure is illuminated through the window at sunrise during the time of the June solstice. A ledge cut into the top of the stone approximately bisects the early rays of the solstice Sun.[40]

---

[39] Paternosto, *The Stone and the Thread: Andean Roots of Abstract Art*, p. 89; Van de Guchte, 'Carving the World: Inca Monumental Sculpture and Landscape', p. 191.
[40] David S. P. Dearborn and Katharina J. Schreiber, 'Here Comes the Sun: The Cusco-Machu Picchu Connection', *Archaeoastronomy* 9 (1986): pp. 15–36.

Below the Torreon is the Royal Mausoleum, a cave, which contains a set of symbolic stairs, niches probably for mummies, and other stonework, and, which opens to June solstice sunrise. The cave gives the impression of a passageway to the underworld and the carved, stepped stone a shamanic stairway of ascent or descent. The major stone lined channel of Machu Picchu makes as sharp turn toward the rock, consistent with the role of water through the process of *camay* in the animation of huacas. Huacas and other forms of sacred architecture were animated by the circulation of running water and the pouring of libations. Most of the major astronomical sites of the Inca were associated with natural or offertory water.[41]

6.2 Intimachay
Also in Machu Picchu's Eastern Urban Sector lies a cave called the Intimachay. Dearborn, Schreiber and White argue that this cave was constructed to observe sunrise at the time of the December solstice and the festival of Capac Raymi.[42] A tunnel, like a horizontal light-tube, was oriented to admit sunlight to the cave for about ten days before and after the solstice. The window did not function to illuminate the cave, but instead was aligned precisely with the December solstice sunrise.[43] The view of the horizon was constrained by an interior stone that limited the field of view to ten arc-minutes. Reconstruction performed in 2006 narrowed the window and presently inhibits direct view of the horizon from the cave's interior.[44] Capac Raymi was a festival celebrated by the nobility that included ceremonies of passage to manhood for young Inca noblemen.[45] A site such as this could have played a role. In 2012 Ziolkowsi, Kosciuk, and Astete confirmed the December solstice orientation of the light tube using 3D laser scanning, although they found

---

[41] J. McKim Malville, 'Machu Picchu', in Ruggles, *Handbook of Archaeoastronomy and Ethnoastronomy*, pp. 879–91.
[42] David S. P. Dearborn, Katharina J. Schreiber and Raymond E. White, 'Intimachay, A December Solstice Observatory', *American Antiquity* 52 (1987): pp. 346–52.
[43] David S. P. Dearborn and Raymond E. White, R., 'Inca Observatories: Their Relation to the Calendar and Ritual', in Anthony Aveni, ed., *World Archaeoastronomy* (Cambridge: Cambridge University Press, 1989), pp. 462–69.
[44] Malville, 'Machu Picchu'.
[45] Bernabe Cobo, *History of the Inca Empire: An Account of the Indians' Customs and Their Origin, Together with a Treatise on Inca Legends, History, and Social Institutions*. 1653. Translated by Roland Hamilton, (Austin: University of Texas Press, 1983).

that a view of December solstice sunrise could only be obtained at the end of the light tube.[46] They also suggest that a northern window of the light tube may have been intended to mark the major lunar standstill. An alternate interpretation is that the northern window may have been a place to insert and offering to the Sun at December solstice. Ziolkowski *et al.* suggest that the light tube reveals that Intimachay was an Inca 'observatory'. The alternate interpretation is that the cave and light tube primarily served ceremonial functions involving entry of the living Sun into the dark cave. The northern window may have been used to place objects to be illuminated by the dawn sun.

6.3 The Temple of the Condor
South of the Intimachay is the Temple of the Condor where designers carved a head in stone while incorporating in-situ rock as wings in the site's overall visual image of a condor, the creature that represented Hanan Pacha, their world above. The site's cosmological significance continued to the underworld through a system of three caves with an entrance below the boulder representing the left wing of the condor. Nonfunctional steps downward are found at the rear of this entrance area (see Fig. 3).[47] James Westerman and Alfredo Valencia explored the caves in 1995, finding such as fissures, stairs, and numerous bones of guinea pigs, animals which were frequently used as food in Inca ceremonies.[48]

Zuidema argues that the Inca were interested in the timing of the region's anti-zenith passages of the Sun, occurring on 26 April and 18 August, the two days that the Sun is at nadir for the latitude of Cusco.[49] The Temple of the Condor's cave is oriented to the anti-zenith sunrise with a true azimuth of 74°, and therefore could have played a role in associated

---

[46] Mariusz. Ziólkowski, Jacek Kościuk, and Fernando Astete, 'Inca Moon : Some Evidence of Lunar Observations in Tahuantinsuyu', in Ruggles, *Handbook of Archaoastronomy and Ethnoastronomy*, pp. 897–912.

[47] Malville, 'Machu Picchu'.

[48] Westerman, James S., 'Inti, the Condor and the Underworld: The Archaeoastronomical Implications of the Newly Discovered Caves at Machu Picchu, Peru', in John W. Fountain and Rolf M. Sinclair, eds., *Current Studies in Archaeoastronomy: Conversations Across Time and Space* (Durham: Carolina Academic Press, 2005), pp. 339–51.

[49] R. Tom Zuidema, 'Inca Observations of the Solar and Lunar Passages Through Zenith and Anti-Zenith at Cuzco', in Ray A. Williamson, ed., *Archaeoastronomy in the Americas* (Los Altos: Ballena Press, 1981), pp. 319–42.

ceremonies if such festivities took place.[50] Westerman relates that as the Sun rises on or near days of the anti-zenith its rays pass between two external structures, illuminate the condor stone, and extend beneath the boulder to the stairs at the cave's entrance. This orientation supports Zuidema's thoughts regarding anti-zenith observations.

Fig. 3 Temple of the Condor cave entrance.

6.4 Huayna Picchu

Framed in the major entrance to Machu Picchu is the peak of Huayna Picchu with two stairways up its southern and northern sides. The southern route, which starts in Machu Picchu, passes through a narrow cave near the summit (see Fig. 4). It would appear to be the paradigmatic exemplar of a liminal passageway between worlds, perhaps the best example we have in the Inca world. On the south-east stairway, one crosses a large observing platform and then enters this narrow, dark tunnel (difficult if one is wearing a backpack) and climb its stairs, moving, as it were, from the mundane world of Machu Picchu to the upper world of the heavens. The

---

[50] Westerman, 'Inti, the Condor and the Underworld'.

northern route starts at the Urubamba River and continues through the largest double-jamb doorway of Machu Picchu, near the cave of the Gran Caverna formerly known as the Temple of the Moon. Double-jamb doorways appear to be markers of liminality, doorways from one realm to another. This large cave contains beautifully formed double-jamb niches, perhaps places for mummies of important ancestors. It is possible that these mummies were carried on the backs of celebrants from the underworld of the cave to the sunrise on upper world of the summit of Huayna Picchu for special celebrations. This combination of stair, cave at the bottom, places for ceremony on the summits makes Huayna Picchu powerfully emblematic of passage between the lowest and the highest worlds. Huayna Picchu's summit is located at S 13° 09.40' W 072° 32.57' and 2698 metres above sea level.

Fig. 4 Cavern passage near the summit of Huayna Picchu.

6.5 Gran Caverna

Low on the northwest face of Huayna Picchu is a shrine referred to as both Gran Caverna and Temple of the Moon. This site includes an upper cave and a lower cave. The upper cave is the larger and within it are five finely constructed double-jambed niches. The double-jambs indicate this site was significant and might have been a place for the storage of mummies. The cave is oriented in the approximate direction of the June solstice sunset and can also be illuminated by the moon. The solstice sunset orientation is only approximate as it differs by approximately 7°. Two routes approach the site. One branches downward from the main trail between Machu Picchu and Huayna Picchu, while the other descends directly from Huayna Picchu's summit. A lower trail passed through a gateway on its way to the river and to the vicinity of the River Intihuatana. The upper cave is located at S 13° 9.09' W 072° 32.78' and 2277 metres above sea level.

Further northeast and lower on the slope is the lower cave which features a constructed doorway bracketed by two windows. As with the upper cave, the lower cave's door and flanking windows are approximately aligned with the June solstice sunset.

6.6 The River Intihuatana

During his exploration of the Vilcabamba, Hiram Bingham located two carved rocks that he identified as intihuatanas. One of these, the Intihuatana of Machu Picchu, is arguably the best-known carved rock of the Inca world. The second intihuatana, lying deep in the Urubamba canyon to the west of Machu Picchu, has been visited far less frequently. When examined this shrine was found to be rich with cosmological symbolism. The River Intihuatana is an important element of the extended ceremonial complex that combines Machu Picchu with sites on the Llactapata ridge. It is located on a hillside between PeruRail switchbacks near a hydroelectric complex.

The principle element of the shrine is a rock carved with steps and tiers. The adjacent upslope section of the sanctuary contains two water basins aligned east-west and has an elaborately engineered water fountain that is situated over a small cave. Eastward of these granite carvings are the remains of several support structures and a tower attached to a large boulder with a second cave beneath. The area exhibits agricultural terraces, but they are presently engulfed by trees.

The significance of the River Intihuatana has become clearer since the rediscovery of the Llactapata Sun Temple in 2003. The site can now be identified as a major shrine (a huaca sanctuary) connected to Machu

Picchu by two intersecting sightlines or ceques from the Llactapata ridge. The concentration of symbolic motifs suggests ceremonial significance at the site.

The site's primary feature is the Intihuatana, a somewhat worn, but finely carved stone situated at the sanctuary's western boundary. Its dimensions overall are 4.27 meters along the flat northern face by 3.20 meters wide. The tiers get increasingly smaller as they rise. The middle tier measures 2.17 meters by 2.14 meters and the top tier 1.50 meters by 1.70 meters. On the east side of the top tier is an intermediate level measuring 48 cm by 1.70 meters and both are adjoined by a set of descending steps too small to serve any necessary function as they are situated. There appears at present to be three symbolic steps, but the stone has been subject to enough erosion to make the original number uncertain.

One of the more intriguing areas found on the site is a complex incorporating several common huaca motifs: a fountain, two basins and a cave. The fountain structure is situated 3.84 meters upslope from the Intihuatana and spans 5.60 meters at its extremes. The face of the fountain points approximately north, is oriented approximately east-west, and was designed to receive water from the east into the channel. A ledge was carved 1.56 meters below the top of the fountain and worn examples of sculpted seats or shelves remain to the west. The channel was engineered to distribute water to each of the four outlets on the fountain's face. The outlets measure 16.5 cm by 9 cm and are spaced 61 cm apart. Within the channel a small baffle was constructed at each outlet to enhance an even diversion of water flow through that opening. The fountain is now dry, but would have once been fed by an upslope spring.

A small cave also exists within the fountain-basin complex. This orifice extends beneath the fountain with its opening situated between the fountain and the boulder with the basins. The cave has enough space for an attendant or priest to function and could possibly have been used for mummy storage. Above the cave entrance and carved into the boulder is a set of three symbolic stairs (see Fig. 5), in this case perhaps representing transition between the underworld and the world of the here and now. The Incas felt caves to express deep connections with the forces of nature.[51]

---

[51] Steven R. Gullberg and J. McKim Malville, 'The River Intihuatana: Huaca Sanctuary on the Urubamba', *Mediterranean Archaeology and Archaeometry* 14, no. 3 (2014): pp. 179–87.

Fig. 5 Cave entrance beneath the fountain at the River Intihuatana.

The property owner provided insight into a second cave that exists beneath the base of the boulder that forms part of the tower constructed on the sanctuary hillside. He stated that the cave was deep enough to have entrapped animals. A circular tower reaching to the sky while sitting atop a cave may have served to symbolise and facilitate strong connections between this world and those of both the heavens and of the underworld.[52]

7. Conclusion

Caves appear to have played a varied role in the ritual and culture of the Incas. We find no evidence that they were significant representations of the creation story of the Inca. As we have noted the Temple of Three Windows may be associated with that cultural history. They appear to have connected the living with their ancestors as evidenced by large niches used to store mummies at the Royal Mausoleum and the Grant Caverna of Machu Picchu. The Incas were a Sun-worshipping society and many of the caves selected and developed as huacas appear to have been chosen because sunlight entered. The portal of a cave as a metaphorical window

---

[52] Gullberg, 'The River Intihuatana: Huaca sanctuary on the Urubamba'.

for sunlight to penetrate the darkness is the most frequently encountered meaning, such as Kenko, Lacco, Intimachay, Choquequilla, The Royal Mausoleum, and the Temple of the Condor. Such a meeting of sunlight with darkness may be another example of binary opposition of Levi Strauss' theory of structuralism. Some of these caves exhibit intentional efforts to bring the solstice sun in their dark interiors. The cave within Kenko Grande showcases light climbing three symbolic stairs on the side of an altar at the time of the June solstice, while a second altar is illuminated with the December solstice sunrise. Two caves in Lacco incorporate light-tubes directed at altars and the opening of a third is positioned for the rising June solstice sun. Caves as transformative passageways are found at Titikaka of Chinchero and near the summit of Huayna Picchu. The precarious north and south stairways to the summit of Huayna Picchu, both involving caves, qualify as genuinely liminal experiences that carried celebrants from a lower to the upper world. Finally, the geopolitical significance of caves is perhaps evident at the two ends of the empire of Pachacuti. Both Kenko and Lacco were major cave complexes in the Cusco valley. At Machu Picchu there is a plethora of caves associated with both June and December solstices. Many of these caves were living beings, animated by water and sunlight with whom humans could interact and establish complex relationships we are gradually beginning to understand.

Rethinking Nahualac, Iztaccíhuatl, Mexico: Between Animism to Analogism in Mesoamerican Archaeoastronomy

Stanisław Iwaniszewski

Abstract: The site of Nahualac (3890–3920 m asl) is situated on the western slopes of Iztaccíhuatl, a well-known volcano in Central Mexico. It consists of a rectangular stone sanctuary located within the seasonally active small lagoon, and the distinct area where multiple deposits of ritual pottery were found. The piles of stone situated on the borders of the lagoon produce alignments towards the nearby and distant landforms offering broad vistas towards the brilliant white peaks of Iztaccíhuatl in the East and restricting the visibility towards the West. The site belongs to the category of high-mountain cult places functioning during the Early and Late Post-classic periods (900–1521 CE) and is associated with the central Mexican cult of fertility, mountain, and rain.

The ritual and worldview meanings of this site are taken together to discuss the ways of how the Post-classic societies in Central Mexico conceptualised their relationship with their surroundings. Using the layout of Nahualac and its astronomical alignments, I conclude that it exhibited cultural configurations that can be classified as characterising analogism rather than animism.

When it comes to the interpretation of the material evidence of past human practices related to the celestial environment, two persistent tendencies are observed. On the one hand, there is the tendency to embrace interpretations in terms of ourselves at the expense of interpretations made by the anthropological Other.[1] As a result, a good amount of ancient worldviews is lost because archaeoastronomers use their own scientific worldview to explain the past as it really was.[2] On the other hand, there is the tendency

---

[1] Anthropologists usually describe 'others' as different and separate from 'us'. I am using this concept to emphasize the notion of perceived cultural differences between modern and non-modern (indigenous, non-Western or pre-modern) societies.

[2] See, for example, Stanisław Iwaniszewski, 'Looking Through the Eyes of Ancestors: Concepts of the Archaeolastronomical Record', in Mauro Peppino

to retrieve knowledge of an existing worldview from the material record itself. This tendency proposes we can grasp deeper cultural meanings beyond the purely logical meanings of things. Since each society can understand and interpret its social and natural environments only within the context of its own cultural tradition,[3] then our attempts to describe the ways people perceived and conceptualised their celestial environments in the past, should always be compatible with the inferred underlying principles established by their worldviews. A pragmatic view asserts that in some way, both of these tendencies supplement each other. While the former one describes the people's engagements with their celestial environments in terms that are intelligible to us, the latter one attempts to understand the past on its own terms.

Given the potential of archaeoastronomy to yield information about now extinct human perceptions and practices relating to the skies, the obvious question here is how we might archaeoastronomically provide the insights into the ways of conceiving and conceptualising of the celestial environment in the past. In trying to answer this question, I will start from the assumption that the fundamental categories of what people use to construct or describe their lifeworlds may differ significantly. Western philosophy, with its distinction between nature and society (or culture), is just one possibility among others. Consequently, if there are other possibilities of being in the world, and Western ontologies no longer can be taken for granted, we should be able to identify, in the archaeological record, other than Western modes of engagements between people and their celestial environments.

For the aims of this paper, I am defining a worldview as a set of the main proposals (prejudices, beliefs, categories, concepts, and so on) through which we view and generate the world. Its form or shape derives from the principles evolved from the practical (day-to-day) engagements of human communities with their surroundings (human lifeworlds).[4] In many non-western societies, the recognition of celestial cycles forms part of their practical knowledge of the environment, so it is practically impossible to

---

Zedda and Juan Antonio Belmonte, eds., *Lights and Shadows in Cultural Astronomy* (Isili: Associazione Archeofila Sarda, 2007), pp. 11–19.

[3] See Hans Georg Gadamer, *Truth and Method,* translated by G. Barden and W.G. Doepel, (London: Sheed and Ward, 1981).

[4] See Tim Ingold, 'Hunting and Gathering as Ways of Perceiving the Environment', in Tim Ingold, *The Perception of the Environment: Essays in Livelihood, Dwelling and Skill* (London: Routledge, 2000), pp. 40–60.

separate the modes of engagement with celestial bodies from all other relations taking place between human groups and their natural environments. It is not surprising, therefore, that their worldviews usually combine conceptions of social relations and conceptions of cosmos treating the human and nonhuman beings within a single conceptual field. Within such models, the patterns which organise the relationships between celestial bodies cannot be conceived as belonging to a different order than those which characterise the relations between humans and between humans and non-human others.[5] In light of this, it becomes apparent that the knowledge about the celestial environment can no longer be viewed as being constructed exclusively by humans; rather it emerges as a result of human and human-nonhuman interactions. The artifacts which today represent the archaeological record once coexisted with other entities like humans, plants, animals, landscape features, meteorological and astronomical phenomena, immaterial beings, sharing the world inhabited by humans.[6] Thus in a broader perspective, past human perceptions and practices related to celestial phenomena are the result of the dialectic between human communities and their material and immaterial contexts, of the web of diverse relationships between humans, components of their natural and celestial environments, other animal and plant species, human-made objects, and non-human others.

To sum up, though Western philosophy sees humans as distinct from their natural surroundings, we do not need to take this pattern as universally shared. On the contrary, our study of the past cultural practices related to the skies should demand a recognition of humans as part of the environment. Assuming that the Western concept of nature/culture dichotomy is just one form of being-in-the-world, we should consider other ways of interpreting the archaeological record. The recent shift observed in archaeological approaches to the past material record consists of the idea that peoples' relationships and engagements with the world they inhabit rely on the recognition that the non-human components of that world might be believed to be animate. In thinking how different peoples developed their relationships with their social and natural environments Philippe Descola developed an inspiring conceptual framework outlining the ways

---

[5] More on this subject, see Stanisław Iwaniszewski, 'Por una astronomía cultural renovada', *Complutum* 20 (2009): pp. 23–37.
[6] See, for example, Bjørnar Olsen, *In Defence of Things: Archaeology and the Ontology of Objects* (Lanham: Altamira Press, 2010).

in which such relationships or engagements might be conceptualized.[7] Descola's proposal focuses on four common types of human interactions with the environment (animism, totemism, naturalism, and analogism). Obviously, peoples' worldviews may show two or three different types of modes of existence, (for example, both animic and analogical, or both analogical and naturalistic) because attributes and properties of nonhuman components of the world may not be regarded as fixed, bounded, autonomous or existing independently of the surrounding world, but remain determined by the relationships between all other components. While in Western societies physical objects and living organisms are just entities regardless of their particular context and the web of relationships within which they are embedded, in non-Western ontologies, their properties are developed by the relationships with all other entities. In other words, the treatment of archaeoastronomical evidence on entirely astronomical grounds may not adequately describe the social reality of past societies. To give an example: celestial alignments may be defined as attributes or properties of specific objects or structures, which are defined by the returning positions of the heavenly bodies, climatic, meteorological, ritual, and other relationships that those structures and objects develop with other entities over time.[8]

As a result of attributing certain human properties to the components of the surrounding world, people establish the rules which serve them to construct and maintain various relationships with them. The diverse forms of cultural conceptualisations of human-environmental relationships rely of course, on their capacity of defining who the human persons are and with whom they maintain interrelationships. As Philippe Descola observes:

> each specific form of cultural conceptualisation also introduces sets of rules governing the use and appropriation of nature, evaluations of technical systems, and beliefs about the structure of the cosmos, the hierarchy of being, and the very principles by which living thing function.[9]

---

[7] Philippe Descola, *Beyond Nature and Culture*, translated by Janet Lloyd, (Chicago: The University of Chicago Press, 2013).
[8] See Ingold, 'Being Alive', pp. 67–75; Nurit Bird-David, '"Animism" Revisited: Personhood, Environment, and Relational Epistemology' (with comments), *Current Anthropology* 40 (supplement), pp. 67–91; Herva, 'Living (with) Things', pp. 389–401.
[9] Philippe Descola, 'Societies of Nature and the Nature of Society', in Adam Kuper, ed., *Conceptualizing Society*, European Society of Social Anthropologists, (Routledge: London, 1992), pp. 107–26.

Mesoamerican ontologies

In pre-Hispanic Mesoamerican worldview, the human being is conceived as a temporary union of different components. Some are seen as material and extracorporeal, others as intangible and tied to the physical human body. Each human person can thus be divided into two building components: a heavy matter which is linked to the earth and a light component of divine origin which formerly is derived from the activities of the gods at the time of the creation of the universe and then repeated in each of the individual creatures. The hard part of each person is considered as a kind of a container or a coverage of intangible essences that give life, intelligence, personality and emotion to the human individual. The physical body supports, houses or wraps three intangible components. Mesoamerican pre-Hispanic societies believed that each newborn child absorbed those three essences, called animated entities. When combined, these enabled human beings to experience emotions, to perform intelligence, and to sustain life. Those three entities were known in Nahuatl as *tonalli* ('heat', 'day name', 'fate'), *teyollia/yolia* ('heart') and *ihiyotl* ('breath', 'blow/puff').[10] The three entities endowed to the person of a specific existence and continuously transformed themselves according to the particular stages of an individual life. In pre-Hispanic Mesoamerican communities, calendar ritual specialists or priests were trained in uncovering the names, character, and other specificities of those entities.

Mesoamerican peoples interacted with a wide spectrum of animate entities. Deities, heavenly bodies, meteorological phenomena, landscape features, animals and plants, and peculiar artifacts were all endowed with some degree of animacy. However, the attribution of animacy to them was neither fixed not static. The attribution of animacy to the whole cosmos, to objects, things, persons, processes, states, etc. was made through a kind of ever-flowing entity known as *teotl*.[11]

---

[10] For the sake of simplicity I am using standard Spanish orthography. In translating the Nahuatl terms I rely on interpretations made by Alfredo López Austin, *Cuerpo humano e ideología. Las concepciones de los antiguos nahuas* (México: Instituto de Investigaciones Antropológicas Universidad Nacional Autónoma de México, 1980) and Roberto Martínez González, *El nahualismo* (México: Instituto de Investigaciones Históricas, Universidad Nacional Autónoma de México, 2009).

[11] *Teotl* has been customarily translated as 'god' or 'deity'. These concepts are misguiding since they seem to represent European concepts of god and divinity rather than those of Native Mesoamericans.

Nahualac – description of its basic features

The site of Nahualac (3890–3920 m asl) is located on the western slopes of Iztaccihuatl, a famous volcano in Central Mexico. The name Nahualac means 'in the river (or spring) of the sorcerers',[12] although it remains unclear whether it was the pre-Hispanic name for this location. The site occupies two neighboring yet remaining divided places, located in the bottom of the oval valley which is shaped by the last glaciation and is connected through narrow passes with similar valleys above. The valley itself is bounded by minor elevations in the south and west that form a kind of ridge culminating with the peak of Mt. Nahualac (3930 m asl) from which it takes the name. To the north lies Nahualac Glen. The Valley, especially its western part, offers wide-ranging views over the summits of Iztaccihuatl (Fig. 1). From the valley eastward, the slopes descending from Iztaccihuatl can be seen. The highest elevations of this snow-capped mountain visible from the valley form an undulating skyline at a distance of about 2 km. The name of the mountain Iztaccihuatl ('White Woman') affects the cultural perception of the peaks since people today commonly affirm that the whole ridge of mountains represents a lying down woman. Thus, the highest peaks receive their names from the parts of the human body. Looking from the north to the south there are: La Cabeza ('the head'), El Cuello ('the neck'), El Pecho ('the breast'), La Panza ('the belly'), Rodillas ('the knees'), Pies ('the feet'), etc. (Fig. 2 and see Fig. 1). In the pre-Hispanic past the mountain received the names of Iztactepetl ('White Mountain'), or Tonacatepetl ('Mountain of Our Sustenance').

---

[12] From *nahualli*, 'sorcerer'. I owe this translation to Leopoldo Valiñas.

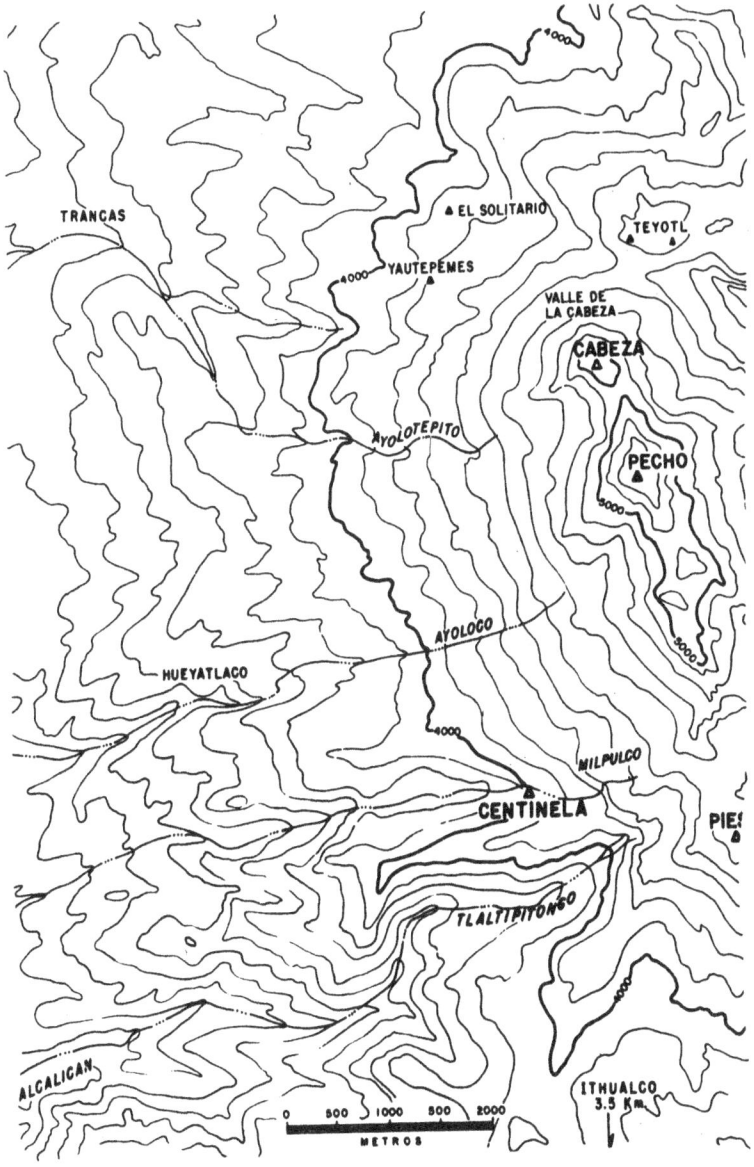

Fig. 1. Western slopes of Iztaccíhuatl

222 Rethinking Nahualac, Iztaccíhuatl, Mexico: Between Animism to
Analogism in Mesoamerican Archaeoastronomy

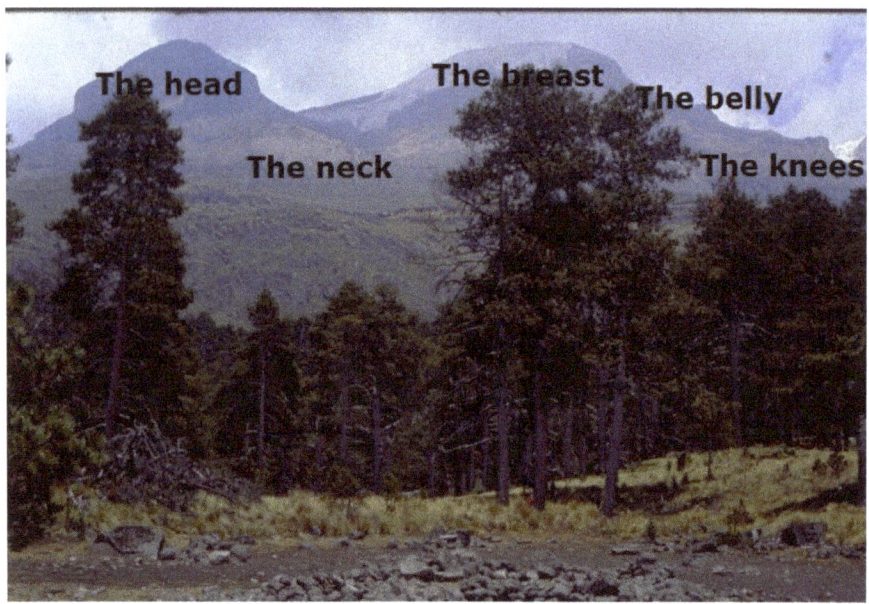

Fig. 2. The view of Iztaccihuatl

The site was first examined by Desiré Charnay who, in 1857 and 1880, excavated archaeological sites on the slopes of Popocatepetl and Iztaccihuatl.[13] He found about 800 items related to the cult of Tlaloc, the Aztec (Mexica) god of rains and mountains. The site was re-discovered in 1956 by José Luis Lorenzo who recollected the surface material in the area of ritual deposits.[14] Between 1984 and 1986 the site was visited by the author who in the company of Arturo Montero did excavations in the area of ritual deposits.[15] Site chronology is based on the pottery typology and indicates the site was visited on different occasions during the Early Post-

[13] See, Desiré Charnay, *The Ancient Cities of the New World Being Voyages and Explorations in Mexico and Central America from 1858 till 1882*. Vol. 10 Antiquities of the New World, Early Explorations in Archaeology. AMS Press, Inc. New York for Peabody Museum of Archaeology and Ethnology (New York: Harvard University, 1973). Originally published in French in 1867.
[14] José Luis Lorenzo, *Las zonas arqueológicas de los volcanes Iztaccíhuatl y Popocatépetl* (México: Instituto Nacional de Antropología e Historia, 1957).
[15] Consult Ismael Arturo Montero García, *Iztaccíhuatl, Arqueología den Alta Montaña*. Tesis para optar por el título de licenciado en Arqueología, (México: Escuela Nacional de Antropología e Historia, 1988).

Culture and Cosmos

classic Period (ninth to eleventh centuries CE, and probably also in the Late Post-classic Period (thirteenth to sixteenth centuries CE). Finally, the site was also archaeoastronomically examined by the author and Arturo Ponce de León.[16]

As stated above, the archaeological site of Nahualac consists of two locations separated by a distance of about 120 metres. The first one is composed of a rectangular stone sanctuary placed within the seasonally active small lagoon or pond and occupies the isolated, lowest and northernmost extension of the valley at approximately 3890 m asl elevation. The pond is enclosed on three sides by slopes descending from higher elevations and slightly opens towards the north where the Nahualac Gorge precipitates. The site lies within a light pine forest. Nevertheless, the highest peaks of Iztaccihuatl ('the head', 'the neck', 'the breast', and 'the knees') are visible to the east behind the pine trees when one is standing in the front of the structure, near to its entrance. A small elevation restricts the view towards the west, but the forest here is not very dense enabling the sun to be seen behind the trees. This view looks uphill behind the trees, but it is possible to see the distant skyline in the background if one moves from the lagoon area. The view towards the south is restricted by the adjacent part of the valley while looking north, the view is of more immediate mountain ridge which descends from the Pulpito del Coyote ('Coyote's Pulpit') summit, just on the opposite side of the Nahualac Gorge. The peaks of Mt. Telapon and Mt. Tlaloc would be visible to the NNW, although they are presently blocked from view by pine trees.

The center of the lagoon is occupied by a small structure made of stone (Figure 3). The structure is roughly rectangular, measuring 7.4 m x 6 m, and today its walls rise to 30–40 cm. It contains two interior spaces and a doorway facing west. The small lagoon is made by water coming from the source at the Chalchoapan[17] lagoon, located some 700 m above, which in turn receives water from the melting snows from the Ayolotepito[18] glacier

---

[16] Stanisław Iwaniszewski, 'De Nahualac al Cerro Ehecatl. Una tradición prehispánica más en Petlacala', in *Primer Coloquio de Arqueología y Etnohistoria del Estado de Guerrero*, (México: Instituto Nacional de Antropología e Historia and Gobierno del Estado de Guerrero, 1986), pp. 497–518; Arturo Ponce de León, 'Propiedades geométrico-astronómicas en la arquitectura prehispánica', in Johanna Broda, Stanislaw Iwaniszewski and Lucrecia Maupomé, eds., *Arqueoastronomía y Etnoastronomía en Mesoamérica* (México: Universidad Nacional Autónoma de México, 1991), pp. 413–46.
[17] From Chalchiuhapan, 'upon green stones water'.
[18] From Ayollotepiton, 'in the small heart of water'.

(today almost entirely vanished) located between the lagoon and the summit of Iztaccihuatl (known as El Pecho, or 'the breast') and slopes of La Cabeza ('the head').

Fig. 3. The pond of Nahualac.

The placement of the entrance implies that the visitors were approaching the precinct from the west facing the main and highest peak of Iztaccihuatl known as El Pecho ('the breast'). The axis of the structure is aligned with the southern edge of the breast, between Ordóñez and Aguilera crags. The orientation of this structure yields the mean azimuth of 107° 03'–107° 15' (the structure's walls are only roughly linear), belonging to one of the most widespread alignments' groups in Mesoamerica.[19] For the latitude of 19° N, azimuths around 105°–107°/285°–287° usually record the sunrise/sunset dates around February 12 and October 30, and April 30 and August 13. These four dates establish intervals of 260 days (from February 12 to October 30 and from August 13 to April 30) being pivoted either on the winter or the summer solstice. It has been argued that these

---

[19] Anthony F. Aveni, *Skywatchers: A Revised and Updated Version of Skywatchers of Ancient Mexico* (Austin: University of Texas Press, 2001), p. 234.

dates marked four critical moments in maize agriculture.[20] However, due to the significant horizon altitude, instead of February 12 and October 30 at Nahualac we can observe sunrises on February 20/21 and October 21. These appear to divide the solar year into two-thirds and one-third (243 and 122 days). With adjustments for horizon elevations the sunset dates corresponding to the same orientation refer to the days of May 3 and August 9, dividing the solar year into the periods of 100 and 265 (approximately nine lunations) days. Interestingly, as viewed from the precincts' entrance, the sun rises over La Cabeza ('the head') on May 3 and August 9.

In the center of the pond a ritual precinct was built and around its edge, were raised small bases or piles of stones which today are almost totally destroyed. They are described by Charnay as 'smaller monuments, pedestals, altars, or chapels, bearing the statue of Tlaloc'.[21] Lorenzo calls them 'small bases quite regularly grouped on the central construction'.[22] I described them as 'piles of stones'.[23] These descriptions only witness the advanced deterioration of the monument. The number of them varies between nine[24] and ten.[25] While the picture of the Nahualac pond depicts them as regularly shaped rectangular stone basements, it is impossible to conclude how many of them were initially erected.[26] The alignments carried out from the precinct entrance displayed a radial pattern. Those alignments laid out over the surrounding prominent landforms, allowed me to attempt to identify the basements with some specific places. In my opinion, this represented the local representation of the Mexica cosmological system, in its spatial, temporal and hierarchical aspects.

---

[20] Iwaniszewski, 'De Nahualac al Cerro Ehécatl', p. 515; Stanisław Iwaniszewski, 'La arqueología y la astronomía en Teotihuacan', in Johanna Broda, Stanislaw Iwaniszewski and Lucrecia Maupomé, eds., *Arqueoastronomía y Etnoastronomía en Mesoamérica*, (México: Universidad Nacional Autónoma de México, 1991), pp. 269–90; Ivan Šprajc, 'Astronomical Alignments at Teotihuacan', *Latin American Antiquity* 11, no. 4 (2000), pp. 403–15; Ivan Šprajc, *Orientaciones astronómicas en la arquitectura prehispánica del centro de México*, Colección Científica, 427, (México: Instituto Nacional de Antropología e Historia, 2001), pp. 79–88, 107–20.
[21] Charnay, 'The Ancient Cities', p. 180.
[22] Lorenzo, 'Las zonas arqueológicas', p. 20.
[23] Iwaniszewski, 'De Nahualac al Cerro Ehecatl', p. 502.
[24] Iwaniszewski, 'De Nahualac al Cerro Ehecatl', p. 502.
[25] Lorenzo, 'Las zonas arqueológicas', p. 20.
[26] Charnay, 'The Ancient Cities', p. 182.

About 20–30 m above the lagoon, on a wide hillock (about 3910–3915 m asl) adjacent to the peak of Mt. Nahualac, at a distance of 120 m, there is an extensive area where multiple deposits of ritual pottery and offerings were found. The structural remains at this site are so ambiguous that it can only be described as the offering place. The site was the subject of intense illicit digging. The landscape setting around this site is quite spectacular. This area affords excellent views of all the major snow-capped peaks of Iztaccihuatl. Also Popocatepetl is clearly visible on the SSE horizon. The view towards the Valley of Mexico is obscured by the slopes of Mt. Nahualac. This site seems to have been located specifically so that Popocatepetl and other mountains in the south can be observed. On the other hand, this site is located so as to be visible from the area adjacent to another rectangular precinct, called El Caracol which and situated about 500 m above.[27]

Of particular importance for my interpretation seems to be the position of El Cuello ('the neck'). Today, this feature is part of the anthropomorphised sierra, but it is not known whether the same visual perception of the lying female was known in Pre-Hispanic times. Avoiding possibly anthropomorphic connotations, I will start with a simple observation that today is called the neck and creates a form known as a saddle among mountaineers. Now, moving from this less anthropomorphic feature, we can notice that the saddle may also be transformed into a cleft between two higher elevations, those of La Cabeza ('the head') and El Pecho ('the breast'). Drawing on these preliminary observations, I can now propose that the undulating profile displayed by Iztaccihuatl may in fact represent a mountain with a deep cleft. As observed from the Nahualac pond, the sun rises over the saddle during the last days of April, coinciding with the onset of the rainy season and the ripening of maize plants. Keeping in mind that one of Iztaccihuatl sixteenth-century names was Tonacatepetl, we can further propose that the cleft-like profile exhibited by the mountain (see Fig. 2) may refer to the well-known pan-Mesoamerican myth in which Tlaloque breaking the huge Tonacatepetl mountain opened a way to a cave containing the life-giving plants (including maize). This symbolism may be reinforced by the fact that the Chalchoapan lagoon is located just below the saddle, and gathers waters that flow towards

---

[27] The exact location of the Nahualac pond cannot be established from the distance due to the forest cover.

Nahualac and fill the empty and dry space around the precinct at the onset of the rainy season just bringing the lagoon back to life. Similar landscape properties have been studied recently by García Zambrano who found many more examples of alignments linking monumental architecture with mountain saddles or passes.[28]

From Nahualac to Tlalocan to Analogism

Agricultural affinities of the site are reassessed by the pots representing Tlaloc's face. Tlaloc is the Mexica (Aztec) god of rain, lightning, and earth. During the Post-classic period, he was believed to dwell in mountains and caves. Tlaloc represented the male aspect of falling waters while his consort Chalchiuhtlicue, or Matlalcueye, was considered as patroness of flowing or static waters (rivers, lakes).[29] In fact, Tlaloc, Chalchiuhtlicue, Matlalcueye and the innumerable assistants called *tlaloque* were treated as mountain and water embodiments.

Tlaloc embodies the Mesoamerican concept of earth and the natural environment. According to Sullivan, his name translates as 'he who has the quality of earth', 'he who is made of earth', 'he who is the embodiment of the earth'.[30] Tlaloc appeared in the Late Pre-classic and was popular at Teotihuacan (Classic Period 1–650 CE). As mentioned, he was one of the most popular gods at Tenochtitlan, the Aztec capital. Therefore, ritual structures and activities at Nahualac may reveal the process through which the relationship between the humans and their surrounding environment was brought into being.

Tlaloc was believed to dwell in a place called Tlalocan, or 'The Place of Tlaloc'. Several mountains around the Valley of Mexico housed shrines where rain-bringing ceremonies, propitiatory, and thanksgiving rituals took

---

[28] Consult Ángel Julián García Zambrabo, 'Transference of Primordial Threshold Crossings onto the Geomorphology of Mesoamerican Foundational Landscapes', in Amos Megged and Stephanie Wood, eds., *Mesoamerican Memory. Enduring Systems of Rememberance* (Norman: University of Oklahoma Press, 2012), pp. 215–28.

[29] Stanisław Iwaniszewski, 'Y las montañas tienen género. Apuntes para el análisis de los sitios rituales de la Iztaccihuatl y el Popocatepetl', in Johanna Broda, Stanislaw Iwaniszewski and Arturo Montero, eds., *La Montaña en el Paisaje Ritual*, (México: IIH UNAM-CONACULTA-INAH-BUAP, 2001), pp.113–47.

[30] Sullivan Thelma Sullivan, 'Tlaloc: A New Etymological Interpretation of the God's Name and What It Reveals of His Essence and Nature', in *Proceedings of the 40th International Congress of Americanists* (Genoa: Tilgher, 1974), Vol. 2, pp. 213–19.

place. They represented *ayaucalli* (Houses of Mist) where Tlaloc with tlaloque (the Tlalocs, rain gods) was believed to reside.[31] Tlalocan was a mythical and watery dwelling of abundance. In the *Historia de los mexicanos por sus pinturas* Tlaloc's abode is described as being four-sided occupied by four tlaloque (the Tlalocs, rain gods) who sent four kinds of rain.[32] In the *Leyenda de los Soles* four tlaloque (blue, white, yellow, and red) steal the staple foods from the Mountain of Sustenance.[33] Classic and Post-classic artifacts (Tizapan Box, Mixtec panel with four Cosijos from Amparo Museum, Zapotec Four-vessel with Cosijo appliqué, Codex Borgia 27–28) represent four Tlaloque associated with four basic directions, kinds of rain and colors. In thinking how rain can be construed as a person, I propose to view Tlaloc as a dividual being, composed of different kinds of rain. In turn, the four Tlaloque beings, or the dwarfish Tlaloc's assistants, can be interpreted as entities embodying a singular aspect of this rain deity. The same may be said of Tlaloc in his aspect embodying a mountain. Tlaloc could be seen as a type of a partible person whose parts are identified with the countless and dwarfish tlaloque beings associated with a singular mountain or a cliff. (see below).

The four-sided world, with the pivotal world axis, is the most common and widespread cosmovisional model in ancient Mesoamerica. It appeared in the Middle Pre-classic period (900–500 BCE) when maize agriculture became a basic staple of Mesoamerican subsistence. The four-sided world probably represented a maize field.[34] In later times, the maize field, house, village, and the created worlds were represented as four-sided. The orderly world was often contrasted with the wild forest. The world was divided

---

[31] Elena Mazzetto, 'Las ayauhcalli en el ciclo de las veintenas del año solar. Funciones y ubicación de las casas de niebla y sus relaciones con la liturgia del maíz', *Estudios de Cultura Nahuatl* 48 (2014): pp. 135–75.
[32] 'Historia de los mexicanos por sus pinturas', in A. Ma. Garibay K., ed., *Teogonía e Historia de los mexicanos,* Sepan cuantos… 37, (México: Porrúa, 2005), pp. 23–87.
[33] 'Leyenda de los Soles', in *Códice Chimalpopoca*, (México: Universidad nacional Autónoma de México, 1975), pp. 119–42.
[34] Karl Taube, '2000 Lightning Celts and Corn Fetishes: The Formative Olmec and the Development of Maize Symbolism in Mesoamerica and the American Southwest', in J.E. Clark and M. Pye, eds., *Olmec Art and Archaeology: Social Complexity in the Formative Period,* Studies in the History of Art 58 (Washington, DC: National Gallery of Art, 2000), p. 303.

Stanisław Iwaniszewski 229

into four directions and at each corner were sacred trees which held up the sky.

I think these examples clearly show that in many aspects the Aztec and Mesoamerican model of the world was based on analogism. All components of the world: *tlaloque*, world directions, kinds of rain, and distinct colors are ontologically different, so it is necessary to find stable correspondences between them. Tlaloc (vertical waters, falling waters) represents the male principle, while his divine consorts, Chalchiuhtlicue and Matlalcueye, represent static or horizontal waters. The principle of analogism is, therefore, achieved through the model of the four-sided world and hot and cold (male and female) classification.[35]

Now as Strathern already demonstrated, the notion of a person as a whole and independent being, enclosed within a skin and possessing one soul or mind is a modern Western conception.[36] As is known, the Aztec understanding of human beings was quite distinct. The human beings consisted of a heavy substance – the physical body – and three different kinds of vital entities (see above). These animate entities depended on exterior components (deities, other animate entities, particular human agents, etc.) so individual destinies might have been affected by exterior influences. Moreover, there were many other animated entities embodied in objects, ritual tools, offices, gesture, vestments, etc. that could confirm the individuality of an individual. Therefore, a human person was derived from the network of all external objects and relationships. In other words, humans had similar physicalities but different interiorities.

Following Strathern typology, in Mesoamerica, human persons may be defined as 'dividuals' in contrast to the 'individuals' of the West.[37] However, taking into account that many Mesoamerican persons are composed of relations that apparently extend beyond the skin boundary, to include objects, other persons, and relationships, they can be defined as partible persons. They absorb parts of other persons and objects and are absorbed by others. On the other hand, Gillespie took another point of view, relying on the concept of networking and corporation.[38] The former

---

[35] Alfredo López Austin, *Tamoanchan y Tlalocan* (México: Fondo de Cultura Económica, 1994).

[36] Marylin Strathern, *The Gender of the Gift: Problems with Women and Problems with Society in Melanesia* (Berkeley: University of California Press, 1988).

[37] Strathern, *The Gender of the Gift*, p. 185.

[38] Susan D. Gillespie, 'Aspectos corporativos de la persona (personhood) y la encarnación (embodiment) entre los mayas del periodo clásico', *Estudios de Cultura Maya* 31 (2008): pp. 65–89.

230 Rethinking Nahualac, Iztaccíhuatl, Mexico: Between Animism to
Analogism in Mesoamerican Archaeoastronomy

view conceptualises a human person as being composed of diverse relationships; the latter one defines a person as being part of a corporate institution.

Now, following the relationship that Tlaloc maintains with the tlaloque, who assist him in his duties (thunder- or rain-making, gathering water and humidity inside mountains), it is possible to define the god either as being partible, because tlaloque might be 'distributed Tlalocs', or as being corporative, because he shared his office with the tlaloque. This structure is represented by an architectural pattern of Nahualac: the rectangular structure surrounded by separated piles of stones (see Figures 4 and 5).

The piles of stones, the ritual precinct and the water in a seasonal pond thus embody the animate entities and divine forces symbolized by Tlaloc and his assistants. They all constitute a miniaturized surface on which rituals were performed to produce effects in the human lifeworld. In performing the rain-bringing ceremony, the ritual specialists pursued to synchronize the entities embodying particular topographical features both located in the Itaccihuatl range and in the piles of stone with the activities of Tlaloc who was supposed to send rain in a proper time.

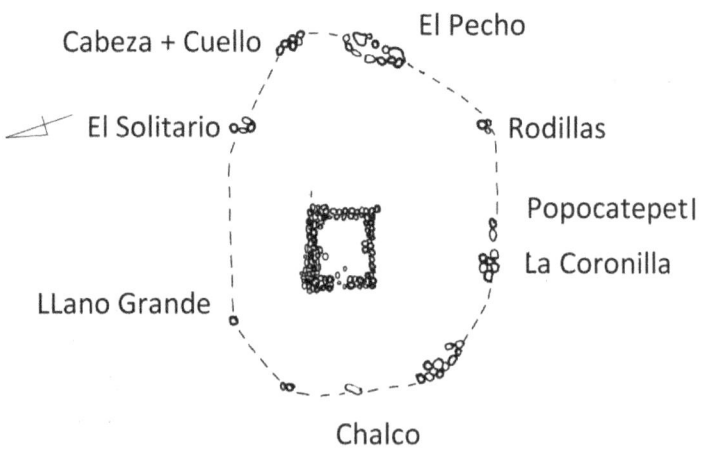

Fig. 4 The piles of stones aligned.

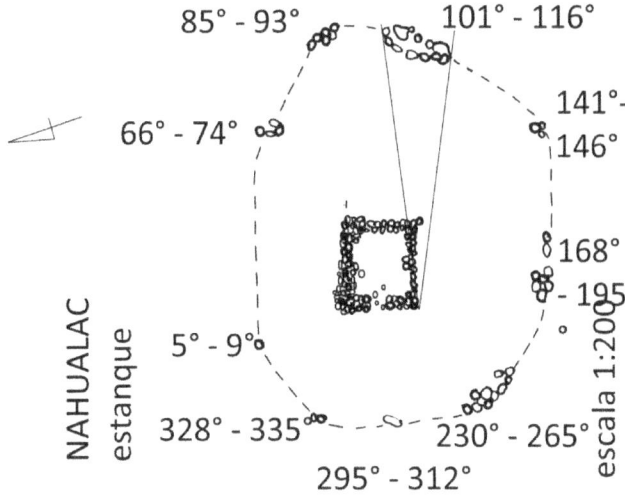

Fig. 5 Tentative identifications of the piles of stone with landscape prominences.

Conclusions
It may be concluded that the ritual precinct is representing the indigenous ontology. This ritual site is a way of world-making not just world-mirroring.[39] In this way, archaeoastronomers may reaffirm a symbolic relationship between monumental ceremonial structures, the offerings and rituals taking place in those structures, and the universe.

The Mesoamerican ontology described in this paper as the combination of animism and analogism is creating a more complex picture than the model of four ontologies proposed by Descola. My example also shows that animic ontology cannot be taken for granted.

In this paper, I explored the utility of archaeoastronomy for producing insights into the worldviews of non-Western peoples. Apart from the 'pure' archaeoastronomical investigation, I widely utilized both archaeological and ethnohistorical evidence. The case described here enables me to suggest that careful archaeoastronomical research may offer a window onto past and non-Western ontologies.

---

[39] I am following W.J.T. Mitchell, *What Do Pictures Want? The Lives and Loves of Images* (Chicago: The University of Chicago Press), pp. xiv–xv.

Evolution of Arabic Astronomy in Relation with the Translation Movement in the Early Abbasid Era

Nasser B. Ayash

Abstract: In this work, the emergence of the Arabic astronomy in the early Abbasid era will be presented as an amalgam of various traditions. For this to be illustrated, a close analysis of the development of the cultural and political aspects in the Middle East will be presented that eventually allowed for the so-called Arabic or Islamic culture to flourish. The translation movement will be discussed briefly in order for various aspects of this period to be shown, emphasizing the duality of tradition and innovation. These aspects will be followed more closely in the field of astronomy, illustrating the various tendencies especially in the case of incorporation of Greek Uranography, and the relation between the Lunar Mansions and the Anwa. Political, religious and cultural changes left their traces on the accepted Academic tendencies of the period. For a better understanding the astronomical view at the dawn of the Abbasid era, a close look on the work of Ibn Qutaiba will take place, depicting the transitional period in which he lived.

The so-called Arabic or Islamic Culture in the period from the seventh until the eleventh century CE adapted quickly to the cultural environment of its surrounding and preceding cultures. A main turning point was the rise of the Abbasid dynasty to power in the mid-eighth century CE, putting a stop to the policy of conquering new lands and focusing on internal organization. This was possibly due to a sense of security due to military successes and the internal conflicts of enemy states. In this period a flourishing of culture and science was observed, and it is even today considered as a golden era.[1]

One of the main problems the Abbasids had to face was the issue of rivalry between the Arabic party on the one hand and the populations of the conquered areas converted to Islam, such as Persians and Syrians, on

---

[1] George Saliba, *A History of Arabic Astronomy: Planetary Theories During the Golden Age of Islam* (New York: NYU Press, 1994), pp. 245–57; P. Adler and R. Pouwels, *World Civilizations* (Boston: Cengage Learning, 2014), p. 214.

Nasser B. Ayash, 'Evolution of Arabic Astronomy in Relation with the Translation Movement in the Early Abbasid Era', *The Marriage of Astronomy and Culture,* a special issue of *Culture and Cosmos*, Vol. 21, nos. 1 and 2, 2017, pp. 233–47.
www.CultureAndCosmos.org

the other.² These had long pressed for equal rights, which the previous dynasty, the Umayyads had to some extent denied, a fact that contributed to their fall in 750 CE.³ The Abbasid Khalifs who rose to power afterwards, made an effort to establish themselves as universal rulers, above nation-divisions, and more importantly above the religious law and its limitations.⁴ This step would put the Khalif on a collision course with the scholars of the Arab fraction and for this debate, a philosophical dialectic was useful, which explains the Khalif's interest for purely philosophical works.⁵

The translations of the Abbasid period, which showed their peak in the ninth century CE, offer a way of understanding the tendencies of this period, as the translations had, apart from the scientific importance, also a fundamental political and religious role in the society of the time. In the previous translation period, during the Umayyads, the focus of the translation projects was primarily bureaucratic and practical.⁶ The State Archives were written in Greek until the shift towards Arabic was performed by 'Abd al-Malik. Also private collectors sought practical knowledge such as the medical treatises of Galen.⁷ After the middle of the eighth century, the Abbasid Khalif, showed interest in acquiring rhetorical and philosophical works and so began ordering translated works, such as the 'Topics of Aristotle' by Khalif al-Mahdi which was one of the first commissions emerging from the need of the Khalif to insert philosophy in the debates of the period.⁸

The books, gathered allegedly in the House of Wisdom, were written in a variety of languages including Greek, Sanskrit, Middle Persian and

---

² Elisabeth Urban, 'The early Islamic mawālī: A window onto processes of identity construction and social change' (PhD Thesis, The University of Chicago, 2012), pp. 86–93.; H.A.R Gibb, *The Arab Conquests in Central Asia*, (London: The Royal Asiatic Society, 1923), p. 10.
³ A. Khanbaghi, *The Fire, the Star and the Cross: Minority Religions in Medieval and Early Modern Iran* (London: I.B. Tauris, 2006), p. 19; Albert Hourani, *A History of the Arab Peoples* (London: Bloomsbury, 1991), pp. 30–32.
⁴ D. Gutas, *Greek Thought, Arabic Culture*, (London: Routledge, 1998) (trans. M. Makri Greek edition: Periplous, 2001), pp. 112–13, 145.
⁵ Gutas, *Greek Thought, Arabic Culture*, pp. 86–95.
⁶ Gutas, *Greek Thought, Arabic Culture*, pp. 32–39.
⁷ I. Giannakis, *The Greek Thought in the Court of the Khalifs* (Greece: Ioannina University: 2000), p. 9.
⁸ Gutas, *Greek Thought, Arabic Culture*, pp. 96–97.

Aramaic.[9] This situation gave birth to an institutionalised translation movement, whose organisation and standards rose rapidly, to cope with the high number of books to be translated. The gradual shift of interest to scientific works was a by-product of the tendency of the time, which provides an insight to the priorities and the necessities of cultural thought and political agendas of the period. It is not a coincidence that in the translations the minorities such as the Nestorians played a major role.[10] Also, the Zoroastrian tradition of the Iranians manifested itself in the importance of astrology, hence the endorsement of works related to astronomy in the early Abbasid era.[11] A. Beinorius argues that translations of astrological works from Pahlavi to Arabic may constitute the earliest scientific texts in Arabic.[12] Sassanid astrology was by itself syncretistic, allowing for a blend of various elements including Greek and Indian, a fact that contributed in shaping many factors of Islamic astronomy, such as the final form of the Anwa system.[13]

Before expanding the gathered knowledge by early scientists of the ninth century, such as Al-Huwarismi and the Banu Musa brothers, the first stage was to translate the texts gathered to the lingua franca of the time, which was Arabic. The importance of the translators can be seen from their fee that could reach 500 golden dinars per month (a currency that had common value as the Byzantine solidus).[14] An interesting aspect of the translation movement was the methodology followed, which focused on overcoming the difficulty of expressing complicated terms, as the Arabic language did not yet contain the necessary vocabulary. Complicated by the fact that many of the scientists and translators working in Arabic were not native speakers.[15] The search for adequate expression shows similarities with the same process that took place in other cultures. For example, in Rome Latinised Greek words were used to express philosophical concepts

---

[9] Gutas, *Greek Thought, Arabic Culture*, pp. 75–85.
[10] D. Hill, *Islamic Science and Engineering*, (Edinburgh University, 1993), p. 4.
[11] Gutas, *Greek Thought, Arabic Culture*, pp. 64–68.
[12] A. Beinorius, 'On the Intercourse between Indian and the Arabic/Persian Astrologies', in F. Pimenta, N. Ribeiro, F. Silva, N. Campion, A. Joaquinito, L. Tirapicos, eds., *Stars and Stones: Voyages in Archaeoastronomy and Cultural Astronomy*, Proceedings of the 2011 SEAC Conference (UK: BAR, 2015), p. 133.
[13] Beinorius, 'On the Intercourse between Indian and the Arabic/Persian Astrologies', p. 134.
[14] Gutas, *Greek Thought, Arabic Culture*, p. 197.
[15] Gutas, *Greek Thought, Arabic Culture*, pp. 193–94.

whereas others, like Cicero, preferred using modified Latin words.[16] Similarly in Baghdad different solutions where followed in the translations performed by different schools that flourished in the ninth century such as the Hunayn Ibn Ishaq school, the Al-Kindi school, and the so called Harran school. Of those, the Harran school was associated strongly with mathematics and astronomy, both in translation and in research. Amongst others Thabit bin Qurra commented on the translation of Ptolemy's Almagest and the Conics of Apollonius.[17] Regarding astronomy, words like 'astrolabe' were transliterated, whereas other words such as the constellation names were, as will be demonstrated, descriptive.

Of great renown to the West are the translation movements of later periods, such as those in the post Mongol-invasions era (after the thirteenth century). Indeed, these revisited works seemed to be more creative and with a tendency towards critical perception and innovation, while the early Abbasid period was greatly reliant on the translation of texts, sometimes with no effort for critical analysis of those translated texts. But it is in this early stage that the Arabic academic language first emerged as a language capable of dealing with philosophical and scientific terms, as it tackled for the first time scientific and philosophical vocabulary and expression. Indeed a great effort was put into determining the appropriate Arabic terms for expressing new terms and expressions. Indeed, a translation of Ptolemy's Almagest in 829 shows a high level of sophistication in the translation method.[18]

Another interesting aspect of the literature of this period is the gradual acceptance of foreign systems that replaced pre-existing Arabic traditions. This will be shown in the case of astronomy, as Greek Iconography of the asterisms prevailed and replaced the pre-existing Arabic names that derived from their pre-Islamic period. But first the role of astronomy in general must be addressed.

---

[16] Baltussen H. 'Cicero's Translation of Greek Philosophy: Personal Mission or Public Service?', in S. McElduff & E. Sciarrino, eds., *A Sea of Languages: Rethinking the History of Western Translation* (Manchester: St Jerome Publishing, 2011), pp. 37–47.

[17] G. Endress, 'The Circle of al-Kindi', in G. Endress and R. Kruk, eds., *The Ancient Tradition in Christian and Islamic Hellenism*, Leiden: Research School CNWS, 1997), pp. 43–76.

[18] George Saliva, *Islamic Science and the Making of the European Renaissance*, (Cambridge, MA: MIT Press, 2007), p. 83.

Why astronomy

The research of astronomical issues and later developing the most advanced astronomical records and instruments of that time such as astrolabes, had a practical background related to religion and state policy. The determination of the exact date of the new moon, was an important and prestigious state function, linked predominantly with religious functions such as the determination of specific calendar dates, such as the starting date of Ramadan.[19] Also, finding the exact direction towards Mecca was necessary for prayer or when building a new mosque.

The position of the astrologers, was often an important one even for high ranking officials, showing a persistence of the pagan practices of the Sassanid empire and their policy that were noticeable in the Muslim Iranian Dynasties.[20] The father of the inventors, Banu Musa, was an astrologer of the Khalif, and the city of the Harran, home to the translators and mathematicians of Harran school, (the so called Sabeans) practiced planet and star worship openly up until the ninth century, if we follow al-Mas'udi description.[21] Finally, the Arabic maritime developed by Khalif Muawiya for war purposes, made astronomical navigation essential.[22]

From Arabic to Greek Uranography

In this period there was a shift towards the Greek system of sky tradition. There was a gradual abandoning of the pre-Islamic Arabic names and sky perception in favour of the concept of the zodiac.[23] Namely the constellations were re-identified and receive their names based on the Greek iconography, which could be studied thanks to the translation movement. The zodiac was also associated with other traditions, but as al-Marzuqi states, the knowledge of the Greeks in matters of astronomy

---

[19] D. A. King, *In Synchrony with the Heavens, Studies in Astronomical Timekeeping and Instrumentation in Medieval Islamic Civilization: Instruments of Mass Calculation* (Leiden: Brill, 2005), p. 17.
[20] Khanbaghi, *The Fire, the Star and the Cross*, pp. 27–29.
[21] Al-Mas'udi, *Kitab muruj-al-Dhahab* (*Meadows of Gold and Mines of Gems*), see chapter, 'On the Earth Surface and Planets'.
[22] G. Ostrogorsky, *History of the Byzantine State* (New Brunswick, NJ: Rutgers University Press, 1969), p. 116.
[23] Daniel M. Varisco, 'The Origin of the Anwa' in Arab Tradition', *Studia Islamica* 74 (1991): p. 7.

surpassed the others.²⁴ Secondly the solar aspect of astrology, and the importance of the zodiac replaced partially the importance of the 'Lunar Mansions' (Manazil al-Qamar) and the Anwa. These were the pre-Islamic constellations that are on or near the ecliptic, delineating the moon's path. This transition was also linked with contemporary political tendencies, most notably the vision of the Abbasids to continue the Sassanid astrological ideology that had close relations with the Greek system.²⁵

As presented, astrology was an important aspect of this tradition which the Abbasids tried to incorporate to the new faith.²⁶ An example of this is the work of the astronomer and translator of the Kindi School Muhammad al-Tajjib al-Sarahsi (d. 899) who found himself in the circle of Khalif al-Mu'tadid. He investigated elements of Ptolemaic geography in the Quran, in a manner similar to what the Mu'tazilits tried to do in their teachings with the allegorical interpretation of the Quran.²⁷ In this way the truth of the Ptolemaic system was seen as absolute and could therefore be found in the holy texts.

The Anwa System
Various bright stars were used in the culture of pre-Islamic Arabia as markers of the periods of the tropical year. An example of this is when the heliacal rising or setting of a star would coincide with a rain or a wind season. These stars were mostly located near the ecliptic (such as Aldebaran and Antares) but could be far from it, (such as Vega and Canopus). However in later periods, the stars of the Anwa would be often replaced by those of Lunar Mansions. This rendering for determining the time periods is usually called the Anwa system.

There were religious reasons for putting aside the Arabic star system, associated with the Anwa, as these were deeply rooted in the folk culture of the desert nomads, and with a history of their veneration, which linked them with heathen and therefore unwanted practices. This was recognized by the astronomers of the Abbasid era, such as al Sufi. The polemic against

---

²⁴ Al-Marzuqi, 'Kitab al-Azmina wa-l-amkina', in Fuat Sezgin, ed., *Natural Sciences in Islam* 53, (Frankfurt am Main: Institute for the History of Arabic-Islamic Science at the Johann Wolfgang Goethe University, 2001), p. 171.
²⁵ Gutas, *Greek thought, Arabic Culture*, pp. 49–64.
²⁶ Gutas, *Greek thought, Arabic Culture*, pp. 106–17.
²⁷ Fuat Sezgin, ed., *Abu Yusuf Ya'qub ibn Ishaq al-Kindi (256/870): Texts and Studies* (Frankfurt am Main: Institute for the History of Arabic-Islamic Science at the Johann Wolfgang Goethe University, 1999), pp. 206–10.

the star worship and the Anwa stems from the Islamic religious texts. This is linked with a broader phenomenon, of investigating pre-Islamic beliefs, which will be presented later in this paper.

As for the emergence of the Anwa cult, one can find statements in the early Abbasid era, not only by astronomers but also by scholars emphasizing the existing knowledge in Arabia. As stated by the scholar al-Jahiz who in the ninth century was trying to promote the value of the Arab culture:

> the Arabs knew the Anwa and the stars for orientation [...] because no matter the situation the sky is always visible, and a man can see the planets and their relation to the fixed stars.[28]

The nomads of the desert recognised the brightest of the stars and gave them names based on their everyday lives. They noticed that some annual changes related to weather or agriculture coincided with the heliacal rising or setting of some of these stars. They further noticed that these stars formed pairs, since a rising star in the morning in the east is considered coupled with a setting star in the west.[29] The rising one is called the observer (Raqib) as it observes the setting star which is named (Naw), i.e., the Falling one, or the leaning star.[30] The plural of Naw is Anwa' which gives the term by which the astronomical system of the Arabs is known, although even amongst the early Arabic scientists there was a controversy regarding the etymology of the term.[31]

It is important to note that the system of Anwa, was not directly connected to the Lunar Mansions, as some of the Anwa were associated with stars which fall outside the Moon's path, such as the significant and anticipated Naw' (heliacal setting) of Sirius and Canopus.[32] These 'constellations' did not have the same width in the pre-Islamic era, as the

---

[28] Al-Jahiz, *Kitabul-Haywan*, part 6.
[29] Ibn Qutaiba, *Kitab al-Anwa*, in M. Mamidullah and C. Pellat, eds., *Natural Sciences in Islam* vol. 52, (Frankfurt am Main, Germany: Institute for the History of Arabic-Islamic Science at the Johann Wolfgang Goethe University, 2001), p.8.
[30] Ibn Qutaiba, *Kitab al-Anwa*, p. 10.
[31] Daniel M. Varisco, 'The Rain Periods in Pre-Islamic Arabia', *Arabica* 34, no. 2 (1987): pp. 251–52; Al-Marzuqi, *Kitab al-Azminawa-l-amkina*, p. 310.
[32] C. Nallino, *Ilm al-falak: ta'rikhuhu 'inda l-'Arab fi l-qurun al-wusta*, edited by Fuat Sezgin (Frankfurt am Main: Institute for the History of Arabic-Islamic Science at the Johann Wolfgang Goethe University, 1999), p.117.

people at that time did not make use of astronomical equipment.[33] The system of the Lunar Mansions seems to have been imported from India at a later time and mingled with the existing Anwa system before Islam.[34] Initially, a number of 27 Mansions was used instead of 28, with the arc of each Mansion measured on the ecliptic being 13 1/3°, or 2¼ zodiacal signs.[35] But later with the addition of the Naw' of Zubana (α,β Librae) the 28 mansions were not of the same width leading to some confusion amongst the scientists of the Abbasid era.[36] This led to the existence of a second Anwa list, such as the one preserved by Abu Zayd Sa'id al-Ansari, which did not correspond to the 28 Lunar Mansions. This list appears to represent 'an indigenous Arab rendering'.[37]

The astrologers however, in their effort to evenly distribute the 28 Lunar Mansions in the ecliptic, and to match them with the zodiacal signs, divided 360° by 28, yielding 12° 51.4' for each Lunar Mansion and 2*13* Mansions per zodiac.[38] The number of 2*13* Mansions per zodiac became the most referenced analogy, as seen in the texts, for example, of al-Marzuqi.[39]

Religious texts as testimonies for astral worship in pre-Islamic Arabia

Up to the present time the pagan sources were scarce, the religious texts were used to retrieve information for the pre-Islamic period. In the religious texts, various legends that existed in the time of Mohamed were either dismissed as pagan or they were adopted from the new religion. In the Quran there is a typical formula for this kind of statements, indicating that his contemporaries were asking him on various legends. If the person or thing they asked Mohamed about was condemned, it is very hard to find other material regarding it, as reproducing or discussing it would be considered heretic.

An example of this scenario was the condemnation of the worship of the star Sirius (Shiara) which led to the conclusion that this star was

---

[33] Nallino, *Ilm al-falak: ta'rikhuhu 'inda l-'Arab fi l-qurun al-wusta*, p. 117.
[34] Ibn Qutaiba, *Kitab al-Anwa*, pp.29–31.; A. Beinorius, 'On the Intercourse between Indian and the Arabic/Persian Astrologies', p. 133.
[35] Nallino, *Ilm al-falak: ta'rikhuhu 'inda l-'Arab fi l-qurun al-wusta*, pp. 117–18.
[36] Ibn Qutaiba, *Kitab al-Anwa*, p. 12; Nallino, *Ilm al-falak: ta'rikhuhu 'inda l-'Arab fi l-qurun al-wusta*, pp. 117–18.
[37] Varisco, 'The Rain Periods in pre-Islamic Arabia', p. 254.
[38] Nallino, *Ilm al-falak: ta'rikhuhu 'inda l-'Arab fi l-qurun al-wusta*, pp. 117–18.
[39] Al-Marzuqi, *Kitab al-Azminawa-l-amkina*, p. 215.

worshiped.[40] An indirect example is the mentioning of the temporary worship of Abraham towards the planet Venus. The Lunar mansions were mentioned in the Hadith as part of heathen practices, but in the Quran only twice, and with no specific emphasis.[41] As al-Sufi mentioned, this was a method of putting aside the behaviour of worshipping the lunar mansions.[42] On the other hand four times the name Buruj, corresponding to the Solar/Greek constellations, appeared.[43] They held prominent position. For example, Allah swore by them at the beginning of a Surah.[44]

In the Quran and the Hadith there are many references to the stars, indicating their significance. Sometimes the stars appear with no connection to the subject of the next verses, for example the evocation "I swear by the position of the stars" which is indicative of the importance of the fixed position of the stars, referring here possibly to the constellations (Buruj).[45] This mentioning could be perceived as contrary to the popularity of the pagan Arabia Anwa system for time measurement.

However, the term Naw' or Anwa is never mentioned in the Quran, although they are mentioned in the Hadith as an example of 'false idols'.[46] Since stars rise helically at the same time within the solar year, they served as indicators of weather and seasonal changes.[47] This was common in many cultures of the past, as is seen for example in Hesiod's 'Works and Days' where the heliacal rising or heliacal setting of stars or star groups like the Pleiades, Orion or the star Spica helped people to coordinate various agricultural activities, or to anticipate the cold season.[48] Sometimes it was common to attribute the observable phenomenon to the stars that indicated them, as is known for example from Egypt where the star Sirius was venerated as it was considered connected with the flooding of the Nile.[49]

---

[40] Quran, Surah *Alnajm*, 49.
[41] Varisco, 'The Origin of the Anwa' in Arab Tradition', p. 7; Quran, Surah 10, 5; Surah 36, 39.
[42] 'Abd al-Rahman al-Sufi, *The Book of Constellations-Kitab suwar al-kawakib*, ed. Fuat Sezgin (Frankfurt am Main: Institute for the History of Arabic-Islamic Science at the Johann Wolfgang Goethe University, 1986), pp.11–12.
[43] Ibn Qutaiba, *Kitab al-Anwa*, p. 9.
[44] Quran, Surah 85, 1.
[45] Quran, Surah 56, 75.
[46] Ibn Qutaiba, *Kitab al-Anwa*, p.10.
[47] Al-Marzuqi, *Kitab al-Azminawa-l-amkina*, Book 9.
[48] Hesiod, *Works and Days*, pp. 599–622.
[49] E. C. Krupp, 'Astronomers, Pyramids, and Priests', in E. C. Krupp, ed., *In Search of Ancient Astronomies* (London: Chatto and Windus, 1979), p. 189.

In a similar fashion, the Arabic tribes used to associate the annual weather periods, like the wind or the rain seasons, with the heliacal rising or setting of stars that coincided with them.[50] For example the hot rain period of Hammim in summer was thought to be caused by the rise of Aldebaran.[51] This would not be acceptable from the new view of the world, where all weather changes are attributed to God.[52]

This tendency was considered heretic to the new religion as is apparent from the Hadith. In one Hadith Mohammed says:

> Whoever claims that the rain is caused by the good will of Allah and his benevolence, then he is a believer in me and does not believe (worships) the star. But whoever claims that the rain is caused by a certain Naw' (Heliacal Rise or Setting of a star) he is an infidel and a believer (worshiper) of the star.[53]

The rain cult was predominant before Islam, as in the Kaaba idols were erected specifically to pray for rain.[54] Distancing from the rain cult was important, and this is evident in the absence of the term 'Anwa' from the Quran. In the light of the above it is clear why the writers that make reference to the Anwa tend to make their position clear on their adoration. Thus, the writer al-Marzuqi tried to justify the ignorance of the people of Arabia before Islam, and their worship of the Anwa stars, as he believed that the Arabs did not have enough information on them.[55] Ibn-Qutaiba distinguishes between those who use the Anwa as season indicators and those who believe that the season change is bought by the Anwa (stars) and he condemns them as unbelievers.[56]

Furthermore, the Lunar Mansions seem to have had a relation to the Indian Naxatra system, which also contributed in putting them out of favour. It is interesting that these reasons where identified by the Islamic Astronomers, such as al-Sufi, in the tenth century.[57]

---

[50] Al-Marzuqi, *Kitab al-Azminawa-l-amkina*, Book 9.
[51] Varisco, 'The rain periods in pre-Islamic Arabia', pp. 256–57.
[52] Al-Marzuqi, *Kitab al-Azminawa-l-amkina*, pp. 167–68.
[53] Hadith, *Sahih Muslim, Book of Faith*, 74 (author's translation).
[54] Varisco, 'The Rain Periods in Pre-Islamic Arabia', p. 252.
[55] Al-Marzuqi, *Kitab al-Azminawa-l-amkina*, p. 167.
[56] Ibn Qutaiba, *Kitab al-Anwa*, pp. 13–14.
[57] Ibn Qutaiba, *Kitab al-Anwa*, pp. 9–10; Al-Sufi, *The Book of Constellations–Kitab suwar al-kawakib*, pp.11–12.

Lastly, the non-Arabic descent population, including the people of Persian and especially those of Aramaic descent, would favour a system that they were accustomed with and that signified their cultural independency from the Arab conquerors.

Legacy
Indeed the cultural conflict of conquered and conquerors was depicted in the adoption or not of Greek philosophers and scientists, who were linked with the Abbasid policy presented above. The anti-Greek fraction was best represented by the scholar Al-Jahiz, who emphasized the superiority of Arabic thought nevertheless the Abbasid dynasty put effort in advancing the adoption of the foreign way of thinking. In the case of the lunar Mansions both sides agreed: the Arabic Islamic academia for religious reason as well as the nonArab side for political reasons agreed on putting them aside. This created a vacuum that was occupied by the Greek Astronomical System.

This is a part of this period's legacy for todays' astronomy. There is an interesting aspect in analysing how the Arabic names were finalised and then how they were transferred some centuries later to western Europe. The way they were transliterated caused distortion of the names, producing inconsistencies in the modern sky map. For example, some common words such as head, tail or leg appear in different ways in the Latin transliteration. The word for leg which is Rijel denotes Rigel in beta Orionis and Rigilkent in alpha Centauri, meaning the leg of the centaur. Even within the same constellation inconsistences may appear. In the Draco constellation the words Thu'ban (big snake) and Tannin (dragon), which were two different descriptions of the constellation, both appear in the constellation's star names: Thuban (alpha Draconis), Rastaban (beta Draconis) meaning head of the snake and Eltanin (Gamma Draconis) meaning the Tannin-dragon. Additionally, in the first two stars of Draco, the same Arabic word Thu'ban, appears differently as Thuban and Taban respectively.

The Arabic names were also often description of the Greek images, rather than translation of the names, with no evident knowledge of the original mythological meaning of the depictions. For example, Orion is described as the Giant – Al-Jabar. The star Algol (beta Persei) in the Greek depiction was illustrated as the head of the Medusa. So the star was named 'Head of the monster', Ras-al-Goul from which the modern name Algol derives. Similarly, Andromeda was described as a tied woman, Al-musalsala. Sometimes the pre-Islamic Arabic names survive in separate stars within a constellation, while the other star names of the same constellations are description of the Greek image. Such is the case with the house or armpit of the Jauza (a pre-Islamic deity and constellation in the area of Orion) which survives in the name of the star Betelgeuse in Orion. Similarly the star Vega in Lyra, whose name is derived from the pre-Islamic Arabic constellation of the falling falcon.

As stated previously religious and political reasons favoured the abandoning of the Anwa system whose study has of itself great interest as it shows the occupation with astronomy in the pre-Islamic times, is a gate to the folklore and also hints to influences of other cultures, most importantly the Indian influence.

The Anwa went out of favour of the academia but remained in folk tradition for a long time. Indeed, in the non-formal rural life, poems and proverbs survived linking a Lunar Mansion with weather changes, not as a lore but more as a practical guide. An analogy with Works and Days of Hesiod can be made. These poems survived largely due to the documentation of the early Islamic astronomers such as Ibn Qutaiba.

Ibn Qutaiba

Ibn Qutaiba is a notable figure in this translation period, who lived and wrote in the ninth century (died 889 CE). He wrote one of the oldest surviving astronomical treatises in Arabic. Indeed from twenty-four writers of books on the subject of the Anwa before the eleventh century CE, the only surviving work is the one by Ibn Qutaiba.[58] His work was influential for the most known Arabic manuscripts in the West, as the works of Al-Sufi and al-Marzuqi. In regard to the many works on the Lunar Mansions, Al-Sufi recognized as valuable only the works of Ibn Qutaiba, as the others are literature and philology oriented with little astronomical understanding.[59] Furthermore Ibn Qutaiba contributed in documenting the Arabic pre-Islamic folklore of Anwa' that were preserved in the rhymed prose (Saja') associated with this phenomenon[60]. This form of Saja' was used by the Arabs to orally immortalise traditions and practices, often providing information on the folklore.

A further importance of Ibn Qutaiba was his testimony of the transition period in which he lived. Firstly, the style of Ibn Qutaiba was influenced

---

[58] Ibn Qutaiba, *Kitab al-Anwa*, p. 19.
[59] Al-Sufi, *The book of Constellations–Kitab suwar al-kawakib*, pp. 1–2.
[60] Varisco, 'The Rain Periods in Pre-Islamic Arabia', p. 251.

by both the traditional Arabic as well as by the Greek astronomical rendering, as he wrote within one generation or less after the translation of the Ptolemy's Almagest. Despite the short period from this translation, he showed great familiarity with the Greek names. At the same time, he was still attached to the Anwa and the Lunar Mansions. Indeed, he dedicated a whole book to them called Kitabu-l Anwa, where he incorporated the Anwa to the Greek astronomical approach.

For example, the fact that the first Lunar Mansion is Sharatan, which is in Aries, reflects a correlation to the Hellenistic model of the zodiac, starting the list of the Anwa at the spring equinox.[61] Ibn Qutaiba mentioned that 'The two stars of Sharatan (β,γ Arietis) are the first stars of Spring'.[62]

Another example of this point is the comparison of the presentation of Orion in the books of Ibn Qutaiba and al-Sufi. Ibn Qutaiba refered to Orion primarily as the Jauza stars, although providing at the same time the Greek depiction as Orion. Namely, he states:

> The Jauza is amongst the southern constellations. She is called 'The Giant' (Al-Jabar) due to similarity with the king. Because she is depicted as a crowned man on a chair.[63]

It is clear that Orion for Ibn Qutaiba is essentially identified with the pre-Islamic female figure 'Jauza', rather than the 'Giant, al-Jabar' which is a depiction used for her stars. Interestingly, the feminine pronoun is used even when the subject in the sentence is the King or the Jabar. Furthermore, the name of the related chapter is 'On the stars of Jauza' and he further identified the stars based heavily on the Jauza depiction, as the name Jauza occurs nine times in the first introductory paragraph.[64]

Al-Sufi on the other hand, almost a century later, introduces the constellation of Orion as follows:

> Constellation of Jabar, and he is the Jauza: His stars that form his depiction are 38, and it is a picture of a man standing south of the Sun's path and he looks like a human...[65]

Al-Sufi uses the masculine pronoun, and the name Jauza appears less prominent than in the text by Ibn Qutaiba. This is an example of the pre-

---

[61] Varisco, 'The Rain Periods in Pre-Islamic Arabia', p. 253.
[62] Ibn Qutaiba, *Kitab al-Anwa*, p. 18.
[63] Ibn Qutaiba, *Kitab al-Anwa*, p. 45.
[64] Ibn Qutaiba, *Kitab al-Anwa*, p. 45.
[65] Al-Sufi, *The Book of Constellations–Kitab suwar al-kawakib*, p. 321.

Islamic depiction that becomes a secondary alternative name or in other cases even omitted.

Secondly, stylistic analysis of his writings and content offers a precious gateway into understanding the developments of this period. He makes use of the newly emerged academic Arabic language, while still keeping a relaxed style in his writing, which imitates a friendly discussion rather than austere academic style, as is seen in other works of later periods, such as in Averroes. Most notably, he speaks directly to the reader, as is seen in the excerpt below.

Some of the observations above can be understood through Ibn Qutaiba's original text. In the translated section from the third chapter from Ibn Qutaiba's book on the Anwa. An effort is made to illustrate the spirit and writing style of the original. Of interest is the fact that he uses the term Lunar Mansions and speaks of their role as Anwa, depicting the merge that has taken place by his time.

> About the mechanism of Rising and Setting: The sun reaches at a certain dawn one of the Lunar Mansions, and it covers (with its light) the Lunar Mansion that it has reached and it also covers the previous Lunar Mansion. Now in this morning you can see the Mansion that lies before these two. This visible one is called 'the one who comes out/rises'. And this is the Mansion the people mean when they say 'if this Mansion comes out, this happens' [....] The sun stays in one Mansion for a period of 13 days before it moves to the next one. Therefore each Mansion that receives the sun will come out/rise and will be visible again at dawn after 26 days. This is the period of time between the sun's arrival to it and its exit from it. I will provide now for you, O reader, an example of what I said to help your understanding: When the sun arrives to Thurayya (Pleiades) at dawn, it therefore covers Thurayya (Pleiades) and Butain (Aries $\varepsilon,\delta,\pi,\rho$) which lies behind it. Therefore the rising Lunar Mansion is Sharatan (Aries α,β,γ). [.] The sun stays at Thurayya (Pleiades) for 13 days before moving to Aldebaran and covering it, while still covering Thurayya (Pleiades), since the sun covers the Mansion where it resides and the previous one, as I taught you. So the sun stays for 13 days at Aldebaran and then moves to Haq'a ($\lambda,\varphi1,\varphi2$ Orionis) causing the Pleiades to be visible again after 26 days.[66]

Conclusions

When today Arabic or Islamic astronomy is discussed the focus is the star catalogues or the innovations of the middle or late Abbasid period, up to

---

[66] Ibn Qutaiba, *Kitab al-Anwa*, pp. 9–10.

the Turkish Era, from the ninth up to the fourteenth century. For example, the names of prominent scientists are in the centre of interest, as well as the influence of the Arabic astronomy in the West. In this paper however, the scope is rather an earlier stage, at the beginning of the Abbasid era in the mid-eighth to mid-ninth century, and the focus is the examination of the reasons behind the developments that shaped the astronomical landscape of that period.

It can be said that reaching for knowledge by pushing forward the translation movement was a part of state policy and thus to a large extent an intentional move. The first steps had included gathering of information via translation and adapting them, giving attention to verbiage and style, to serve the various needs of the political and religious dialogue.

The impact of these translations on the astronomy of the time, was a gradual transition from the pre-Islamic to the Greek and Persian perception of the sky. This affected many aspects of the sky Uranography, primarily the embedding of the Greek depictions of the constellations and the accepting of the importance of the zodiac. The interaction of the pre-Islamic Anwa and the Lunar Mansions are central in this process. In this scope, the work of Ibn Qutaiba relating to the Anwa was emphasized, as a fine example of this transition period, and a valuable source of ethnographic and scientific information.

Metrology, Memory and Long-Term Landscape Inhabitation: Evidence for the Septarian Package on the Atlantic Façade

Roslyn M. Frank

Abstract: All along the Atlantic façade there is evidence that distinctive septarian units of measurement were employed, suggesting a continuity of metrological practice and more particularly the association of the Septarian Package with agro-pastoral practices. The goal of this paper is to show how this metrological *Sprachbund* with its geographical diffusion and memory traces can be brought into play to examine cultural conceptualisations and practices that might have been associated with megalithic structures found along the Atlantic façade.

Methodology
The findings presented in this paper are the result of nearly forty years of research, based on a mix of three approaches to data collection.[1] Initially, I spent many years conducting fieldwork in the Basque Country, including interviewing elderly native informants in rural settings. At the same time, I was inspecting archives, digging out old documents that had not been examined previously. All of this was complemented by access to the extensive ethnographic work published previously by others. The process of data collection was greatly facilitated by the fact that I had learned Basque. This gave me a real advantage in carrying out my fieldwork, talking to native informants and gaining access to materials written only in Basque. In addition, over the years I have benefited greatly from the many conversations I have had with Basque colleagues, such as Luis Mari Zaldua Etxabe and others, who are pursuing similar research objectives.

---

[1] See, Roslyn M. Frank, 'Basque Stone Circles and Geometry', *Archaeoastronomy: Bulletin of the Center for Archaeoastronomy* 3, no. 1 (Winter 1980): pp. 28–31; 'The Basque Nautical League and Terrestrial Geometry', *Archaeoastronomy: Bulletin for the Center for Archaeoastronomy* 5, no. 1 (Winter 1982): pp. 24–29.

Roslyn M. Frank, 'Metrology, Memory and Long-Term Landscape Inhabitation: Evidence for the Septarian Package on the Atlantic Façade', *The Marriage of Astronomy and Culture,* a special issue of *Culture and Cosmos*, Vol. 21, nos. 1 and 2, 2017, pp. 249–67.
www.CultureAndCosmos.org

The Septarian Package

The term Septarian Package refers to a pre-decimal metric metrological system characterised by septarian units of measure, as well as to the material and immaterial instantiations of these units in the way they were manipulated and integrated into social practice. As such, memory traces of these units point to a type of long-term landscape inhabitation, guided by cognitive supports and affordances intrinsic to the Septarian Package.[2] More specifically, we are talking about a cultural complex that assigned numerical importance to septarian units of measurement; units consisting of seven subunits or multiples of seven, and consequently to the use of measuring devices such as a 7 ft. staff or rod, a 21 ft. pole and a 49 ft. unit known as a 'knot'.[3] Taken collectively, these units were conceptual devices which allowed users to configure the landscape in a distinctive fashion.

While the use of septarian units of measurement is particularly well documented in Euskal Herria (the Basque Country), there is evidence that similar septarian units were employed along the Atlantic façade, suggesting a continuity of metrological practice and more particularly the association of the Septarian Package with agro-pastoral practices. In Euskal Herria, these septarian units of measure form an integral part of pastoral sociocultural practices that include the Basque *sarobe*, known as a 'stone octagon' in English: an irregular octagonal structure laid out, using septarian units around a centre stone with eight uprights on the perimeter, aligned to the cardinal and intercardinal directions.[4] As will be demonstrated, units based on seven and its multiples also structured the laying out of land holdings along the Atlantic façade, pointing to the potential longevity of metrological practices in this zone and their role in social memory. As Bloch observed, using Durkeim's terminology, the

---

[2] See, Adrian Chadwick and Catriona D. Gibson, '"Do you remember the first time?" A Preamble through Memory, Myth and Place', in Adrian Chadwick and Catriona D. Gibson, eds., *Memory, Myth and Long-term Landscape Inhabitation* (Oxford: Oxbow, 2013), pp. 1–31.

[3] See Roslyn M. Frank, 'An essay in European Ethnomathematics: The Basque septuagesimal system. Part I,' in Arnold Lebeuf and Mariusz S. Ziólkowski, eds., *Actes de la Vème Conférence Annuelle de la SEAC* (Gdansk: Département d'Anthropologie Historique, Institut d'Archéologie de l'Université de Varsovie-Musée Maritime Central, 1999), pp. 119–42.

[4] Roslyn M. Frank and Jon D. Patrick, 'The Geometry of Pastoral Stone Octagons: The Basque *Sarobe*', in C.L.N. Ruggles, ed., *Archaeoastronomy in the 1990s* (Loughborough: Group D. Publications, 1993), pp. 77–91.

persistence of measures is closely bound up with the questions of communal memory (*mémoire collectif*).[5]

That elements from the Septarian Package have survived in this zone raises the question whether they are an indication that the Septarian Package might have a much deeper time depth and, as is the case of the Basque Country, that it might be a survival from even earlier efforts to cognitively configure the landscape. When highly distinctive units belonging to a well-structured metrological system are encountered in contiguous geographical locations, this often implies the existence of a type of metrological *Sprachbund*, a zone in which social collectives are dependent on a particular system of communication which permits the recording, storing and sharing of information in a consistent and meaningful fashion. In short, the employment of a unified system of metrology involves a tacit agreement on the part of its users, much in the same way that a common language does, and hence confers significant communicative advantages and benefits to the social groups involved.[6]

In sum, the term Septarian Package brings together intangible and tangible artefacts into a single distributed cognitive system. In this sense, it is an approach that subscribes to a broad socioculturally situated view of the component parts of the cultural complex in question. Furthermore, the approach affords a means of studying the evolution of discrete elements across time, tracing given lineages of cognitive and material artefacts, for example, specific septarian units of measurement, that make up the Septarian Package, how they relate to each other, and the permutations that they have undergone across time.

Survival of the 7 ft. unit of measurement along the Atlantic façade
In order to bring into clearer focus the time-depth that might be assigned to the Septarian Package as well as the way in which units from it have evolved, we will begin by charting the use and geographical diffusion of a key septarian unit of measurement, that of 7 ft. In Euskal Herria this 7 ft. unit was and is called a *gizabete* which translates as 'full-man' or 'whole-man'. It consisted of a pre-measured rod or staff, often carried by shepherds while tending their flocks as well as by officials. In the rest of the Iberian Peninsula a unit set at the same 7 ft. length went by the

---

[5] Marc Bloch, 'Le témoignage des anciennes mesures agraires,' *Annales d'Histoire économique et sociale* 6, no. 27 (1934) : p. 280.
[6] See Witold Kula, *Measures and Men* (Princeton, NJ: Princeton University Press, 1986).

following names: *estado*, *brazada* and *braza*.[7] As for the 'feet' in question, they are defined as 'common feet of Castile' and measured .278 m, based on the feet of the 'vara de Burgos' standard which eventually came to be used in all of Spain. That bar standard is identical in length to the one used in Castile and the two Basque provinces of Gipuzkoa and Bizkaia.[8]

In France another standard unit of measure was employed which once formed part of the Septarian Package. The unit in question, known as a *toise*, corresponds to the *gizabete*, that is, it was equal to '7 common feet of Castile'.[9] Although much has been written about the *toise*, this correspondence has gone relatively unnoticed and consequently its underlying septarian nature has not been taken into consideration. Whereas the *toise* bar standard is regularly defined as being composed of 6 *pieds du roi* or *pieds de Paris*, when examined through the lens of the Septarian Package we find a 7:6 ratio holding between the 'feet' of the Basque *gizabete* and those of the French *toise*. That is, both standards are defined as equivalent to 1.949 m. Moreover, the terms used to refer to the 'feet' of the *toise* standard, namely, *pieds du-roi* and *pieds de Paris*, suggest that a recalibration of their length took place in the early Middle Ages, possibly by means of a royal decree coming out of Paris.[10]

Whereas much has been written on the vicissitudes of the *toise* bar standard and its role in the development of the decimal metric system, the genealogy of the *toise* itself and its linkage to the Septarian Package has not caught the attention of researchers.[11] Nevertheless, in the eighteenth

---

[7] Vicente Martínez Gómez, *Manual de comercio en que se halla la descripción de las monedas, pesas, y medidas que se usan en los reinos de España*, 2nd ed. (1795; repr. Madrid: En la Imprenta de la Viuda de Barco, 1816).
[8] See Roslyn M. Frank. 'An Essay in European Ethnomathematics: The Social and Cultural Bases of the *vara de Burgos* and its Relationship to the Basque Septuagesimal System', *Zentralblatt für Didaktik der Mathematik* 31, no. 2 (1999), pp. 59–65, http://emis.impa.br/EMIS/journals/ZDM/zdm992a.html [accessed 20 Dec. 2016].
[9] Fermín Leizaola, *Euskalerriko Artzaiak* (Donostia: Etor, 1977), p. 125; Ronald E. Zupko, *French Weights and Measures before the Revolution: A Dictionary of Provincial and Local Units* (Bloomington: Indiana University Press, 1978), pp. 175–76.
[10] Bloch, 'Le témoignage', pp. 280–82.
[11] See Georg Strasser, 'The Toise, the Yard and the Metre: The Struggle for the Universal Unit of Length', *Surveying and Mapping* 35 (1975): pp. 25–46.

century the 7:6 ratio holding between the two standards was common knowledge and is well documented in the historical record.[12]

In short, the geographical extent of the 7 ft. unit is significant. We find that in Euskal Herria the unit known as a *gizabete* has its equivalent in the rest of the Iberian Peninsula, such as the *estado*, *brazada* and *braza*. At times the term *toesa* was used in Spanish to refer 7 ft. unit, that is, as a translation of the French word *toise*. The *toise* was the official pre-decimal metric standard of France, and by the middle of the eighteenth century it was functioning as the linear measure most commonly used by the royal armies and navies as well as in mathematical works in Spain, Italy, and a large part of the rest of Europe.[13]

In other words, because of its widespread usage, we may state that in practice the *toise* functioned as the pre-decimal metric standard for most of Europe. Moreover, in the eighteenth century when astronomers from various nations undertook surveys to determine the length of 1° of meridional arc at different latitudes, their astronomical calculations were carried out using the *toise* bar standard.[14] Consequently, the French bar standard was a key component in the calculations carried out to establish the modern decimal metric system and the length of the new universal standard, the *metre*. Yet the *toise* unit itself appears to have evolved from an earlier septarian-based unit set at 7 'common feet'. As such the *toise* appears to be a deeply entrenched cognitive and material instrument found in metrological practices along the Atlantic façade and should be understood as forming an integral part of the Septarian Package.[15]

---

[12] Pedro Bernardo Villarreal de Berriz, *Máquinas hidráulicas de molinos y herrerías y gobierno de los árboles y montes de Vizcaya* (1736; repr. San Sebastián: Sociedad Guipuzcoana, 1973), pp. 116–17.

[13] Salvador García Franco, *La legua náutica en la Edad Media* (Madrid: Consejo Superior de Investigaciones Científicas, Instituto de la Marina, 1957), p. 76.

[14] See Paul Guilhiermoz, 'De l'équivalence des anciennes mesures: A propos d'une publication récente,' *Biblioteque de l'École des Chartes* 74, no. 1 (1913): pp. 267–328, http://www.persee.fr/doc/bec_03736237_1913_num_74_1_448498 [accessed 27 Dec. 2016].

[15] Roslyn M. Frank, 'Saroiak eta Euskal Sistema tradizional zazpitarra,' in Fernando Bustillo, Roslyn M. Frank and Alfonso Mtz. Lizarduikoa, eds., *Antzinako Euskal Matematikaz Zeinbait Burutazio* (Donostia: Gaiak, 2008), pp. 169–235.

254 Metrology, Memory and Long-Term Landscape Inhabitation: Evidence for the Septarian Package on the Atlantic Façade

Survival of other septarian units of measurement along the Atlantic façade

In Euskal Herria, the *hamalauoin* unit, a pole of 21 'common feet' was regularly employed. The *pértiga* unit used for laying out land in the Iberian Peninsula was 21 'common feet' in length. Then there is the *perche* unit used for laying out land in France, set originally at 21 ft. and later converted to 18 *pieds du roi*; the latter being feet of .325 m and hence somewhat larger than the 'common feet of Castile' of .278 m. As Grierson points out, in France after literally centuries of debate over the correct length of the *perche* for arable land, it was finally settled at 18 *pieds du roi*.[16] And once again the 7:6 ratio was retained: 21 'common feet' = 18 *pieds du roi*. In addition, there is evidence that earlier in the British Isles a 21 ft. pole was also used to lay out land holdings.

The use of 21 ft. poles along the Atlantic façade is a topic taken up by Garnier who investigated the antiquity of the commonplace use of a 160 sq. *perche* land division (70,560 sq. ft.) in Normandy. He concluded that 'feet' and poles of 21 ft. were the two base unit types of the Merovingian and Carolingian Gaul agrarian measures. Garnier believed that they dated back at least to the Gallo-Roman period. He suggested a Celtic origin for the 160 sq. *perche* land division, called *acre* in French.[17]

When Garnier suggested a Celtic origin for these continental units, he was unaware of the fact that a land division of 70,560 sq. ft. also shows up across the channel as the 'old Irish acre' and that there, too, poles of 21 ft. were used to lay out land holdings.[18] Hence, we see evidence for the use of a land division of 70,560 sq. ft., an area equal to 160 sq. *perche*, documented both in Normandy and in Ireland.[19] The fact that this land division was laid out using poles of 21 ft. suggests that at that point, at

---

[16] Philip Grierson, *English Linear Measures: An Essay in Origins* (Reading, UK: University of Reading, 1972), p. 19.

[17] Bernard Garnier, 'Sur quelques recherches métrologiques françaises,' in Harold H. Witthöft, Günter Binging, Franz Irsigler, Ivo Schneider and Albert Zimmermann, eds., *Die historische Metrologie in den Wissenschaften* (St. Katharinen: Scripta Mercaturae Verlag, 1986), p. 266.

[18] Andrew McKerral, 'Ancient Denominations of Agricultural Land: Ancient Denominations of Agricultural Land in Scotland: A Summary of Recorded Opinions, with Some Notes, Observations and References', *Proceedings of the Society of Antiquaries of Scotland* 78 (1943–1944), p. 47.

[19] See Andrew Jones, 'Land Measurement in England, 1150–1350', *The Agricultural Historical Review* 27 (1979): p. 13.

least in Normandy, the 'feet' used in the *perche* unit utilised were still 'common feet', rather than *pieds du roi* or *pieds de Paris*.[20] What is certain is that in both locations a septarian unit of measurement was a key component in laying out land holdings and that there is evidence for a continuity of metrological traditions stretching along the Atlantic façade. This geographical extension could imply a time-depth for the Septarian Package which reaches beyond the Gallo-Roman period.

Basque stone octagons
References to the Basque stone octagons are found in the earliest medieval records while radiocarbon dating of the sites takes us back to the Late Bronze Age. Given their employment by transhumant pastoral groups to set out clearings within common woodlands and pastures, the Basque stone octagons may date back to even earlier pastoral practices.

Fig. 1 Aerial view of stone octagon sites in Dima, Bizkaia. Source: Eusko Jaurlaritza / Gobierno Vasco. geoEuskadi.

---

[20] See Gérard d'Arandel de Condé, 'Les anciennes mesures agraires de Haute-Normandie,' *Annales de Normandie* 18, no. 1 (1968): p. 6, http://www.persee.fr/doc/annor_0003-4134_1968_num_0018_0001_6277 [accessed 27 Dec. 2016].

Likewise, there is the numerical importance assigned to septarian units of measurement, such as the use of a 7 ft. staff, a 49 ft. 'knot' and a 21 ft pole to lay them out. Thus, the sociocultural emplacement of stone octagons along with the septarian units of measurement that characterise their design and construction, provides a kind of cognitive template that can be explored to extract, albeit tentatively, the knowledge base that informed social practices existing in this zone in times past.[21]

In the case of the Basque data, that such a metrological tradition has been preserved so effectively over such an extended period of time affirms the remarkable resilience of the Septarian Package. The continuity associated with Basque cultural assemblage is testimony of the staying power of metrological practices, including those grounded primarily in oral tradition. Undoubtedly one of the major factors contributing to the perpetuation of these septarian units of measurement in Euskal Herria has been the fact that until quite recently this type of traditional mathematical knowledge has been transmitted orally and, hence, through embedded mechanisms, 'hidden' or 'frozen' in the artefacts utilised.[22] Another key factor contributing to the stability of the cultural complex in question is the linguistic continuity of the zone which has facilitated analyses of the interaction between cognitive and material residues left by past activities. In short, the octagons, their design and associated functions, serve as memory traces of actions carried out earlier by pastoral collectives from the same region.[23]

With respect to the Basque data we need to remember that rather than being codified explicitly in written treatises, Basque metrological traditions

---

[21] See Luis Mari Zaldua Etxabe, *Saroeak Urnietan / Seles en Urnieta / Stone Octagons in Urnieta* (Urnieta, Gipuzkoa: Kulturnieta, S.A., 1996). http://j.tinyurl.com/stone-octagons-Urnietan [accessed 22 Dec. 2016]; *Saroiak eta Kortak: Mendialdeko Antzinako Gizartearen Oinordeak*. (Urnieta, 2006). http://www.euskomedia.org/PDFAnlt/mono/saroiak/saroiak001218.pdf [accessed 22 Dec. 2016]; *Saroiak eta Kortak: Abeltzaintza-sareko lotuneak / Los seles: Nodos de la red pastoril* (Donostia: Gipuzkoako Parketxe Sarea / Fundación Gipuzkoako Parketxe Sarea, 2010).

[22] Paul Gerdes, 'On Culture, Geometrical Thinking and Mathematics Education', in Arthur B. Powell and Marilyn Frankenstein, eds., *Ethnomathematics: Challenging Eurocentrism in Mathematics Education* (Albany, NY: SUNY Press, 1997), pp. 223–47.

[23] See Julio Caro Baroja, *Estudios vascos* (San Sebastián: Editorial Txertoa, 1973), p. 180; G. Adriano García-Lomas, *Los pasiegos: Estudio crítico, etnográfico y pintoresco (Años 1011 a 1960)* (Santander, 1960), pp. 273–88.

represent a form of 'distributed cognition'; collective knowledge manifested in this case in the architectural design of the stone octagons and in the cultural conceptualisations that supported them, as well as in the units of measurement and devices used to lay them out. This non-codified form of embodied metrological knowledge and practice can be recovered through the careful examination of the historically attested material artifacts and social practices that have acted to preserve these earlier mathematical understandings. The stone octagons themselves act as material anchors and represent examples of the off-loading of collective understandings held at the group level in times past whereas the units of measurement composing the Septarian Package are additional evidence of these earlier understandings.

Pastoral usages of the stone octagon sites
According to custom and later written law codes, during the daylight hours the animals and their keepers had access to pastures belonging to the common lands; each night the shepherd group had to return with its flock to the confines of the group's own stone octagon to bed down.[24] The stone octagons were constructed originally within the limits of the common lands and represented a multipurpose space within which the shepherd collectives (*olhak*) were required to build their huts and corrals.[25]

Initially, the grazing lands were open and the stone octagon sites located on these lands were held in usufruct by each shepherd collective. Later many of the octagons passed into the hands of ecclesiastical authorities while many others evolved into privately held land holdings. Traditionally, each stone octagon has had its own distinctive name which aids in their identification in archival sources and locating them on maps.[26]

---

[24] Gurutze Arregi, 'Auzo,' in Enrique Ayerbe, ed., *La Etnia Vasca: Euskaldunak*. Vol. 3 (Donostia: Etor, 1980), pp. 601–56.
[25] Arantza Gogeascoechea Arrien, Joseba Juaristi Linacero and Iñaki Moro Deordal, 'Las formas de propiedad de los seles en Bizkaia', *Lurralde: Investigación y Espacio* 34 (2011): pp. 151–88. http://www.ingeba.org/lurralde/ [accessed 27 Dec. 2016].
[26] Arantza Gogeascoechea Arrien, Joseba Juaristi Linacero and Iñaki Moro Deordal, 'Del uso común del monte a la propiedad privada: Introducción al estudio de los seles en Bizkaia', *Lurralde: Investigación y Espacio* 32 (2009) 15–46. http://www.ingeba.org/lurralde/ [accessed 22 Dec. 2016]; Daniel Rementeria Arruza and Robert Quintana Peña, *Los seles de Busturialdea-Urdaibai. Paisajes, cultura y etnografiía* (Lorra Kultur Elkartea, 2010). http://www.urdaibai.org/eu/etnografia/seles.pdf [accessed 21 Dec. 2016].

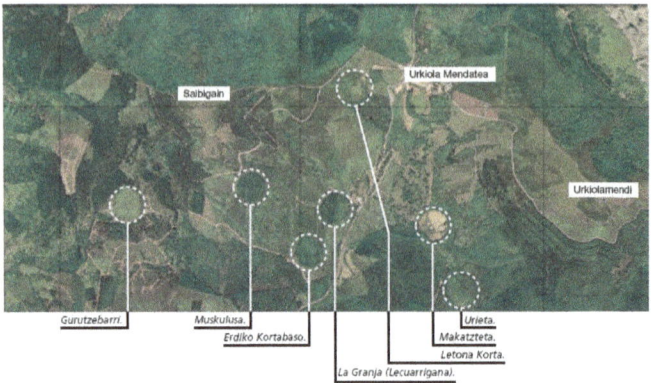

Fig. 2 Stone octagons located inside the Parque Natural de Urkiola in Bizkaia. © Foru Aldundia Bizkaia.

At this point some 500 sites have been identified. Over time many stone octagons have evolved into farmsteads and become private property while others remain as part of common lands controlled by municipalities.[27]

Fig. 3 Farmstead occupying the stone octagon of Akola, near Hernani. © Luis Mari Zaldua Etxabe.

---

[27] Ariznabarreta Zubero, Abel, Kepa Velasco Irazabal and Zefe Ziarrusta Artabe, 'Kortabasoak: Lurraren jabetza eta erabilpenak mendialdean (Dima, Bizkaia)', *Zainak. Cuadernos de Anthropología-Etnografía. Mendia, Gizartea eta Kultura / Montaña, Sociedad y Cultura / Montagne, Société et Culture* 17 (1998): pp. 33–43; Vicario De la Peña, 'De los seles,' in Vicario De la Peña, ed., *Derecho consuetudinario de Vizcaya* (Madrid: Asilo de los Huérfanos del Sagrado Corazón de Jesús, 1901), pp. 92–95.

Roslyn M. Frank 259

As a result, in many instances, the land enclosed by a stone octagon is recognised judicially today, a fact which has contributed to the maintenance of boundary markers and the survival of centre stones, although certainly many sites have lost one or more of their stones. In fact, the vast majority of the extant centre stones as well as the uprights on the periphery are markers that several centuries ago were carefully set in place to replace earlier ones. Meanwhile the names of the sites can often be traced back in the written record to the early Middle Ages. As will be discussed, in some cases they have been radiocarbon-dated to much earlier epochs.

Design of the standard stone octagon
At this point we can turn our attention to the design and symbolic architecture of sites. Rather than being conceptualised geometrically as a regular octagon, that is, in such a way that the distance between the centre stone and each of the eight stones on the perimeter was always the same, as is often popularly assumed, instructions found in Basque law codes indicate otherwise. In reality we are talking about a design that produced an irregular octagon, even though to the untrained eye the area would appear to be circular. The law codes are quite explicit concerning the way that the sites were to be laid out as well as the instruments used to calculate their dimensions.[28] For example, distances were measured using a cord marked off with knots indicating units of 49 ft. called *gorapila*. In the case of the standard stone octagon, known as a 'sel común' in Spanish, the radius is set at 12 *gorapila* units which give a radius of 588 g. ft. or 164.64 m. To give the reader an idea of the size of the standard stone octagon, its radius of 164.64 m can be compared to that of the Avebury circle, estimated by Alexander Thom to be 165.8 m, although we need to keep in mind that it is not a perfect circle.[29]

The construction process, as laid out in the law codes, begins with the placement of four outer stones at the prescribed distance of 12 *gorapila* from the centre stone so that they marked the four cardinal directions. Next, four additional uprights were placed on the perimeter to mark the intercardinal positions so that the straight-line distance between each of them was 9 *gorapila*. The result was an irregular octagon with a ratio between the perimeter and the radius of 72:12. Since that 6:1 ratio is only

---

[28] Frank and Patrick, 'The Geometry of Pastoral Stone Octagons: The Basque *sarobe*', pp. 77–91.
[29] See Alexander Thom, *Megalithic Sites in Britain* (Oxford: Oxford University Press, 1967), pp. 89–91.

geometrically possible for a six-sided figure, a regular hexagon, the result is an irregular octagon. At the same time because of the attention paid to marking the cardinal and intercardinal directions, the overall design ends up resembling an eight-point compass-rose.

It was Dr. James Van Allen of the Department of Physics and Astronomy of the University of Iowa who brought these facts to my attention in the spring of 1991. Earlier I had given him a copy of a draft article I had written on the stone octagons which I had assumed were circular configurations laid out with eight uprights on their perimeter. We discussed my material and he explained the reason for his concern, namely, that the only figure whose perimeter has a 6:1 ratio with its radius is a six-sided one, a regular hexagon. Up to that point, no one had noticed this anomaly in the data.

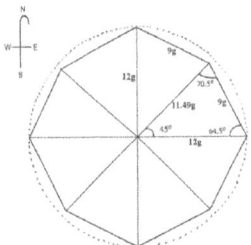

Fig. 4. Design of the standard stone octagon.

Dating the stone octagon sites

Of the eight attempts to use Carbon-14 dating techniques to determine the time depth associated with the octagons, only three have been successful.[30] Radiocarbon analysis of pieces of charcoal found at the base of the centre stones gave the following dates: fourth to sixth century CE (Mendabio); second to third century CE (Gorostarbe); and ninth century BCE (Pikuetaondarra).[31] The Late Bronze Age date of Pikuetaondarra emphasises the stability of the architectural design over nearly 3000 years.

---

[30] Juan José Agirre Mauleón and Alex Ibáñez Etxeberria, 'Gorostarbeko saroea (Urnieta)', *Arkeoikuska* (1995): pp. 211–14; Juan José Agirre Mauleón, María Flores and Sonia San José Santamarta, 'Mendabioko saroea (Urnieta)', *Arkeoikuska* (1996): pp. 142–44; Jaione Agirre García, Alfredo Moraza Barea, José Antonio Mujika Alustiza and Luis Mari Zaldua Etxabe, 'Aralar mendialdea – Sierra de Aralar', *Arkeoikuska* (2009): pp. 404–7.

[31] See Luis Mari Zaldua Etxabe, 'Basque *saroiak*', trans. from Basque by Roslyn M. Frank, in C.L.N. Ruggles, ed., *The Handbook of Archaeoastronomy and Ethnoastronomy* (Berlin: Springer, 2014), pp. 1187–96.

Fig. 5. Pikuetaondarra centre stone.
© Luis Mari Zaldua Etxabe.

The significance of the range of dates resulting from these excavations, from the Late Bronze Age to the Iron Age, will become clearer as we turn our attention to the cromlechs often located near the stone octagons. As for their composition, the circles formed by the cromlech enclosures tend to be quite small, most of them falling into the range of 3 m to 9 m in diameter although there are notable exceptions with diameters that reach up to 20 m. The stones themselves tend to be small when compared to the tall uprights of other zones along the Atlantic façade even though it is not unusual for a Pyrenean cromlech to include a tall standing stone, one that would be visible from a significant distance. The circles tend to be situated on or near crests and, therefore, along routes that would have formed an integral part of the communication networks associated with the transhumant practices of the time.[32] Stated differently, the emplacement of cromlechs coincides closely with locations where we find stone octagons. In short, both phenomena appear to be related to earlier pastoral social practices.

The assumption is that the shepherd collectives who occupied the stone octagons might well have been involved in constructing the cromlech circles nearby. Hence, the activities carried out at the cromlech sites would have formed a regular part of pastoral sociocultural practices and responded in some way to the needs of this particular collective. At present, 853 extant cromlechs have been identified inside Euskal Herria itself. These include

---

[32] Xabier Peñalver Iribarren, *Mairubaratzak: Pirinioetako Harrespilak* (Donostia: Aranzadi, 2004), p. 227, http://www.euskomedia.org/PDFAnlt/munibe/2004015271.pdf [accessed 22 Dec. 2016].

sites with and without uprights while the cromlechs themselves can be located alone or in groups. The range of architectural styles is quite varied, from minimalist to extravagant and, as has been noted, stone octagons are regularly found in the vicinity of such megalithic complexes.

In 2007 Zaldua studied the distance between 70 megalithic sites and the stone octagons located nearby. Bearing in mind the large size of the standard stone octagon itself – it has a diameter of 327 m – the results of his study are noteworthy: 30% of the megalithic sites are 0-300 m from a stone octagon; 60% of the megalithic sites are within 500 m or closer; and taken altogether, 81% of the megalithic sites are within 0-700 m from a stone octagon.[33] Consequently, when examining Zaldua's statistical results concerning the possible linkage between the stone octagon sites and the megalithic structures often encountered nearby, we need to keep in mind that in Euskal Herria cromlechs were still being built into the Late Iron Age while we have the Carbon-14 dating of the Pikuetaondarra site that takes us back to the Late Bronze Age. Hence, the time frame for the two types of structures appears to overlap.

In a listing of cromlech sites compiled by Peñalver as part of his exhaustive study of 1104 cromlechs from throughout the Pyrenean zone, we find only twenty that have been excavated and dated, specifically from the Late Bronze Age (1300 to 700 BCE) and Iron Age (800 BCE to 50 CE).[34] And, of the twenty, nearly a third fall roughly within 500 years or less of the second to third century CE dating for Gorostarbe, while the rest date back further in time. Even more important is the fact that the Late Bronze Age cromlechs overlap closely in time with the results of the excavation of the centre stone of Pikuetaondarra, dated to the ninth century BCE. So it, too, goes back to the Late Bronze Age. This suggests that the two phenomena were co-terminus. Consequently, it is highly likely that groups who were involved in laying out stone octagons were also actively involved in the construction of the small stone circles located nearby. In other words, it is quite probable that the two entities were linked in some fashion, in terms of the functions they fulfilled for the population in question and furthermore, that these functions were closely tied to the sociocultural needs of a herding community.

---

[33] Luis Mari Zaldua Etxabe, 'Seles en Legazpi: Sobre la antigua organización de territorio', *Euskonews & Media* 457 (2008).
http://www.euskonews.com/0457zbk/gaia45702es.html [accessed 29 Dec. 2016].
[34] Peñalver Iribarren, *Mairubaratzak: Pirinioetako Harrespilak.*

Roslyn M. Frank 263

Chronological overlaps
At this juncture we can appreciate that the cromlechs are structures that form part of a set of well-documented pastoral sociocultural practices which also include the Basque stone octagons. Whereas there is inferential evidence these small stone circles located in the immediate vicinity of Basque stone octagons might have been assembly sites with socio-economic and political significance, in Euskal Herria no large-scale ritual complexes have been identified, such as, for example, on the scale of Avebury. Nonetheless, the design of the stone octagons suggests they originally had a ritual component, in light of the concern shown with respect to the precise orientation of the stones and the marking of cardinality and intercardinality. In short, there is evidence that those who came up with the architectural design of the octagons were preoccupied with metrological precision unrelated to the pastoral activity at the site so that therefore the design had some symbolic meaning.

Although outside the scope of this study, analogies might be drawn between the design of the stone octagons and the cosmic geometry of built sites in other parts of the world. For example, Malville in his discussion of the sacred geography of temple building in India, states that in

> every Hindu city the most accessible demonstration of the merging of the macrocosm and microcosm is the temple. The initial phase in its construction involves the ritual of measurement during the laying-out [of] the *vastupurusha mandala*. This ritual is understood to be a reenactment of *cosmogony*, i.e., a repetition of the primordial creative act of measurement of space.[35]

Likewise, given that acts of measurement can have a cosmological dimension, it is significant that the stone octagon has to be oriented to the true cardinal and intercardinal directions and as such it has characteristics that sets that space apart from the rest of the landscape. This interpretation would assign a powerful symbolic value to the precise measurements and accurate cardinality of its architectural design. At the same time, having these alignments built into the design of the sites could have responded to quite practical needs. They certainly acted to orient the users correctly in the landscape and consequently could have provided a helpful means of mapping features visible in the surrounding landscape and perhaps even celestial phenomena.

---

[35] J. McKim Malville, 'Astronomy at Vijayanagara: Sacred Geography Confronts the Cosmos', *National Geographical Journal of India* 40 (1994): p. 172.

Discussion of results: Archaeology ethnography, ethnocultural substrate and social memory

In other parts of the Atlantic façade, we have found clear evidence that septarian units were traditionally used, suggesting a continuity of metrological practice across this zone and particularly its association with pastoral practices. It follows, therefore, that septarian units, such as the 21 ft. *perch*, may have structured agro-pastoral practices and the laying out of traditional land holdings across much of entire Atlantic façade. In sum, in Euskal Herria the stone octagons are intimately linked to transhumant practices of the Basque-speaking shepherd collectives while the absolute *terminus ante quem* date of the design of the octagons is quite uncertain. Similarly, there is no way of determining when these septarian units of measure first came into use along the Atlantic façade.

As is well recognised, when a unified metrological system is identified and its geographical reach documented, the results provide a mechanism for reconstituting, albeit always in a tentative fashion, pre-existing social conditions that must have been in place in order to give rise to the development, instantiation and diffusion of the system itself. In the case of the Septarian Package, there is reason to believe that it came into being at a point in time for which there is little or no access to the type of written documentation that we associate with literate societies. Moreover, not only its origins, but also its subsequent transmission from one generation to the next was achieved, in all likelihood, through mechanisms typical of orality. That fact, however, does not preclude the possibility that members of the social collective utilised types of recording devices that did not depend directly upon the written word, for example, knots and tallies. However, these, too, are perishable and would leave behind little or no material evidence.

Nevertheless, even though we lack material attestations for the use and storage of these septarian units, due to the highly entrenched nature of the Septarian Package, units of septarian measure have managed to survive along the Atlantic façade and these, in turn, point to a metrological *Sprachbund*, namely, to a zone in which a common metrological system was in use, one that provided its users with communicative and technological advantages. At the same time the presence of substantial memory traces of the Septarian Package seems to countenance the earlier cultural and social contacts that must have taken place in this geographical region and which contributed to the establishment and transmission of the system itself from one generation to the next.

However, the processes that led to the geographical diffusion of these septarian units are far from clear and equally obscure is the directionality of the propagation of the cultural complex. At this stage, what can be determined is that the densest network of extant data is concentrated in the Basque-speaking zone where cross-generational continuity is most apparent and where there is no evidence of linguistic rupture. In addition, it should be noted that this is the only zone in western Europe where a pre-Indo-European language has continued to be spoken. Unquestionably, linguistic rupture can contribute to the fracturing of cultural conceptualisations and complicate the integral transmission of the cultural heritage from one generation to the next.

As Bahn noted in his extensive palaeoeconomic study of Pyrenean transhumant pastoralism, throughout the study area it is evident that megalithic monuments are intimately associated with highland pastures and the practice of local transhumance, given that they cluster along well-travelled highland tracks.[36] However, in the case of western European megalithic sites less attention has been paid to the analysis of palaeoeconomic data which can give crucial clues about the important constants and long-term factors governing human behaviour. Inherent in such investigations has been the assumption that for the rest of western Europe, because of cultural and linguistic rupture brought about by the incursions of Indo-European speakers into the zone, no traces of earlier indigenous pre-Indo-European cultural assemblages have survived in the present-day practices and knowledge base of populations inhabiting areas where megalithic monuments are found. This position was summarised by Ruggles:

> In some countries, archaeological studies benefit from evidence other than just the present disposition of archaeological remains: in Mesoamerica, for instance ... there exists ethnohistoric evidence (accounts by Spanish invaders of practices current when the sites were in use), ethnographic material which is clearly relevant (present-day practices by descendants of the groups being studied) and first-hand accounts (Maya and Aztec codices). However, none of these other sources is available in British work.[37]

---

[36] Paul G. Bahn, *Pyrenean Prehistory: A Palaeoeconomic Survey of the French Sites* (Warminster, Wiltshire: Aris & Phillips, Ltd., 1983).
[37] C.L.N. Ruggles, *Megalithic Astronomy: A New Archaeological and Statistical Study of 300 Western Scottish Sites* (Oxford: British Archaeological Reports, 1984), p. 14.

266 Metrology, Memory and Long-Term Landscape Inhabitation: Evidence for the Septarian Package on the Atlantic Façade

The implications of the Basque stone octagons and septarian units of measure have not been ignored by researchers outside the Basque Country.[38] For instance, a decade after declaring that no ethnohistorical evidence was available for British work, Ruggles had the opportunity to do fieldwork at various stone octagon sites in the Basque Country. In 1996, he came back to the question of whether we can use analogy with other cultures to gain insights into prehistoric astronomical practices, saying that an analogy of great potential interest has emerged recently, from far south in the Basque Country. Here there are many examples of what appear to be eight-stone rings.[39]

Ruggles goes on to note that in some instances these *sarobe*, constructed by transhumant shepherding collectives, continued to be utilised into the twenitieth century which means that we have both first-hand accounts and extensive documentary evidence relating to their purpose and function. Furthermore, their design, construction and celestial orientation suggest that they 'functioned within a cosmological network of social practices and beliefs rather than merely at an instrumental level'.[40]

He concludes his discussion of the Basque stone octagons by pointing out their value as a heuristic, a means of rethinking the functions that might have been associated with similar configurations in other parts of Europe, such as Avebury. At this stage there is no proof of direct links between the cultural practices documented for the Basque Country and those in the Neolithic and Bronze Age of the British Isles; 'they do, however, provide a strong motivation for studying the Basque Country further as a useful analogy for ancient cultural practice elsewhere in Europe, and such investigations are well under way'.[41]

While Ruggles' commentary clearly emphasises the possible heuristic value of the Basque materials, it was published before much of the new evidence for the existence of the Septarian Package in Atlantic Europe, discussed above, was brought forward. At this stage, I am confident that investigations currently underway on this topic will new shed light on the time depth that should be assigned to the Septarian Package and provide additional evidence for socioculturally entrenched practices involving

---

[38] Evan Hadingham, 'Europe's Mystery People,' *World Monitor* 5, no. 9 (1992): pp. 34–42.
[39] C.L.N. Ruggles, 'Archaeoastronomy in Europe', in Christopher Walker, ed., *Astronomy before the Telescope* (London: British Museum Press, 1996), p. 25.
[40] Ruggles, 'Archaeoastronomy in Europe', p. 25.
[41] Ruggles, 'Archaeoastronomy in Europe', p. 25.

septarian units of measure not only in the British Isles but also in the Franco-Cantabrian zone itself.

Conclusions
To conclude, in the case of investigations that have been carried out on the stone octagons of the Basque Country, data collection has been facilitated by the wide range of resources available to the researcher, from the written evidence encountered in archives, interviews with informants, radiocarbon results from archaeological excavations, all supplemented by access to extensive ethnographic materials collected by other Basques researchers. Moreover, until quite recently, the shepherds of this Pyrenean zone were still transmitting their traditional repertoire of social practices primarily by means of mechanisms typical of oral cultures. And the transmission process, from one generation to the next, was done almost exclusively in Euskara, a language recognised as the only surviving pre-Indo-European linguistic system in western Europe. In other words, Euskara is considered the indigenous language of this part of Europe while there is increasing evidence that a Basque-like ethnocultural substrate may have existed in other parts of western Europe. What is still not known with certainty is the nature and diffusion of the languages spoken along the Atlantic Façade during the Late Bronze Age.

On the other hand, as has been demonstrated in this study, there is ample evidence that units of measurement from the Septarian Package survived along the Atlantic Façade and have left an indelible mark in the historical record. Further investigations into this topic might reveal whether, as appears to be the case in the Basque Country, units and conceptual frames of reference emanating from the Septarian Package were employed in the design and construction of megalithic sites in other parts of the Atlantic façade. In the Pyrenean zone, there is every reason to believe that the Septarian Package with its associated knowledge base and constructional technologies was recruited in laying out the stone octagons and probably played some role in the conceptualisation and laying out of megalithic sites located in close proximity to them. The question remaining is whether these survivals can be traced further back in the ethnocultural record for the rest of the Atlantic Façade.

The Skyscape Planetarium

Georg Zotti, Florian Schaukowitsch and Michael Wimmer[1]

Abstract: Communicating scientific topics in state of the art exhibitions frequently involves the creation of impressive visual installations. In the exhibition 'STONEHENGE. A Hidden Landscape.' in the MAMUZ museum for prehistory in Mistelbach, Lower Austria, LBI ArchPro presents recent research results from the Stonehenge Hidden Landscape Project. A central element of the exhibition which extends over two floors connected with open staircases is an assembly of original-sized replica of several stones of the central trilithon horseshoe which is seen from both floors. In the upper floor, visitors are at eye level with the lintels, and on a huge curved projection screen which extends along the long wall of the hall they can experience the view out over the Sarsen circle into the surrounding landscape. This paper describes the planning and creation of this part of the exhibition, and some first impressions after opening.

Introduction

The classical projection planetarium has over decades been regarded as optimal environment for the communication of fundamental concepts in astronomical phenomenology. With a dedicated opto-mechanical projection system, the entire sky can be projected onto the hemispherical dome which spans over the visitors, providing an immersive experience of being under the starry night sky. The lower limit of the dome representing the mathematical horizon is sometimes decorated with cut-out city skylines, in other planetaria the horizon is decorated with various projected panoramas. However, explanations of architectural alignments with features on the landscape horizon are not correctly possible when part of the view of interest lies on the ground and thus below the mathematical horizon. Therefore systems involving computer graphics solutions seem more useful for the purposes of visualizing horizon alignments.[2]

---

[1] Georg Zotti: Ludwig Boltzmann Institute for Archaeological Prospection and Virtual Archaeology; Florian Schaukowitsch and Michael Wimmer: Institute of Computer Graphics and Algorithms, TU Wien.

[2] G. Zotti, A. Wilkie, W. Purgathofer, 'Using Virtual Reconstructions in a Planetarium for Demonstrations in Archaeo-Astronomy', in Cecilia Sik Lanyi ed.,

In the last decade or so the use of desktop planetarium programs running on conventional computers has become almost commonplace. A few of these programs allow the inclusion of a user-created panorama photograph to be shown in the sky/ground boundary zone, and can also show parts of the visual environment which are located below the mathematical horizon. When the panorama has been properly aligned to a surveyed horizon line, the combined display of horizon panorama and sky simulation can act as proper visual representation of such land-and-sky environments for which the term 'skyscape' has recently been introduced[3] and allow orientation studies almost like on-site, as long as three-dimensional surveys or motions around small objects like stone circles are not required. One of these programs is the well-known open-source program Stellarium,[4] which has grown into one of the favourite simulation applications used by observing amateur and even professional astronomers and also many researchers in cultural astronomy.[5] The freely available source code allows code review and addition of features as required.

Third Central European Multimedia and Virtual Reality Conference (Proc. CEMVRC2006) (Pannonian University Press, 2006), pp. 43–51.
[3] See, for example, Fabio Silva and Nicolas Campion, eds., *Skyscapes: The Role and Importance of the Sky in Archaeology* (Oxford: Oxbow, 2015).
[4] http://stellarium.org [accessed 27 Dec. 2016].
[5] Applications seem almost as numerous as cultural-astronomy research topics. For example, 9 of 70 papers from the SEAC 2015 proceedings published in *Mediterranean Archaeology and Archaeometry* 16, no. 4 and available online at http://www.maajournal.com/Issues2016d.php mention the use of Stellarium, from orientation research to ethno-astronomical studies. Skyscape experience in particular is discussed in that volume by R. Mukundu, W. Ktorides and D. Brown, 'Skyscapes of Clifton', *Mediterranean Archaeology and Archaeometry* 16, no. 4 (2016): pp. 33–39. Applications of Stellarium towards light pollution education include G. Zotti and G. Wuchterl, 'Raising Awareness of Light Pollution by Simulation of Nocturnal Light of Astronomical Cultural Heritage Sites', in Fabio Silva, Kim Malville, Tore Lomsdalen and Frank Ventura, eds., *The Materiality of the Sky: Proceedings of the 22nd Annual SEAC Conference 2014* (Ceredigion, Wales: Sophia Centre Press, 2016): pp. 197–203. Stellarium's hitherto unique 3D mode has been used for orientation research, for example, in B. Frischer, G. Zotti, Z. Mari, G. Capriotti Vittozzi, 'Archaeoastronomical Experiments Supported by Virtual Simulation Environments: Celestial Alignments in the Antinoeion at Hadrian's Villa (Tivoli, Italy)', *Digital Applications in Archaeology and Cultural Heritage* 3 (2016): pp. 55–79.

The Stonehenge exhibition

The Austrian MAMUZ museum of prehistory is located in Asparn/Zaya and Mistelbach, about 50 km north of Vienna. The Mistelbach site has been added a few years ago as location for annually changing exhibitions. For the 2016 season, a team around Wolfgang Neubauer, director of LBI ArchPro, has curated an exhibition[6] around new results of the Stonehenge Hidden Landscape Project (SHLP)[7] in which LBI ArchPro had performed the large-scale motorized geophysical prospection which had detected several hitherto unknown monuments. The high exhibition hall, previously a factory for agriculture machinery, offered a unique opportunity: it provided enough space to set up original-size replica of large parts of the central trilithon horseshoe. A view which included the impression of the outer Sarsen circle should be added with some visual means. The rectangular layout and size of the hall precluded an all-encompassing circular projection, so quite early in the exhibition planning phase the idea came up to have the open side of the replica horseshoe's main axis (north-east) point towards a wall and show the view out of the stones as projection on the upper floor of the hall. The visitor would be surrounded by the trilithons, but at eye level with the lintels, an elevated position which cannot be experienced on site. The view direction obviously invited development of a splendid view of the summer solstice sunrise over the heelstone, which can however only be observed from the ground. So quite early it was clear that we would switch through vistas of several places on the projection wall, and that we should indeed tell a story of Stonehenge also with views of important locations in its vicinity. The projection should fill the visitors' field of view, so the exhibition team decided to develop a 4 m high projection area of 20 m or more width, which can only be achieved by installing several projectors.

Movie or real-time graphics?

The exhibition team was investigating the options for simulating views of Stonehenge and other sites in the area and their development over the course of millennia. These included the production of a pre-rendered movie, or a simulation involving a scriptable panorama viewing program. Given that Stellarium had scripting capability, can show high-resolution

---

[6] MAMUZ, *Stonehenge – Verborgene Landschaft. Begleitheft durch die Ausstellung* (Guide brochure through the exhibition), (2016).
[7] C. Gaffney *et al.*, 'The Stonehenge Hidden Landscapes Project', *Archaeological Prospection* 19 (2012): pp.147–55.

panorama views as 'landscapes' and already provided a multitude of projection options including cylindrical or Mercator projections, it seemed better-suited for the wide screen envisioned than most other panorama viewing applications which are mostly limited to a standard perspective projection.

Using Stellarium promised the added benefit of using the projection system for special occasions where pre-produced content is not applicable. Given that Stellarium is first and foremost an astronomical simulation program, it would enable us to demonstrate the astronomical elements in landscape and architecture which have been identified around Stonehenge, and astronomical facts like the slow changes of the ecliptic obliquity.

Fig. 1. Sunrise in Late Neolithic Stonehenge. Panorama by LBI ArchPro's media partner 7reasons based on LBI ArchPro's image-based model. Simulation with Stellarium 0.15.0.

Developing the Skyscape Planetarium

In the year before the exhibition opened (March 2016) we had time to make Stellarium ready for running a show on such a big screen, which brought several useful additions to the program in general. At first we thought about using five synchronized small computers (one per projector), but soon found that a single modern PC equipped with a graphics card that can drive several screens or projectors would be far easier to handle. The raw pixel count of 5x 1920x1200 pixels gives a viewport size of

9600x1200 pixels, which was tested to run sufficiently fast. The 5 projector wall was now redesigned into a partly curved screen which required edge blending, and given the 25x4 m screen area results in a net display area of 100 m² with 7500 (equivalent width) x1200 pixels. The placement of the two rightmost projectors turned out to be challenging: they had to be mounted so that their diverging projection beams avoid one of the trilithon replica and still overlap and seamlessly blend on the screen. Using this additional projector, as seen from the inner area of the trilithon horseshow, the screen extends behind the stones, which adds to the immersive experience of being part of this skyscape, but requires that no important or detailed content is presented in that section of the screen.

We had planned to develop a narrated, media-rich show in the months before the opening on a computer identical to the show PC, but only with regular screens attached. Panorama renderings of several sites in the landscape, based on a terrestrial laser scan taken by LBI ArchPro during the SHLP, from which we had erased all modern constructions like road dams or houses, were prepared by our media partner 7reasons. With this setup we could run a few performance tests and landscape pans, and look for any rendering artefacts which may not be apparent on a single screen.

The best projection usable on the wide screen is the cylindrical equirectangular projection, where the mathematical horizon forms a straight horizontal line on the screen. Centring the horizon in the viewport would put it in 2 m height on the wall, and ground and sky would get equal share of screen space. However, depending on narrative context, the horizon should be allowed to shift up (to show more of the ground in aerial views, e.g., of the Cursus stretched out in all its length) or down (to show ground views with the Sarsen circle surrounding the visitor), which required the implementation of such an off-center projection mode that also had been on Stellarium's public feature wishlist for years. When the horizon line is moved to the bottom of the screen, and we zoom out to show 180° of the horizon, the vertical view angle is about 30°. From the viewpoint in the centre of the stone circle, the ring of lintels lies at about 10° altitude, but the horseshoe lintels just extend to about 30°.

Fig. 2. The Skyscape Planetarium, a 25 m wide partly curved screen with 4 m height, providing a view out of Stonehenge from an elevated position within the horseshoe. An automated tour takes the visitor to several sites in the area and explains the landscape development from the Mesolithic to the Bronze Age. The horizon line can move up or down to show more of the foreground or sky, respectively.

The astronomical simulation had to become more accurate than what was available in version 0.13.3 to properly simulate the sky of the Neolithic if not earlier. Stellarium had been developed mostly for contemporary amateur astronomers, with planetary positions from the analytical VSOP87 model which is recommended for use only for years −4000 to +8000, with degraded (undefined) accuracy outside this range. Also, precession had been implemented in a very simple way, and ecliptic obliquity was assumed constant, so that the slight but noticeable changes in solstitial sunrise positions between the time of construction and today could not be demonstrated.[8] The biggest astronomical improvement was therefore the implementation of a recent proper long-time model of precession of the equinoxes and changes in the obliquity of the ecliptic.[9] Also, the long-time

---

[8] See a discussion about skyscape experiences and about a few well-known limitations of an even older version of Stellarium in D. Brown, 'Exploring Skyscape in Stellarium', *Journal of Skyscape Archaeology* 1, no. 1 (2015): pp. 93–111.

[9] J. Vondrák, N. Capitaine and P. Wallace: 'New Precession Expressions, Valid for Long Time Intervals', *Astronomy & Astrophysics* 534 (2011): p. A22.

planetary position ephemeris DE431 was made accessible to the program.[10] This can be installed as an optional data file and provides NASA/JPL's results of a numerical simulation of planet positions for the year range −13000…+17000, so that Stellarium should now be able to simulate planetary phenomena even over a Mesolithic landscape.[11] (Note that without an alternative way of computation, we cannot give estimates of total accuracy at the moment and would invite further investigation by other researchers.)

For several years, Stellarium has been able to run pre-programmed scripts, which allows the development of complete astronomy shows. The scenes and narration should also be enriched with image and movie inserts, which necessitated the re-activation and enhancements of Stellarium's multimedia extensions that had been disabled years ago due to changes in the internal software structure.

The RemoteControl plugin
Running the shows in a completely automated manner with occasional manual access required the development of another new extension (plugin): a remote control interface which allows sending commands via network. In addition to triggering the beginning of the next show run, it should also allow the manual control of the application on a separate handheld device, in case a museum guide or lecturer wants to take over the screen and show some specific views without also showing the graphical user interface (GUI) on the projected wall.

A thorough rewiring of Stellarium's internal modules has allowed us to access their properties and settings in a much more consistent manner than before. In addition, the new plugin provides an HTTP network interface which fulfils both purposes: On access with a web browser, a mixed HTML/JavaScript interface which replicates most tabs and buttons of the usual Stellarium GUI is presented to the user. Actions which are commonly performed with mouse and keyboard (e.g., panning or zooming) had to be implemented with a few additional intuitive GUI elements. Accessing the HTTP interface with particular command-line programs allows triggering of single commands or the start of scripts.

---

[10] W.M. Folkner, et al., 'The Planetary and Lunar Ephemerides DE430 and DE431', *IPN Progress Report* 42–196 (Feb. 2014), available from URL: https://ipnpr.jpl.nasa.gov/progress_report/42-196/196C.pdf
[11] G. Zotti, 'Open-Source Virtual Archaeoastronomy,' *Mediterranean Archaeology and Archaeometry* 16, no. 4 (2016): pp. 17–23.

To complete the requirements of a fully automated setup which should be easy to operate for the museum staff, our media partner 7reasons developed an operation panel based on a tiny Raspberry Pi single-board computer with touch screen which allows the museum operators to switch on the PC and 5 projectors with a single keypress on the touch display. The Raspberry Pi in addition sends the required script trigger command every 25 minutes during daytimes via cronjob. To start shows after hours, a manual trigger is available on the control panel. If required, an operator in front of the wall can access almost all of Stellarium's controls using a web browser on a handheld tablet computer connected via WLAN.

We have already seen that the web browser based control interface is welcome also in other settings like presentations at amateur astronomers' meetings when operators or presenters want to control the scene displayed on the screen without exposing graphical user interface panels to the audience. Not every functionality of Stellarium was deemed a requirement for this new interface, configuration options, optional downloads or also features of the telescope control or ocular plugins have not been exposed. However, the webpage which comes with the installation can be copied and adapted to other users' requirements by adding or removing functionality, changing screen layouts etc. A more compact layout for a 7 inch tablet which has been developed for use in the museum is provided as example.

The final weeks
With software and the hardware installation at the museum almost ready, we finally received our narrative text, 'The Story of Stonehenge', by the exhibition's co-curator Julian Richards to base the show on.[12] We selected the fitting existing horizon panoramas but found we need a few more created by our partner 7reasons (about twenty in total; creating high-quality renders take many days of computation time, so these should be decided and produced early in the show development), picked additional material to use as inserts from our material collection and found we had to make or find a few more. Less than two weeks before opening the bilingual narration recordings became available, crucial for timing the transitions and pans over the landscapes. Only then a few more issues with automated pans and transitions became apparent which could not be tested before, but which required a few ad-hoc amendments to Stellarium's script capabilities. The opening week approached, and final adjustments could

---

[12] Based on Julian Richards' *Stonehenge: The Story So Far* (English Heritage, 2007), but with additions resulting from the SHLP.

only be made on-site. While the showcases were filled with exhibits in bright working light conditions, we were desperately trying to calibrate the blending zones in the panorama wall. With work lights switched off later that day, some illuminated showcases were identified which caused severe light spill onto the screen. The worst cases could be fixed before the exhibition opened, but of course exhibition designers should take such light spill on open screens or even glaring show case illumination into account. Finally it was possible to fine-tune the script to the wide screen, a process which lasted literally until minutes before the first pre-opening tour was given to selected guests.

The system has run successfully six days per week during the eight-month exhibition season. A few minor improvements to the show were installed in this time.

Fig. 3. A fisheye view of the exhibition's upper floor. Visitors are sitting on the bench circled by the Horseshoe replica, watching the automated show. The Skyscape Planetarium shows a view into the landscape with several inserts explaining building methods.

Reactions

The exhibition topic 'Stonehenge' is attractive to a wide audience, and the exhibition has been the most successful at the MAMUZ Mistelbach site so far, so that it will stay for the 2017 season. The combination of full-size replica of the horseshoe sarsens with the huge screen makes a big impression to even expert visitors, and reactions have been throughout positive. Visitors can sit on the resting bench and let the narration take them from the Mesolithic hunting site of Blick Mead through the development of long barrows, causewayed enclosures, Cursus monuments, Durrington Walls, Woodhenge and finally Stonehenge in its various phases, ending with the burial mounds in the Bronze Age, recapitulating the main stations of the exhibition which they had seen on the ground floor. Headphones can be used to listen to Julian Richard's English narration instead of the German translation coming from the loudspeakers. Some sites are shown in several phases of use centuries apart. On the other hand, given the ongoing discussion and apparent current inability to identify one stratigraphically correct order of construction and reconstruction phases of Sarsens and Bluestones,[13] we do not show views of Stonehenge 'under construction' but only the finished monument, based on LBI ArchPro's image-based model of the site and reconstruction of obviously missing parts. To connect the visitor to the prospection research done by the LBI ArchPro, the final scene shows our motorized sensor systems rushing through the Bronze Age landscape, detecting hidden monuments.

The mostly archaeological content and rather fast pace of the narration provide only very few moments where Stellarium's astronomical capabilities are exploited. These include a reference to the newly found 'Cursus pits' possibly marking the summer solstice sunrise and sunset seen from the heelstone and just a few seconds of summer solstice sunrise.[14] During the breaks between two shows, the system repeats the famous scene of the majestic summer solstice sun rising and slowly moving over the heelstone (Fig. 1). This silent scene, which runs for about three minutes, was intended to provide enough time for visitors to relax, talk, enjoy the view, experience and make themselves part of this skyscape, or take photographs of themselves with that 'epic' sunrise in their backs. However, just these few minutes of silent, narration-free sunrise, which was also not accompanied by any of the well-known 'sunrise' music pieces from the movie industry or any other soundscape, reportedly caused confusion and made some visitors report to the museum staff that the system was 'stuck'. It appears that in our time of ever-louder media overkill, a few moments of

---

[13] T. Darvill, P. Marshall, M. Parker Pearson and G. Wainwright, 'Stonehenge remodeled', *Antiquity* 86 (2012): pp. 1021–40.

[14] Gaffney *et al.*, 'The Stonehenge Hidden Landscapes Project'.

silence and just moderate time-lapse come all too unexpected. (A sunrise in natural speed had immediately been considered too slow.) This was finally remedied by adding the Museum's mascot, a little mammoth named Zotti (unrelated to the first author), to the screen, explaining what is seen on display and announcing that the next narrated presentation would play soon.

Night at the Museum
On a few occasions, the museum provides special evening tours for elementary school children. The lights are switched off, and the young visitors can explore the exhibition during a guided tour using flashlights. Eventually they wake up a friendly 'man from the past', an actor explaining some exhibits using an unknown language and acting, and performing some music or dances with the children. In the lower floor, the connection of monuments like timber circles to the course of the sun and seasonal changes, with the solstices as turning points, are introduced to the children. Ascending to the upper floor, the group enters a nightly scene of Stonehenge, with the stars twinkling through the Sarsen circle. Only here the astronomical capabilities of the Skyscape Planetarium simulating a starry sky together with landscape panorama and ground are utilized, but serve to create the nightly ambient only, unfortunately without being part of the didactic narrative. After explanation of the exhibited beaker pottery and bronze artefacts under flashlight conditions, morning twilight sets in (manually controlled using the tablet computer), and the summer solstice sunrise is greeted with a little music and dance performance by the 'man from the past', soon to be joined by the children, close to the end of their excursion into the past.

Fig. 4. For the Flashlight Tours, the Skyscape Planetarium can deliver stunning night ambient settings.

Final thoughts

The huge screen lets visitors almost immerse themselves into the landscape, even though the panorama renderings are only two-dimensional images pre-rendered in a given lighting setting. An interesting observation during the simulation from sunrise to afternoon was an almost three-dimensional appearance of the Sarsens caused by the independent and changing light brightness mixing of sky hues and landscape panorama.

We have not tried Stellarium's Scenery3D mode on the big screen yet, which would really allow a virtual walk through the landscape. The Stonehenge landscape model would have to be simplified (especially cleared from highly detailed vegetation only possible to show in the pre-rendered landscapes) to allow such an interactive mode.

Adding the astronomical simulation quality of Stellarium, such a system appears much better suited than the classical half-domed planetarium for research and demonstration of prominent skyscapes in the sense of combination of landscape, built monuments and the celestial experience,[15] although in the course of this museum installation such further application is currently not exploited. The current exhibition's topics concentrate on the archaeology and do not highlight the various

---

[15] Silva and Campion, *Skyscapes*.

archaeoastronomical theories around Stonehenge in any detail, and therefore the exhibition show uses mostly its aspect of 'moving panorama display', and does not exploit the full astronomical capabilities of Stellarium. A similar setup with more astronomical content should be replicable at other sites, though.

Acknowledgements
The Skyscape Planetarium has been developed as part of the exhibition 'STONEHENGE. A Hidden Landscape.' at MAMUZ museum Mistelbach, Austria, which has provided considerable developing time for several new and improved features which have been released in Stellarium version 0.15.0 (released 31 July 2016).

In the weeks before opening, we were supported also by Stellarium maintainer Alexander Wolf (Barnaul, Russia) who even created a new customized installer package with the final changes built overnight just in time for setup at the museum two days before opening.

Florian's work on the RemoteControl plugin, and the DE431 integration, were supported by the ESA Summer of Code in Space programme 2015.

The Ludwig Boltzmann Institute for Archaeological Prospection and Virtual Archaeology (http://archpro.lbg.ac.at) is based on an international cooperation of the Ludwig Boltzmann Gesellschaft (A), the University of Vienna (A), the Vienna University of Technology (A), ZAMG-the Austrian Central Institute for Meteorology and Geodynamics (A), the Province of Lower Austria (A), Airborne Technologies (A), 7reasons (A), the Austrian Academy of Sciences (A), the Austrian Archaeological Institute (A), RGZM-the Roman-Germanic Central Museum Mainz (D), the National Historical Museums – Contract Archaeology Service (S), the University of Birmingham (GB), the Vestfold County Council (N) and NIKU-the Norwegian Institute for Cultural Heritage Research (N).

Edward Burne-Jones's *The Planets*: *Luna, A Celestial Sphere*

Liana De Girolami Cheney

Abstract: Edward Burne-Jones (1833–98), a Pre-Raphaelite painter, was fascinated with astronomy as noted in his memorials and accounts. In 1879 he executed cartoon drawings for a cycle on the planets for the artisans of the William Morris firm, who would transform them into stained-glass windows. The commission was for the decoration of Woodlands, the Victorian home of Baron Angus Holden (1833–1912), a mayor of Bradford. Presently, seven of the cartoons – *The Moon* (*Luna*), *Earth* (*Terra*), *Sol* (*Apollo*), *Venus*, *Jupiter*, *Saturn*, and *Evening Star*) – are in the Torre Abbey Museum in Torquay, UK, while the cartoon for *Mars* is part of the collection of drawings at the Birmingham Museum of Art, UK, and the drawing *Morning Star* is located at Lady Margaret Hall in Oxford, UK. In the creation of the *Planets* cycle, Burne-Jones was inspired by cultural events of the time, such as British scientific astronomical discoveries and British and Italian humanistic sources in literature and visual arts portraying astronomy. This essay examines – art historically and iconographically – only one of the eight planets, the cartoon of *Luna* (The Moon) as an astral planetary formation and a celestial sphere. This study is composed of two sections. The first section discusses the history of the artistic commission and the second section explains some of Burne-Jones's cultural sources for the *Planets* cycle and the Moon, both of which partake of heavenly and terrestrial realms.

History of the Commission

In 1879, Burne-Jones and William Morris (1834–96) received a new commission from Baron Angus Holden (1833–1912), a woollen manufacturer and mayor of Bradford.[1] The project consisted of a glass decorative cycle about the planets for a large, twenty feet high, semi-circular window in Holden's Music Room in his Victorian mansion, called Woodlands, outside of Bradford. The Music Room was built in 1866 on the north end of the house as part of the Victorian Gothic wing, which was

---

[1] See Archives of Torrey Abbey Museum. Correspondence of 11 June 1992, between D. J. R. Green of Gloucester and L. Retallack, former Curator of the Torre Abbey Museum.

an addition to the mansion. Burne-Jones completed nine cartoons for the Morris firm, which transformed them into stained-glass panels in 1879. Before Holden's commission, Edward Burne-Jones and William Morris were involved in earlier projects on this subject. In early 1858, the two men decorated with stained glass the church windows of All Saints at Selsley in Gloucestershire. The rose window illustrates God's creation from the Book of Genesis. One roundel represents the planets, the stars, the Moon, and the Sun. A few years later, in 1861–65, Morris and Burne-Jones engaged in another astronomical project, using the signs of the zodiac in the decoration of the Green Dining Room at South Kensington Museum in London, now part of the Victoria and Albert Museum.

According to an essay by Michael Rhodes, former curator of the Torre Abbey Museum, Angus Holden purchased Woodlands to display his famous collection of paintings and works of art. This area was described by E. Healey as a 'large and lofty apartment, lighted partially from the room and partially from the semi-circular window. This large bay window contained in the upper compartment nine stained glass figures, a work by the celebrated artist Burne-Jones'.[2] A note from Rhodes's research indicates the purpose of Lord Holden's commission of the new wing at the Woodlands; this architectural addition functioned as a music room. The incorporation of stained-glass windows with the theme of the planets, the celestial spheres, is traditionally associated with the music of the spheres; hence a fitting topic for the site.[3]

Twenty years later, the nine cartoons and stained-glass panels encountered a series of mishaps. In 1890, the Holden family left Woodlands and moved to Nun Appleton Hall, near York. Woodlands was demolished in 1899. It is uncertain whether Burne-Jones and Morris's windows were still in the house at the time or if they were transferred to the new mansion in Nun Appleton. Regrettably, Lord Holden died in 1912 and, in his account of property sale occurred years later in 1917, there is no mention of the stained-glass windows or panes. Between 1919 and 1920, Sir Benjamin Dawson (1878–1966), 1st Baronet and Bradford textile manufacturer, purchased Nun Appleton Hall and lived there for more than

---

[2] See Michael Rhodes's essay dated 17 November 1997, part of the Archives of the Torre Abbey Museum. Here Rhodes cites an article by E. Healey, 'Woodlands, Bradford'. No references as to when it was published.

[3] See Jamie James, *The Music of the Spheres: Music, Science and The Natural Order of the Universe* (New York: Copernicus, 1993); and Marc Lachièze-Rey and Jean-Pierre Luminet, *Figures du Ciel: de l'armonie des spheres à la conquête spatiale* (Seuil: Bibliothèque nationale de Frances, 1988), pp. 58–62.

forty years. Two articles on this mansion appeared in *Yorkshire Life* (March 1955 and March 1960); neither of them mentioned Burne-Jones's glass panels.[4] But the *Yorkshire Life* of March 1955 noted that when the Dawson family moved into Nun Appleton Hall, they renovated the Victorian house 'by demolishing the incongruous Gothic wing on the west side';[5] perhaps the glass-stained panels were located in this area. Hence, a large part of the Nun Appleton Hull mansion was demolished in 1920 and totally destroyed before 1935. Thus, since 1899, there is no clear trace of Burne-Jones and Morris's stained-glass windows. This essay will not examine the complicated history of Burne-Jones and Morris's glass panels but will focus on the cartoon drawings (compare Fig. 1 and 2).

A possible arrangement of the cartoons suggests this presentation at the Woodlands window: *Luna, Earth (Terra), Sol (Apollo), Mars, Morning Star, Venus, Jupiter, Saturn,* and *Evening Star*. Auctioneer Gabreal Franklin proposes an alternate format for the surviving stained-glass panels: *Morning Star, Evening Star, Jupiter, Sol (Apollo), Mars, Venus,* and *Luna*. Unfortunately, *Saturn* and *Earth (Terra)* are missing.[6]

Three entries in Burne-Jones's *Account Books* at the Fitzwilliam Museum record his progress on the project: 1) 18 August 1878: 'First four figures of Planets. Pound 15 ea. viz. Venus, Luna [see Fig. 1], Morning and Evening stars. Pound 60'; 2) 22 August 1878, two more Planets were completed, 'Saturn and Mars'; and 3) 1 November 1878, noting complete 'Figures of Earth, Jupiter and Apollo. Pound 45'.[7]

---

[4] See Archives at the Torre Abbey Museum. Correspondence between D. J. R. Green of Gloucester with Gordon Hand, Librarian of North Yorkshire County and L. Retallack, Curator of the Torre Abbey Museum. The letters are dated 23 June 1993 and 25 June 1993. See also A. C. Sewter, *The Stained Glass of William Morris and his Circle* (New Haven: Yale University Press, 1974–75), p. 208.

[5] Archives at the Torre Abbey Museum. Correspondence between D. J. R. Green of Gloucester with Gordon Hand, Librarian of North Yorkshire County, and L. Retallack, Letter of June 25, 1993. In the literature, there is a discrepancy regarding the location of this music room, whether it was located on the north or west side of the Victorian mansion.

[6] See http://www.allplanet.com/glass/BJ5.htm [access 1 Dec. 2016]. These stained glass windows were remanufactured; it still is unknown when this copy took place.

[7] See Malcolm Bell, *Edward Burne-Jones: A Record and Review* (London: George Bell and Sons, 1892), p. 102ff., for a discussion of four designs on the seasons as conceptions for the *Planets'* cycle.

Fig. 1 Edward Burne-Jones, *Luna (The Moon)*, 1878–1879, watercolour cartoon. Photo by Liana De Girolami Cheney. Courtesy of the Torre Abbey Museum, Torquay, UK.

Fig. 2 Edward Burne-Jones, *Luna (The Moon)*, 1878–1879, stained-glass panel. Photo courtesy of the Auctioneer Gabreal Franklin.[8]

---

[8] See, http://www.allplanet.com/glass/BJ5.htm [accessed 2 Dec. 2016].

The cartoons for the nine planets have suffered a less tragic fate than their companion stained-glass windows. The cartoons were dispersed after being used, but today their location is known. In 1930, Torre Abbey was a historic site with private collections. According to Rhodes, the art collector G. H. Earle of Rocklands, residing at Warren Road in Torquay, purchased Burne-Jones' seven cartoons of the *Planets* – *Venus, Luna, Sol, Jupiter, Saturn, Earth*, and *Evening Star* – from Angus Holden's wife, Lady Holden.[9] In early 1930, Earle donated these cartoons to the Torre Abbey collection, which became a museum in 1936.[10] In honour of Torre Abbey Museum's fiftieth anniversary in 1966, one of the Burne-Jones's stained-glass panels, *Earth*, was once again constructed after the original cartoon in the collection.

The eighth cartoon, *Mars*, was acquired through Christie's in 1898 by the Birmingham Museum and Art Gallery,[11] and the ninth, *Morning Star*, a large drawing, not a cartoon, is presently located at Lady Margaret Hall in Oxford.

Iconography: Edward Burne-Jones's Cultural Sources for the *Planets* Cycle and The Moon
Burne-Jones's writings, such as the records at the Fitzwilliam Museum of Art and the recollections of his wife, Georgiana MacDonald (1840–1920), assist in understanding the scope of his interest in astronomy. The astronomical books that he collected and viewed at the British Museum contributed to his creative quest. He was captivated by the study of astronomy, as indicated by: 1) his personal notations on astronomy; 2) British scientific events (astronomical discoveries at the time); 3) British literary collections on astronomy (poems, books, manuals, celestial maps, playing cards); and 4) Italian literary and visual sources on astronomy and astrology.

---

[9] See Michael Rhodes, *Devon's Torre Abbey: Faith, Politics and Grand Designs* (New York: The History Press, 2015).
[10] After the death of Angus Holden in 1912, his wife and family moved to Torquay. After the death of her mother, Donna Holden lived for a long time in a nursing home. Prior to her death in 1995, the curator of Torre Abbey Museum contacted her, but she provided no information about Burne-Jones' *Planets* cycle. See Archive, Accession Number A-111, Torre Abbey Museum, notations made by Michael Rhodes, former Curator of the Museum before 2000.
[11] See Registrar's notation at the Birmingham Museum and Art Gallery. Rest moved to Acknowledgements at end.

In *Memorials of Edward Burne-Jones*, Georgiana MacDonald recounted how fascinated he was with the study of astronomy. She wrote: 'He kept astronomical books at his bedside, and often turned to them when unable to sleep'.[12] She described him saying: 'I terrified myself in the night with more astronomy'.[13] In another passage, Georgiana repeated that 'Astronomy had a great fascination for him – almost a terrible one'.[14] One year after his trip to Italy in 1871, Burne-Jones composed a list of the paintings he was working on, mentioning that he made 'an oil picture of *Luna*, in tones of blue' (Fig. 3).[15] After the death of his friend and soulmate, Morris, Burne-Jones decided to delve further into astronomy. He stated: 'So I've had to take to my astronomy again'.[16]

---

[12] See Georgiana Burne-Jones, *Memorials of Edward Burne-Jones*, 2 vols (London: Macmillan Company, 1904), Vol. 2, p. 304 [hereafter GBJ, *Memorials*].
[13] GBJ, *Memorials*, 1, p. 58.
[14] GBJ, *Memorials*, 2, p. 304.
[15] GBJ, *Memorials*, 2. p. 30. *Luna* is an oil painting on canvas (101 x 71cm; 39¾ x 28in.) and signed with initials 'EBJ' at the lower left. Alexander Ionides (1840–98) owned the painting. Before his death it was sold anonymously to Christie's London, 1 March 1897, lot 121, then re-acquired by Christie's London in 1898 and sold as lot 21 on 15 May 1902 to Agnew R.H. Benson (1839–1912). The painting was eventually purchased and was part of the Collection Yves Saint Laurent et Pierre Bergé in Paris, who sold it to Christie's (lot/sale 1209), 23–25 February 2009, Paris. See http://www.christies.com/= lotfinder/-paintings/–sir-edward-coley-burne-jones-bart-ara-rws-5157409-details.aspx 2009 lotfinder/-paintings/–sir-edward-coley-burne-jones-bart-ara-rws-5157409-details.aspx 2009. After the 2009 Christie's sale, the painting became part of a private collection. The description of the painting, the size (101 x 71cm. (39¾ x 28in.), materials (oil on canvas) and the signature with the monogram of 'EBJ' on the lower left of the painting attest and coincide with the description of the painting mentioned by GBJ in the *Memorials*, Vol. 2, p. 30.
[16] GBJ, *Memorials*, 2, p. 303.

Fig. 3 Edward Burne-Jones, *Luna*, 1871–72, oil on canvas. Photo credit: CHRISTIE'S IMAGES LTD. 2017

British scientific culture and discoveries provided impetus for Burne-Jones's artistic expression in representing astronomical imagery.[17] In 1781, William Herschel (together with his sister Caroline) discovered Uranus. Both were also musicians; William composed music and conducted the Bath orchestra.[18] In 1822, Alexander Jamieson published a *Celestial Atlas* with G. & W. B. Whittaker and Co. in London. William Lassell made new astronomical discoveries, such as the largest moon of Neptune (Triton) in 1846 and (together with William Cranch Bond and George Phillips Bond) a moon of Saturn (Hyperion) in 1848. In 1851, William Lassell recorded Uranus' moons (Ariel and Umbriel). In 1859, Richard Christopher Carrington and Richard Hodgson (independently) made the first observations of a geomagnetic solar storm. In 1864, William Thomson discovered the thermodynamics of the earth and estimated the earth's age to be about 20 million years. And in 1873, George Biddell Airy explained and improved upon the orbital theory of Venus and the Moon.

---

[17] See Peter Whitfield, *Astrology: A History* (New York: Harry N. Abrams, Inc., 2001), pp. 165–79, for a discussion on astrology in England between 1550–1700.
[18] William Herschel with his sister Caroline lived at New King Street in Bath. Today the residence houses the Herschel Museum of Astronomy. In 1780, He was appointed director of the Bath Orchestra, while his sister performed as a soprano. See Richard Holmes, *The Age of Wonder* (New York: Vintage Books, 2008).

Jamieson's *Celestial Atlas* inspired an important market for cartographic maps and playing cards in London. The Rev. Richard Bloxam popularized Jamieson's atlas by composing a set of 72 playing cards based on constellations, which were printed by Samuel Leigh for The Royal Astronomical Society in London in 1825 under the title of *Urania's Mirror* or *A View of the Heavens*.[19]

Moreover, these British scientific discoveries and media applications provided inspiration for Pre-Raphaelite artistic expressions not just in painting but also in poetry. Alfred Tennyson composed poems on astronomy and cosmology such as *Maud*, *Lucretius*, on Herschel's great star, and *In Memoriam*; Gabriel Dante Rossetti wrote a stanza on *A Match With The Moon*; his sister, Christina Georgina Rossetti, poetically questioned the colours of the moon in *Is the Moon Tired. She Looks so Pale?*; and Morris evoked the beauty of the planet Earth in *Flora* and *Earth the Healer*. Burne-Jones visualised their poetical visions and formulated his own astronomical ideas into a fanciful stellar ensemble in the *Planets*.

During his many journeys to Italy, from 1859 until 1871, Burne-Jones became interested in Italian culture, including literature and visual arts. Travelling through Florence, Padua, Ferrara, Rome, and Venice and their surroundings areas, he experienced the visual presence of the Renaissance's cosmology as depicted in numerous ceilings and walls of churches and palaces. For example: in Florence, the Medicean astrological cupola in the old sacristy of San Lorenzo, 1426; in Padua, the Palazzo della Ragione with its combination of Niccolò Miretto and Stefano da Ferrara's *Labours of the Months*, the seasons and the zodiac signs of 1425–40; and in Rome, Baldassare Peruzzi's ceiling with the family Chigi's astrological chart in the Loggia of Galatea of 1515 at Villa Farnesina. But Burne-Jones must have been absorbed by the astrological and astronomical representations of Francesco Cossa and Cosmé Tura's wall decorations in the Hall of the Months of 1470–84 at the Palazzo Schifanoia in Ferrara.[20] These astrological seasonal cycles, designed for the duke of Ferrara, Borso

---

[19] See facsimile edition of the First American Edition of *Urania's Mirror*, 1832 (New York: Barnes and Noble, 2004) at http://www.atlascoelestis.com/16.htm [accessed 2 Dec. 2016].

[20] See Francesco Cossa and Cosmé Tura, *Luna, The Month of June*, 1470–84. Wall fresco in Hall of the Months, Palazzo Schifanoia, Ferrara, Italy. Photograph by Sailko, at https://commons.wikimedia.org/wiki/File:Palazzo_schifanoia,_salone_dei_mesi,_06_giugno_(maestro_dagli_occhi_spalancati),_cancro_01.JPG [accessed 2 Dec. 2016].

d'Este, were programmed after the ancient astrological manuscript of Hyginus's *Poeticon Astronomicon*, published unillustrated in Ferrara in 1475 and in 1482 with illustrations.[21] As a frequent visitor of the British Museum – and because of his avid interests in astronomy noted in his wife's memoirs – it is indubitable that Burne-Jones was familiar with Hyginus's astrological manuscript (Harley MS 647) housed in the library.[22]

Among the Italian Renaissance literary sources was Francesco Colonna's *Hypnerotomachia Poliphili* (*The Dream of Poliphilo* or *The Strife of Love in a Dream*), published in Venice in 1499 by Aldo Manutius.[23] Burne-Jones considered Colonna's tome the most beautiful printed and illustrated book of the Italian Renaissance. He owned a copy, a gift from Morris, which is now at the Houghton Library in Cambridge, MA.[24]

Colonna explained the function of the planets and their innate qualities of order, governance, and harmony of the spheres.[25] He elaborated upon a system dealing between the eternal and the physical realms as manifested in the text through Poliphilo's journey. He guides 'the reader through an

---

[21] Hyginus, *Poeticon Astronomicon with the illustrations of the constellations and planets*, Jacobus Sentinus and Johannes Lucilius Santritter, eds., (Venice: Erhard Ratdolt, 1482/3).

[22] Probably Burne-Jones saw these manuscripts: London, British Library, Harley MS 647, 820–11th century. This collection contains the following astronomical texts and diagrams: 1) Astronomical texts based on Isidore of Seville, with two short prayers (ff. 1r–2r); 2) Marcus Tullius Cicero, *Aratea*, with 22 constellation figures containing extracts from Hyginus, *Astronomica* (ff. 2v–17v); 3) Excerpts from Pliny, *Natural History*, Macrobius, *Commentary on the Somnium Scipionis*, and Martianus Capella, *De nuptiis Philologiae et Mercurii*, books 8 and 6 (known as *The Seven-Book Computus* (ff. 17v–20r and 16r–16v (margin)); and 4) Diagram of the Northern and Southern Hemispheres (f. 21v). Decoration: 22 full-page representations of the constellations in colours (ff. 2v–6r, 7r–13v); a full-page diagram of the constellations in brown ink (f. 21v); and a large diagram of the solar system in brown and red (f. 19r).

[23] Leonardo Crasso from Verona, who financed the printing, dedicated it to the duke of Urbino, Guido da Montefeltro. See Lilian Armstrong, 'Benedetto Bordon, Aldus Manutius and Lucantonio Giunta: Old and New Links', in Lilian Armstrong, *Studies of Renaissance Miniaturists in Venice*, 2 vols (London: Pindar, 2003), Vol. 2:161–83.

[24] See Mark Samuel-Lasner, 'Note on Burne-Jones' *Hypnerotomachia Poliphili*', *Pre-Raphaelite Review* 1 (1978): p. 110.

[25] See Joscelyn Godwin, *The Pagan Dream of the Renaissance* (Grand Rapids, MI: Phanes Press, 2002), pp. 32 and 36, on the planets and their association with the cult of Venus.

initiation rite composed of different phases and realities, for the purpose of arriving at a consciousness where body and soul are united through the metamorphoses of love between Poliphilo and Polia'.[26]

Moreover, during his numerous trips to Italy, Burne-Jones probably knew the popular and beautiful astrological manuscript of the Italian Renaissance by Cristoforo de Predis (1440–86), *De Sphaera* of 1470, at the Biblioteca Estense Universitaria in Modena (Fig. 4). A visual comparison with some of Predis' imagery assists in decoding many of the symbols in Burne-Jones's *Planets*. These planetary personifications are accompanied by constellations and signs of the zodiac, similarly visualized in Cossa and Tura's seasonal cycles at the Hall of the Months in Palazzo Schifanoia. For example, Burne-Jones's *Venus* holds a mirror as a symbol of light and beauty; this planet rules the houses of Taurus with the bull and Libra with the scale. *Mars*, the red planet, rules the houses of Scorpio and Aries, but Burne-Jones replaced the traditional tame ram as a symbol of Aries with a fierce feline, perhaps a puma, lynx or wolf.[27] Apollo, the sun god, rules the house of Leo with the sign of the Lion. Saturn, considered in Renaissance as the oldest planet, rules the houses of Aquarius and Capricorn with respective signs; but Burne-Jones substitutes Gemini or the twins for Capricorn. Jupiter rules the houses of Pisces and Sagittarius with their signs, but Burne-Jones focused on the cosmic regal power of Jupiter with a thunderbolt. Burne-Jones' creation of the planet Earth (*Terra*), a conflated combination, rules the house of Virgo, trees, Aquarius, water, and Syrus, the sky. And *Luna* (The Moon), on a boat, controls the seas and rules the houses of Cancer and the zodiac sign of the Crab (compare Figs. 1 and 4).[28]

---

[26] See Esteban Alejandro Cruz, *Hypnerotomachia Poliphili: An Architectural Vision*, 2 vols (USA: Xlibris, 2011), Vol.1:201; Godwin, *Pagan Dream of the Renaissance*, p. 30.s

[27] See Julia Cresswell, *Legendary Beasts of Britain* (Oxford: Shire Library, 2013), p. 24. Puma and lynx were seen in Inverness-shire.

[28] See P. G. Maxwell-Steward, *Astrology: From Ancient Babylon to the Present* (Stroud: Amberley, 2010), pp. 2–44; R. Hinckley Allen, *Star Names and Their Meanings* (Glastonbury: The Lost Library, 1899), pp. 10–30; and Giuseppe Maria Sesti, *The Glorious Constellations: History and Mythology* (New York: Harry N. Abrams, Inc., 1991).

Fig. 4. Cristoforo de Predis, *Luna*, 1470–76, miniature. *De Sphaera* (Ms. lat. 209). Biblioteca Estense, Modena.[29]

I would like to suggest as well an iconological signification. Burne-Jones continued the cultural tradition of envisioning the planets as personifications of goodness. These personifications partake of both the spiritual and physical realms. In this manner, in the *Planets*, Burne-Jones appropriated from Michelangelo's Ignudi, Prophets, and Sibyls in the Sistine Ceiling their composition – such as their royal seated posture and peaceful attitude – as well as their meaning by capturing their projected powers of divination and creation. Michelangelo depicted genii as astral sources of inspiration and benevolent guidance, while Burne-Jones drew constellations and zodiac signs to convey the positive destiny of the cosmos.[30] The majestic and voluminous representation of their personification in the *Planets* suggests that Burne-Jones portrayed them as monumental guardians of the celestial sphere. Burne-Jones surrounds their heads with specific allusions to sacred forms or entities, such as a halo (a celestial attribute) or a crown (a regal allusion), with attributes specific attributes to each planet. For example, in the female planets, Luna is decorated with a bluish luminous ring alluding to a full moon; Venus has a golden starry disk, alluding to her coloured light source; and Earth has no

---

[29] https://commons.wikimedia.org/wiki/File:Luna_-_De_Sphaera_-_Biblioteca_Estense_lat209.png [accessed 2 Dec. 2016].

[30] See Valerie Shrimplin, *Sun-symbolism and Cosmology in Michelangelo's Last Judgment* (Kirksville, MO: Truman State University Press, 2000); and Valerie Shrimplin, *Michelangelo, Copernicus and the Sistine Ceiling: The Last Judgment Decoded* (Saarbrucken: Lambert, 2013).

halo, since she is a terrestrial planet. With the male planets, Burne-Jones depicted Mars wearing a warrior helmet surrounded by a solid golden disk while Jupiter, with a celestial aura, wears a royal crown with jewels. Apollo's starry halo transforms into an open sunflower – an allusion to the Sun – while he is crowned with laurel, a poetic touch; and Saturn is portrayed as a Moses-type with a galactic halo, which is marked with large stars, while his crown is composed of tablets, alluding to ancient knowledge and times. These astral rulers are seated on allegorical thrones surrounded by attributes that combine their cosmic powers. These attributes are part of the constellations, an ancient formation for understanding the movement of the spheres and the stars as well as controlling the destiny of terrestrial beings.[31]

Burne-Jones composed a scroll with large Latin inscriptions to accompany each image in the cartoons; these assist in comprehending the planets' role in the firmament. For example, the inscription *Saturn Pallidum sidus* refers to Saturn's colour: a silvery white star like a silver jubilee. The inscription accompanying the Sun (Apollo), *Solis aureu idus*, decrees that the Golden Age is here. Venus's *[S]Tella Candida Venere* refers to Venus as a brilliant star. *Regia Stella Jovis* proclaims Jupiter to be the Regal Star, meaning the ruler of the universe. *Terra Omniparens* alludes to Mother Earth or the Planet Earth, where all terrestrial forms are created.[32] *Stella Matutina* alludes to the early morning soft light while *Stella Vespertina* refers to the luminous Evening Star. *Mars Terreus* signifies 'made of earth colour', alluding to Mars's reddish tones, and Luna's *Stella Mutabilis Lunae* refers to the changeable movements of the Moon. In addition to providing the Latin inscriptions, Burne-Jones also coloured his cartoons according to what he perceived are the planetary attributes, e.g., blue for the Moon, red tones for Mars and pale yellow for Venus.

---

[31] Plato's *Timaeus*, 90a2–d7 and 90d5–7. See D. J. Zeyl, trans, *Plato: Timaeus* (Indianapolis, IN and Cambridge, MA: Hackett Publishing, 2000); *On Plato's Timaeus-Calcidius*, ed. and trans. John Magee (Cambridge, MA: Harvard University Press, 2016), pp. 211–87 and 289–317; and F. M. Cornford, *Plato's Cosmology* (1937; repr. Indianapolis, IN and Cambridge, MA: Hackett Publishing, 1997).

[32] See Theodore Chilb, 'On the Decorations of Pianos', in *Early Journal Content* (1883), p. 167, for a description of Edward Burne-Jones's decoration of a pianoforte for William Graham in which he employed the term Earth as *Terra omniparens*. '[She] is the universal mother'. See also Liana De Girolami Cheney, *Edward Burne-Jones: Mythical Paintings* (London/New York: Peter Lang Publishers, 2013), pp. 107–13.

Burne-Jones's poetical inscriptions and visualisations allude to a *paragone* (comparison) among the sister arts or fine arts as noted by the classical poet Horace, *ut pictura poesis* (as is painting, so is poetry).[33] In this type of poetical combination the text assists the viewer in interpreting the image, and the image in turn poetically makes manifest the inscription, a union between the sister arts. With the personifications of the planets and constellations, Burne-Jones created, in his *Planets* cycle, a stellar tapestry as a spatial background for the cosmic dwelling of the planets, hence providing a heavenly realm for terrestrial beings to contemplate and wonder about the firmament.

Christina Rossetti (1830–94), a Pre-Raphaelite poet and the sister of the Pre-Raphaelite painter and poet Dante Gabriel Rossetti (1828–1882), who was one of Burne-Jones's teachers, also followed Horace's poetical dictum, 'as is poetry so is painting'.[34] She composed a sonnet in honour of the Moon, verbalizing her feelings upon perceiving the Moon and noticing the Moon's pale colorations and oscillations. Her poem, *Is the Moon tired? She Looks so pale*, was published in a collection of *Sing-Song: nursery rhyme book* in 1868–70.[35]

> Is the moon tired? she looks so pale
> Within her misty veil:
> She scales the sky from east to west,
> And takes no rest.
>
> Before the coming of the night
> The moon shows papery white;
> Before the dawning of the day
> She fades away.

---

[33] The phrase originates from Horace's *Arts poetica* (line 361), first century BCE; for further study, see *Horace on Poetry: The Ars Poetica*, ed. C.O. Brink (Cambridge: Cambridge University Press, 1971). See Jean H. Hagstrum, *Sister Arts: The Tradition of Literary Pictorialism and the English Poetry from Dryden to Gray* (Chicago: University of Chicago Press, 1958), p. 59; and Michael J.B. Allen, *Marsilio Ficino and the Phaedran Charioteer* (Los Angeles: University of California Press, 1981), pp. 339–439.
[34] See GBJ, *Memorials*, Vol. I, pp. 1333, 136, 145, and 149.
[35] An original copy, Ashley MS 1371, is at the British Museum Library in London. George Routledge and Sons were the first publishers in 1872. For this edition, Arthur Hughes (1832–1915) and Dalziel Brothers illustrated the text. In 1893, Macmillan and Co. of London reprinted the book.

Burne-Jones's first *paragone* is between the word and the brush inspired by British poetical compositions on the colouration and status of the Moon; for example, Rossetti's *paragone* between seeing what the Moon appears to be and employing words to express the nature of her colouration. Burne-Jones's second *paragone* is between music and painting, the note and the brush. A modification of Horace's dictum *ut pictura poesis* to *ut pictura musica* (as is painting, so is music) relates as well to the sister arts. Holden's commission was for the purpose of decorating the music room with the image of the Planets. The traditional concept of *musica universalis* or Music of the Spheres (Harmony of the Spheres) is an ancient philosophical association related to the movement of the celestial bodies.[36] The movement of these celestial bodies, the Sun, Moon and planets, are revealed metaphysically through mathematical constructions in order to create harmony and spiritual uplift and physically through sounds played with musical instruments. Earlier, Pythagoras, who invented the theory of music and sounds, claimed that the Sun, Moon and planets all emit their celestial sound or orbital resonance based on their orbital revolutions.[37] As Plato noted, 'As the eyes, said I, seem formed for studying astronomy, so do the ears seem formed for harmonious motions: and these seem to be twin sciences to one another, as also the Pythagoreans

---

[36] See Joscelyn Godwin, ed., *The Harmony of the Spheres: A Sourcebook of the Pythagorean Tradition in Music* (Rochester, VT: Inner Traditions International, 1993), pp. 3–9; 21–40, 163–70, and 395–99; and Jamie James, *The Music of the Spheres: Music, Science and The Natural Order of the Universe* (New York: Copernicus, 1993), pp. 41–59.

[37] See W. Stirling, *The Canon: The Pagan Mystery as the Rule of all the Arts* (Glastonbury: The Lost Library, 1897), pp. 260–74, p. 260, for an explanation of Pythagoras's claim that 'the planets in their revolutions around the earth uttered certain sounds differing accounting to their respective magnitude, celebrity and local distance', establishing his harmony canon. See also Pliny the Elder, *Natural History*, Book I and II, II.xviii.xx, trans. H. Rackham (Cambridge, MA: Harvard University Press, 1938), pp. 277–78. See also in the Renaissance, Franchinus Gafurius, *Practica musicae* (Milan: Gulielmum signer Rothomagensem, 1496, and Venice: Agostino Zani, 1512), when an engraving connects Apollo with the Muses and the planetary spheres, e.g., the planet Luna with the muse Clio. See Pliny the Elder, *Natural History*, Book I and II, II.xviii.xx. The translation of this ancient volume by Philemon Holland (London: Adam Firm, 1601) had a special significance for Burne-Jones. During their marriage, his wife read it aloud to him. See GBJ, Memorials, Vol. 2, p. 55; and James, *The Music of the Spheres*, pp. 20–40.

say' (Republic VII.XII).[38] These Platonic notions of associating music with astronomy resurfaced in the eighteenth and nineteenth centuries. For example, the British astronomer William Herschel not only discovered the planet Uranus but also composed and orchestrated music. In the nineteenth century, Burne-Jones combined his artistic talent as a painter in 1870 depicting *Luna* and in 1879 composing a cartoon about *Luna* with his musical inclination in playing the piano; in addition, he was involved in constructing and decorating musical instruments.[39]

Throughout his artistic career, Burne-Jones composed several images depicting the Moon. His earliest known composition was created during his travels in Italy in 1871 (see Fig. 3). In this painting, Burne-Jones depicted a creative image of the Moon hiding from the Sun's emissions. In a blue-starry firmament, a female figure, Luna, rides on a globe. While the tresses of her hair have been turned golden by the Sun's radiation, Luna employs her attire to hide from the Sun's reflections. Her face is veiled with the sleeves of the dress and the folds of her dress train cover the circular form of her planet, as only a small crescent ridge is visible. In this composition, Burne-Jones experimented with a variation of tonalities, employing only the colour blue to form an orchestral symphony of blue tones.[40] He may have been influenced by James Abbott McNeill Whistler's paintings on nocturnes and symphonies, for example, *Nocturne: Blue and Silver–Chelsea* of 1871, now held at the Tate Gallery in London.[41] Burne-Jones's curved lines of the figure, globe, drapery and crescent moon

---

[38] See Henry Davis, *The Republic. The Statesman of Plato* (1901; repr. Nabu Press, 2010), p. 252; and James, *The Music of the Spheres*, p. 41–59.

[39] See Christopher Wood, *Burne-Jones: The Life and Works of Sir Edward Burne-Jones, 1833–1898* (New York: Stewart, Tabori and Chang, 1998), p. 78, for records, which document the early design on this subject, *Le Chant d'Amour*, created for a decoration of a small upright piano made by F. Priestley of Berners Street in London (now in the Victoria and Albert Museum, London). The American walnut piano was given as a wedding present to Burne-Jones in 1860. Burne-Jones paints the lid of the piano in monochrome in 1863. For Francis Homer, daughter of William Graham, an important collector and patron of Burne-Jones, he painted a piano in 1879, with themes on Orpheus, during the same time that he was designing the *Planets* cycle. See also Aymer Vallance, *Sir Edward Burne-Jones Baronet* (London: The Art Journal, 1900), p. 28, Figs. 54 and 55, for Burne-Jones' constructed psaltery and harp, respectively.

[40] See Stirling, *The Canon*, p. 262, where he notes that Pythagoras, 'using the term of music, called the interval between the earth and the moon, a Tone'.

[41] For the image, see http://www.tate.org.uk/art/artworks/whistler-nocturne-blue-and-silver-chelsea-t01571 [accessed 2 Dec. 2016].

provide a rotary movement, alluding to the rotation of the lunar sphere, while the reflections of the sky and the water produce an eerie atmosphere. Inventively, Burne-Jones interplays these with bright light effects, showing how Luna hides from the bright sunlight, demonstrating the astral connection and rivalry between the planets: the Sun and the Moon, signifying time – day and night – and radiance, with light and dark effects. The overall luminosity of the blue light and its reflections augment the understanding of the Moon's colouration (compare Fig. 1 and 3). Luna, whose ancient Roman name *lumina,* signifies light or to illuminate. In the painting, Burne-Jones interplays these meanings of *lumina:* physically alluding to illumination of a visible area as the Moon radiates at night, and metaphysically referring to Luna as the astral planet that not only illuminates through shining stars but also through its mysterious bluish coloration which evokes a mystical atmosphere for the imagination and the intellect of a creative mind.

The compositional design for the cartoon *Luna*, like the other *Planets*, is in arched-shape format. The haloed female figure is identified as a goddess or astral being. The figure of Luna is beautifully dressed in a classical garment with a wet drapery motif recalling Burne-Jones's admiration for the Greek sculptor Phidias (480–430 BCE), whose marbles statues for the frieze and pedimental area of the Parthenon were (and are) in view at the British Museum during his numerous visits. Burne-Jones depicted *Luna* as seated female navigating through the sea in an ancient Roman boat (see Fig. 1). Holding steadily with her left hand a crescent moon as a rudder of a celestial ship, she intensely watches the nautical direction of her physical and metaphysical realms. With her right hand, Luna holds fast to the edge of the boat, adjusting to the physical movements of the sea's waves. In the cartoon, Burne-Jones has skilfully contrasted the astral deity's beautiful hands with tapered fingers and the crab's clumsy regenerative claws, alluding to both – the Moon and the crab – fastening to atmospheric and climatic changes.

Astrologically, Luna rules the sign of Cancer or constellation of Cancer, which is depicted by the zodiac sign of the crab.[42] The animal symbolism of the crab is associated with water movement, as well as illuminating effects. Burne-Jones depicted a crab surfacing from the sea and following the boat's navigation. The nautical movements of the boat are associated with the crab's oscillation, which alludes to the Moon's monthly cycle and the seasonal time of summer (July and August).

---

[42] See Sesti, *The Glorious Constellations: History and Mythology*, pp. 267–71.

A shower of stars decorates the celestial background of the composition for the navigation of the Moon, as indicated by the large scroll with a Latin inscription which reminds the viewer about the mutable astral Moon, *Stella Mu/Tabilis Lunae*.[43] The Moon's mutable transformation oscillates the cycles of the sea, as noted in the cartoon by the blue tonalities and reflections of the sky in the water, and the sound of the wave is paralleled to the sound of the star. Burne-Jones created a physical and metaphysical bond among art, astronomy, music and poetry. *Luna* is an image of a poetical planetary symphony.

Acknowledgements
I express my gratitude to Dr Joseph Harvey, Custodian, Torre Abbey Historic House and Galleries, for his email communications and generous assistance on Edward Burne-Jones's seven cartoon and glass panels on the *Planets* in the collection. My gratitude is further extended to Jane Palmer, Assistant Curator of the Torre Abbey Historic House and Galleries, for her assistance on the archival and photographic documentation. I thank Serena Trowbridge of Birmingham City University, for her generous comments. My thanks also to Gabreal Franklin for his email communications and assistance on the history of the Burne-Jones's *Planets* in stained-glass panels as well as in reproduction of Burne-Jones's Luna in stained-glass panel. A special thanks to Victoria Osborne, Curator (Fine Art) and Curatorial Team Leader, Birmingham Museums Trust at the Birmingham Museum and Art Gallery for her assistance. A shorter version of this study was published in *The Pre-Raphaelite Society* as Liana De Girolami Cheney, 'Edward Burne-Jones's The Planets: The Cartoon of Mars at the Birmingham Museum and Art Gallery', Vol. 24 (Winter 2016): pp. 15–26.

---

[43] See Archives, Accession Number A.113, Torre Abbey Museum. The drawing is made of charcoal and coloured chalks, blues and gold and violets. The drawing was restored and cleaned in 1988 by the Bristol City Art Gallery. In the Entry in the *Catalogue of Designs* of July 1870, Fletcher is named as of the glass painter who composed the panel (2512/35 and 2477/36).

An Examination of the Images of the Sun and the Moon in the Visoki Dečani Monastery in Kosovo

Dragana Van de moortel-Ilić

Abstract: This paper investigates the celestial-religious images in the Visoki Dečani monastery in Kosovo, particularly the tear-shaped paintings with human figures inside located on the left and right side of *The Crucifixion of Christ* fresco. The aim of this research is to put these images into a cultural and contemporary cosmological context. The images in the fresco that contain human figures have been the subject of controversy in the second part of the last century. A highly speculative popular view was put forward, that the images portrayed extra-terrestrials in UFOs. Yet these images have been mostly ignored in academic circles. In this research the images from *The Crucifixion* fresco were compared with similar frescoes from other Serbian medieval churches and with the philosophical thought of the time. The methodology used was, first, a personal visit and observation of the images, including photos taken by a professional photographer. A comparison of those findings was then made with what had been written about those images in the academic literature. The conclusion is that they present personifications of the Sun and the Moon which could be explained by the synergy of Hellenistic and Christian thought.

Introduction

The research is focused on the images of the Sun and the Moon in the fresco *The Crucifixion of Christ* and other celestial-religious images from the Serbian fourteenth-century Visoki Dečani monastery. Located just a few kilometres west of the village Dečani, in the western part of Kosovo, more specifically in the region called Metohija, the monastery is close to the borders of both Albania and Montenegro.

The images of the Sun and the Moon in *The Crucifixion* fresco, containing human figures that are illustrated in Fig. 2, have been the subject of controversy. A highly speculative popular view, supported by the author Erich von Däniken, was put forward, maintaining that the images portrayed extra-terrestrials in UFOs. In the academic world,

---

Dragana Van de moortel-Ilić, 'An Examination of the Images of the Sun and the Moon in the Visoki Dečani Monastery in Kosovo', *The Marriage of Astronomy and Culture,* a special issue of *Culture and Cosmos,* Vol. 21, nos. 1 and 2, 2017, pp. 301–19.
www.CultureAndCosmos.org

however, scholars have for the most part neglected doing any research on this fresco. Yet, celestial-religious images at other Serbian monasteries and churches dating back to the same period have been analysed in the academic literature, in the writings of Branislav Todić and Milka Čanak-Medić, among others.

The aim of this research, therefore, is to put these celestial-religious images into a cultural and contemporary cosmological context by comparing them with other celestial images in Serbian churches, constructed and painted in the same period. Most celestial images used as references are found in the Bogorodica Ljeviška church, located in the town of Prizren, Kosovo, 60 km from the Visoki Dečani monastery, and the Lesnovo monastery, which is between the towns of Zletovo and Kratovo in FYRO Macedonia – 260 km from the Visoki Dečani monastery.

Methodology
With the objective of collecting primary source material, I conducted a site visit of the Visoki Dečani and Lesnovo monasteries, Bogorodica Ljeviška church and other churches in the neighbourhood. In order to take photographs of the interior and religious objects at the sites in Kosovo, which are under the patronage of the Serbian Orthodox church, permission from the church authorities was obtained. My husband, Koen Van de moortel, who is a professional photographer, took photographs of the fresco *The Crucifixion of Christ* and other celestial-religious images in the churches while I observed the exact locations of the images on the interior walls and searched for anything connected to medieval cosmology. The photos were taken with a Canon Eos 6D camera and composed in the High Dynamic Range (HDR) style in order to stress details in the highlights and shadows as well as the usual details from a standard exposure. I could then compare these photos with photos from previous studies of these frescoes, most of which was published in the late twentieth century. My first hand observations and on-site photographs constitute the primary source material for my research.

An assessment of the secondary data sources was realised by means of a literature review. Mirjana Gligorijević-Maksimović, Srđan Đurić, Branislav Todić and Milka Čanak-Medić are some of the scholars who investigated the images in medieval Serbian churches. However, they were not engaged with considering these images in a cultural or cosmological context, although Đurić investigated the form of the images and the background of the painters. The celestial images of the Lesnovo monastery were explored in detail by Smiljka Gabelić, though also not completely in a cosmological context. This paper will make a link between observations from the field research and from the literature, and, subsequently, place the findings in the context of medieval cosmology.

Comments on the images from Visoki Dečani from non-academic sources

Aleksandar Paunović, an art student, was observing frescoes in Serbian medieval churches in 1964 when he spotted unusual paintings in the monastery Visoki Dečani. In the fresco *The Crucifixion*, he interpreted two objects on the right and the left side above Christ (Fig. 2) as 'spaceships with a crew' and, subsequently, the French magazine *Spoutnik* published images of them.[1] For a very short time these images were hugely popular and featured in many Internet presentations, television programs and books. One of the films posted on YouTube, featuring dramatic music in the background, claims that the person in the spaceship on the right is observing the other, that the jets are clearly visible, and comments on the presence of spectators (angels from the painting) who appeared 'obviously terrified'.[2] This populist and speculative way of interpreting the paintings was further supported by the Swiss author Erich von Däniken, who became famous writing books about UFOs in ancient civilisations, such as *Chariots of the Gods*, *The Gods were Astronauts* and many others. He raised the popularity of the UFO theory associated with the images of the Visoki Dečani monastery, voicing the idea that this was one of the most obvious examples of an alien visit to Earth.[3] On his website, the right image of the fresco is provided as an example of extra-terrestrial visitors, with the explanation 'Curious wall painting from the Middle Ages'.[4] Inspired by the findings of von Däniken, the History Channel US made a series of episodes about the ancient aliens, in which these images were also portrayed.[5]

---

[1] 'Ancient aliens debunked – part 5' at https://illuminutti.com/ancient-aliens-debunked/ancient-aliens-debunked-part-5/ [accessed 10 Oct. 2014].
[2] https://www.youtube.com/watch?v=3yd76pX1MT0 [accessed 25 Oct. 2014].
[3] 'Da li se u manastiru Visoki Dečani nalaze dokazi u vanzemaljcima', Vestinet 15 March 2015 at http://www.vestinet.rs/pogledi/da-li-se-u-manastiru-visoki-decani-nalaze-dokazi-o-vanzemaljcima [accessed 15 March 2015].
[4] 'Pictures' at http://www.daniken.com/e/index.html [accessed 10 October 2014].
[5] 'Ancient Aliens' at http://www.history.com/shows/ancient-aliens [accessed 25 Oct. 2015].

Even if the theory of the space visitors, as thought to be represented in these paintings, was not taken seriously, it inspired many journalists to use the images when referring to something else, for the most part anything of an extra-terrestrial nature. On the Serbian television Pink news broadcast of 7 December 2009, witnesses stated that they spotted unidentified flying objects in the sky above Novi Sad, resembling lights in the shape of triangles.[6] The screen showed images of 'the crew' from the monastery Visoki Dečani while the narrator quoted Miroslav Kostić, the president of the UFO association, claiming that many people saw these lights, yet nobody saw the crew of these flying objects. However, public enthusiasm for the idea that those images represent visitors from space has not yet diminished or disappeared.

Academic work on cosmological elements in the religious images in fourteenth-century Serbian monasteries, with an emphasis on images in the monastery Visoki Dečani

Paintings of the monastery Visoki Dečani that are connected with celestial elements, especially the Sun and the Moon from the fresco *The Crucifixion of Christ*, are barely mentioned in academic literature, mentioned above. However, the images of astrological signs, the Sun, the Moon and the sky in the Lesnovo monastery, located in Serbia at the time of painting in the fourteenth century, have been described as parts of biblical texts by Gabelić.[7] Most scholars agree that the personification of natural and cosmic phenomena in this time period was a common theme in frescoes. As Gligorijević–Maksimović, who observed classical elements in Serbian medieval churches, said, the personification of the Old and New Testaments, the Church, the Synagogue and Kosmos were very popular in the first half of the fourteenth century, but, towards the end of the century, their popularity waned.[8] The form of 'floating' celestial luminaries, as seen in *The Crucifixion* fresco and on the ceiling of the exonarthex of Bogorodica Ljeviška, were, according to Srđan Đurić, 'valid only in theophanic scenes, since, in the composition of the *Journey to Emmaus,* the

---

[6] 'Nacionalni dnevnik', RTV Pink, 7 December 2009 (between 56" and 1' 14" and between 1'39" and 2'03') at
https://www.youtube.com/watch?v=RL4qcgw7TmA [accessed 25 Oct. 2014].
[7] Smiljka Gabelić, *Manastir Lesnovo* (Beograd: Stubovi kulture, 1998), pp. 183–85.
[8] Mirjana Gligorijević-Maksimović, 'Classical Elements in the Serbian Painting of the Fourteenth Century', *Zbornik Radova Vizantološkog Instituta* 44 (2007): pp. 368–69.

sun was depicted in its usual circular shape, with the personification inscribed'.[9] Igor Stojić, Milan S. Dimitrijević, Edi Bon and Vesna Mijatović considered that the teardrop shape of the paintings, which emphasised movement, could have been inspired by the appearance of Halley's comet in 1307.[10] However, scholars were, for the most part, not engaged in placing these images into a cosmological context.

Between folk beliefs and Christianity
In order to understand the background of the Serbian belief system in the fourteenth century, some attention needs to be given to Serbian folk beliefs. According to Nenad Janković, before embracing Christianity, the Serbian people believed that natural phenomena were connected with different gods; these resembled many other Indo-European pre-Christian beliefs.[11] They also believed that some people could speak a 'mute language' which was, as Branislav Rusić said, 'the power of expressing feelings and thoughts to members of animate and inanimate nature'.[12] According to Rusić, this talent belonged to 'saints, tsars or their sons, magnates, eminent and mysterious persons' and was considered powerful not only in the pre-Christian period and medieval Serbia but it even survives to the present day.[13] Serbs started to Christianize from the ninth century onwards, accepting religious art from Byzantium. However, as Janković stated, they also stayed loyal to some parts of the old traditions. One example of this is not believing that the Sun and the Moon were gods by themselves, but rather that the Sun and Moon were aided by 'servants of the God' up to the sky and back on a daily basis, and that these servants also woke up all the stars at night.[14]

---

[9] Srđan Đurić, 'The Representation of sun and moon at Dečani', *Dečani I vizantijska umetnost sredinom XIV veka* (Beograd: SANU XLIX, knjiga 13, 1989), p. 343.
[10] Igor Stojić, Milan S. Dimitrijević, Edi Bon, and Vesna Mijatović, 'Possible Representations of Comets in Serbian Religious Medieval Art', *European Journal of Science and Theology* 12, no. 3 (2016): p. 181.
[11] Nenad Đ. Janković, *Astronomija u predanjima, običajima i umotvorinama Srba* (Beograd: SANU, 1951), p. 4.
[12] Branislav Rusić, 'The Mute Language in the Tradition and Oral Literature of the South Slavs', *The Journal of American Folklore* 69.273, *Slavic Folklore: A Symposium* (1956): p. 304.
[13] Rusić, 'The Mute Language', pp. 305–7.
[14] Janković, *Astronomija u predanjima, običajima i umotvorinama Srba*, p. 4.

From the twelfth to the fourteenth centuries, Serbia expanded its ~~own~~ territory, resulting in a widespread cultural interchange. Monumental religious buildings started to be built and books were translated from Greek to the Old Church Slavonic language. However, there are opinions, such as those espoused by Ihor Ševčenko, that Orthodox Slavs in general 'did not absorb much of the scientific and philosophical literature available in Byzantium'.[15] Whether or not they understood the scientific and philosophical literature from Byzantium and Antiquity, they accepted Orthodox Christian spiritual values, which Atanasije Jeftić explained, was an 'inseparable part of the Serbian national soul, as well as its lifeblood in the early stages in its history'.[16] Some members of the Nemanjić dynasty, which ruled Serbia from the twelfth to the end of the fourteenth century, were not only devoted to Christian values, but left the mundane life and became monks and nuns.

The church of the Visoki Dečani monastery
According to Bratislav Pantelić, King Stefan Uroš III Nemanjić, who was also called Stefan Dečanski, chose a spot on the banks of the Bistrica River to build a temple that would impress all visitors. Together with the archbishop Danilo II, he prayed all night long and, in the morning, decided to dedicate the church to the Ascension of Christ.[17] It seems that the construction of the monastery started in 1327 and was finished by 1331 or, more precisely, by August 1331, as Mirjana Šakota noted, adding that the interior without the wall paintings was completed before 1334.[18] The main architect of the monastery was Fra Vita, a Franciscan monk from Kotor, who let his own name be carved outside the church, above the south portal, together with the names of the king and his son, as mentioned by Todić and Čanak-Medić.[19] They also stated that the wall paintings were

---

[15] Ihor Ševčenko, 'Remarks on the Diffusion of Byzantine Scientific and Pseudo-Scientific Literature among the Orthodox Slavs', *The Slavonic and East European Review* 59, no. 3 (1981): p. 325.
[16] Bishop Maxim (Vasiljević) (ed.), *The Christian Heritage of Kosovo and Metohija, the Historical and Spiritual Heartland of the Serbian People* (Los Angeles: Sebastian Press, 2015), p. 269.
[17] Bratislav Pantelić, *The Architecture of Dečani and the Role of Archbishop Danilo II* (Weisbaden: Reichert Verlag, 2002), p. 22.
[18] Mirjana Šakota, *Dečanska Riznica* (Beograd: Prosveta, 1984), p. 48.
[19] Branislav Todić and Milka Čanak-Medić, *Manastir Dečani* (Beograd: Muzej u Prištini (displaced), Mnemosyne, 2013), pp. 208–9.

completed by the summer of 1348, after the crowning of Dušan, the son of Stefan Dečanski, as the Emperor of the Serbs and Greeks in 1346.[20]

According to the information booklet about the monastery written by the monks of Visoki Dečani in 2009, the church managed to largely preserve the original look of the fourteenth century, even though many valuable artefacts have been stolen over time.[21] Šakota stated that the most valuable pieces were books and crosses decorated with gold and gemstones, rich carpets and clothes that the king Stefan Dečanski and subsequent rulers and aristocrats had bestowed as gifts. Many other valuable golden items were stolen, even the huge church bell, which was stolen by the Bulgarian army during the First World War.[22] The primary source material about those gifts comes from original Charters (Golden bulls or Chrysobulls) written by the King Stefan Dečanski personally, and which are now available in modern Serbian and English translations.[23]

I observed that the frescoes in the central part of the church – in the nave and the altar – are mostly well preserved, but in the narthex there are sections of the walls with damaged frescoes. The gold that covered the floor and some other parts of the church was stolen and only a few remnants of it are left.[24] My first impression was like stepping inside a fairy tale. The more than 3000 very colourful frescoes, literally covering every part of the walls, domes and arches, for me had an enchanting effect.

The central part of the church, or the nave, is particularly impressive. With the central dome reaching up to 29 metres, the church is quite imposing. On the floor, under the central dome, there is a twelve-point ornament or a rosette which, according to Todić and Čanak-Medić, symbolises cosmic cycles, the God of light and Christ as the Sun.[25] I argue that this rosette might represent twelve months, mirroring the movement of the Sun in one year. This ornament used to be covered by gold in the fourteenth century, but now there are just traces of gold remaining. The coffin containing the relics of the king Stefan Dečanski is located on the

---

[20] Todić and Čanak-Medić, *Manastir Dečani*, p. 326.
[21] Monks of the monastery Visoki Dečani, *Manastir Visoki Dečani* (Beograd: Manastir Visoki Dečani, 2009), p. 29.
[22] Šakota, *Dečanska riznica*, pp. 47–48.
[23] Milica Grković, *Prva Hrisovulja Manastira Dečani-The First Charter of the Dečani Monastery* (Beograd: Centar za očuvanje nasleđa Kosova i Metohije–MNEMOSYNE, 2004), p. 88.
[24] Information from the monks of the Monastery, during our visit on 20th and 21st February 2015.
[25] Todić and Čanak-Medić, *Manastir Dečani*, p. 246.

right side of the rosette. This is where believers would crawl, hoping for a miracle.

Frescoes with the Sun and the Moon in the Visoki Dečani monastery

The fresco *The Crucifixion of Christ* is situated high up in the central dome, in the north arch (see Figs. 1 and 2). From the vantage point of a visitor observing this fresco from the ground, its details are difficult to see due to the fact that it is positioned so very high up. However, it is from the vantage point of the king's throne that his fresco could be best observed, which may indicate that the fresco was painted for the king's eyes. The other fresco that was examined in this research study is presented in detail in Fig. 3. Its location is marked by the number 2 on the map in Fig. 1. This fresco is situated in the central part of the west bay of the nave, where it is easier to view from the ground. The plan of the church was taken from the Kosovonet website.[26]

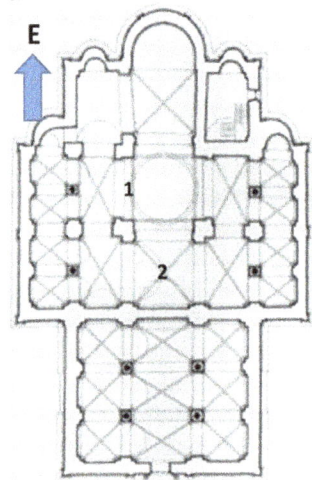

Fig. 1. The plan of the church. 1. The Crucifixion of Christ. 2. The central part of the West Bay of the nave. Image: Kosovonet.

---

[26] 'The Art of Decani Monastery', Visoki Dečani Monastery at http://www.kosovo.net/edecani2b.html [accessed 28 Feb. 2015].

The top part of Fig. 2 depicts *The Crucifixion* fresco, with the enlarged images of the controversial figures below.[27] In *The Crucifixion* fresco, these images are on the left and the right sides above Christ, as tear-shaped forms with people inside. The person on the left side is wearing a cloak around his body and across his left shoulder. This wrap looks like a himation or an ancient Greek cloak and according to Đurić this is the representation of the Sun.[28] On the right side, according to Đurić, a painting of the Moon is portrayed.[29] The person inside this painting is wearing a cloak wrapped around the hips. There are two eight-pointed stars on the painting of the Moon, one in front of the figure and one behind it. The figures are looking at each other, or possibly at angels flying above the Christ. The rays on those paintings could represent the movement from the left side to the right side, since the sharpest point of the tear-shaped form of the Sun is on the left side, and the sharpest point of the Moon is on the right side. The Sun has six rays and the Moon only three.

The other fresco that was examined is located in the central dome of the west bay (see Fig.1, number 2). These paintings, according to Đurić, represent the Sun, the Moon and the Sky, carried by archangels with Christ in the middle.[30] This part of the dome is illustrated in Fig. 3, with enlarged images of the personified Sun (right) and Moon (left).[31] Between these two paintings, the *Christ in the Mandorla* fresco is situated, carried by two archangels.

---

[27] http://www.srpskoblago.org/Archives/Decani/exhibits/Frescoes/Dome/ThirdLevel-2/CX4K1777_l.html [accessed 23 March 2015];
http://www.srpskoblago.org/Archives/Decani/exhibits/Frescoes/Dome/ThirdLevel-2/CX4K1722.html [accessed 23 March 2015];
http://www.srpskoblago.org/Archives/Decani/exhibits/Frescoes/Dome/ThirdLevel-2/CX4K1675.html [accessed 23 March 2015].
[28] Đurić, 'The Representation of sun and moon at Dečani', p. 345.
[29] Đurić, 'The Representation of sun and moon at Dečani', p. 341.
[30] Đurić, 'The Representation of sun and moon at Dečani', p. 343.
[31] https://www.facebook.com/Decani.Manastir/photos/ [accessed 15 Nov. 2014].

310 An Examination of the Images of the Sun and the Moon in the Visoki Dečani Monastery in Kosovo

Fig. 2. Above: *The Crucifixion of Christ* fresco. Below: enlarged images of the personifications of the Sun (left) and the Moon (right). Images: Blago.

Fig. 3. Above: The paintings of the Sun and the Moon in the central dome of the West bay with *The Christ in the Mandorla*. Below: enlarged personifications of the Moon (left) and the Sun (right). Image: FB Visoki Dečani.

The personifications of the Sun and the Moon in both paintings are portrayed as angels wearing fine garments, which is in sharp contrast to the representations of the Sun and the Moon in *The Crucifixion* fresco, where the figures neither have clothes nor wings, merely a himation around the body. The personification of the Sun is depicted as a winged king, clothed with aristocratic golden garments. In the image of the personification of the Moon, a girl with wings and shabbier clothes is portrayed. Both angels in these paintings are inside a circle with a single ray in front of their bodies. The painting of the girl as the personification of the Moon is contained inside the crescent-shaped Moon. In the middle of the fresco of the Moon, two archangels carry something that looks like a canvas with a painting of many small circles and two larger discs, with winged figures inside, partly hidden behind the clouds. Đurić explained that this canvas was the scroll of heaven, as the part of the scene of the Last Judgement, and that the discs with winged figures were personifications of the Sun and the Moon.[32]

Personification of the Sun and the Moon in other Serbian medieval churches and medieval cosmology
The objective for visiting two other churches – specifically the church of the Lesnovo monastery and the Bogorodica Ljeviška church – was to observe celestial elements and Hellenistic motifs in the frescoes. Most details of Hellenistic motifs were found in the exonarthex of Bogorodica Ljeviška church. On the ceiling there are remnants of paintings that appear to have once been very beautiful; however, now they are quite badly damaged. In the remnants, numerous personifications of natural phenomena are depicted. According to Draga Panić and Gordana Babić, the centre of the painting showing Christ inside the Sun was a representation of the Last Judgment.[33] At the top of these images, two angels are holding up the sky, which is presented as a banner on either side of Christ, as explained by Gligorijević-Maksimović.[34] Moreover, she claimed that the personification of the Sun and the Moon are located on the sides of the sky.[35] I observed that the personification of the Moon is located on the right side of Christ, and the personification of the Sun is on the left. Both images are in tear-shaped forms with rays, like the images from the Visoki Dečani monastery. I recognized these personifications by comparing them with the drawn images from Panić and Babić. Without these references, it would be highly difficult to recognize them.

---

[32] Đurić, 'The Representation of sun and moon at Dečani', p. 343.
[33] Draga Panić and Gordana Babić, *Bogorodica Ljeviška* (Beograd: Srpska književna zadruga, 1975), p. 49.
[34] Gligorijević-Maksimović, 'Antički elementi', p. 208.
[35] Gligorijević-Maksimović, 'Antički elementi', p. 208.

Dragana Van de moortel-Ilić 313

Furthermore, Panić and Babić explained that many ideas portrayed in this church originate from the philosophy of Aristotle and Plato and Orphic poems and that these concepts were subsequently integrated with Christian motifs in many medieval Serbian orthodox churches.[36] Indeed, I observed that, in Bogorodica Ljeviška church, despite the detrimental state of its interior, frescoes of the Greek philosophers Plato and Plutarch as well as of Sybille can be seen, which could confirm the theory of the synthesis of ancient philosophy with the vision of Christianity at the time of painting.

The most impressive painting containing cosmological elements in the Lesnovo monastery is the Zodiac, as shown in Fig. 4. On the left side of Fig. 4 there are enlarged images of the personifications of the Moon (above left) and the Sun (below left).[37]

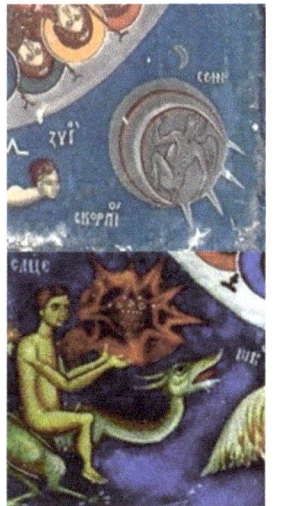

Fig. 4. Right: the zodiac in the South dome of the narthex of the Lesnovo monastery. Left: personification of the Moon (above) and the Sun (below). Image: The Astrologer's Community.

I observed that the head of Christ, which is in the centre of the South dome, is oriented to the centre of the narthex. The signs of the Zodiac are located on the vertical part of the dome, on the East side, partly visible

---

[36] Panić and Babić, *Bogorodica Ljeviška*, p. 71.
[37] The image is downloaded from the website of the forum of the Astrologer's Community, with the theme 'The Zodiac in churches, cathedrals, in Christianity' at http://www.astrologyweekly.com/forum/showthread.php?t=71513 [accessed 29 March 2015].

from the entrance door. Next to every sign or celestial object are words written in the Old Church Slavonic language. These words are difficult to read looking at them with the naked eye inside the church, but the images are clearly visible. Gabelić provided explanations of those words, connecting them with Biblical texts and names of planets and astrological signs, which I could understand and recognize looking at the image from Fig. 4, which is enlarged in the book written by Gabelić.[38] In the first horizontal row, under Christ with angels, there is a naked winged figure, sitting on a bipod beast and carrying the Sun in its hands. Above its head is written 'the Sun', and above the image of the Moon, portrayed with a winged figure inside, is written 'the Moon'. On the right side of the personified Sun there are zodiac signs and personified planets. I argue that the whole fresco with its astrological motifs conveys the message that man could reach God via planets with souls. However, as Gabelić claimed, they might have been painted in order to illustrate the cosmological character of Christ.[39] Above the head of Christ and the circle of angels words can be seen, translated by Gabelić as Psalm 148:1. She wrote that the Psalm continues further, above the astrological signs and other frescoes on the south wall, and ends with Psalm 148:10.[40] Psalm 148:1–3 reads as follows:

> 1 Praise the Lord from the heavens;
>   praise him in the heights!
> 2 Praise him, all his angels;
>   praise him, all his hosts!
> 3 Praise him, sun and moon,
>   praise him, all you shining stars![41]

On the top right side, together with most of the astrological signs, a point with five circles around it is located, containing words written in the Old Church Slavonic language from Psalm 148:3: 'Praise him, sun and moon'; Gabelić identifies this image as the Solar system.[42] The frescoes with cosmological elements in the mentioned Serbian medieval churches are, on

---

[38] Gabelić, *Manastir Lesnovo*, pp. 183–85.
[39] Gabelić, *Manastir Lesnovo*, p. 184.
[40] Gabelić, *Manastir Lesnovo*, pp. 184–85.
[41] https://www.biblegateway.com/passage/?search=Psalm+148&version=ESV [accessed 15 May 2015].
[42] https://www.biblegateway.com/passage/?search=Psalm+148 [accessed 15 May 2015].

the one hand, connected with the Christian cosmology, praising the God, as written above. On the other hand, one can consider that the influence of the Hellenistic cosmology can be observed in these frescoes. John Meyendorff writes, 'unlike their Latin contemporaries, who "discovered" Greek philosophy… the Byzantines had never forgotten Plato or Aristotle, who represent their own Greek cultural past and were always accessible to them in the original Greek texts'.[43] However, according to Meyendorff, metaphysical and religious truths were, especially in the monastic circles, solely connected with the Christian revelation and Plato and the Neoplatonists were looked at with suspicion.[44] Orthodox theologian Thomas Hopko explained that some Greek philosophers were considered to be 'enlightened by God' and this was why their teaching was considered to be a valuable part of the roots of the Orthodox doctrine.[45] The bishop of Thessaloniki from the fourteenth century, Gregory Palamas, who was in his youth educated in Aristotelian philosophy, wrote that 'the mind of demons, created by God, possesses by nature its faculty of reason' but 'the intellect of pagan philosophers is likewise a divine gift' which could be used for its 'therapeutic values even in substances obtained from the flesh of serpents'.[46] The painter of the Bogorodica Ljeviška church and many other Serbian medieval churches was Astrapa, whose origin was Greek, more specifically, according to Đurić, he was Thessalonican, like Palamas.[47] Astrapa belonged to the Macedonian school of painters, which reflected the theological and philosophical thought of that time. Nevertheless, the fact that in several Serbian medieval churches the paintings of the philosopher Plato and other Greek philosophers are to be found indicates the possibility that their teaching might have be respected and somehow integrated in the Orthodox exegesis in this period.

The cosmology of Plato and Aristotle permeated the universe with the divine and the soul. Plato wrote that the God created 'the soul in origin and excellence prior to and older than the body… to whom the body was to be subject'.[48] Furthermore, according to Plato, the soul 'began a divine

---

[43] Gregory Palamas, *The Triads,* ed. John Meyendorff, trans. Nicholas Gendle (New Jersey: Paulist Press, 1983), p. 10.
[44] Gregory Palamas, *The Triads,* pp. 10–11.
[45] Thomas Hopko, *Dogmatische theologie: een inleiding,* trans. Patrick Pauwels (Gent: Orthodox vormingcentrum 'Heilige Johannes de Theoloog', 2003), p. 2.
[46] Gregory Palamas, *The Triads,* pp. 27–28.
[47] Đurić, 'The Representation of Sun and Moon at Dečani', p. 344.
[48] Plato, *Timaeus*, trans. Benjamin Jowett (1959) at http://classics.mit.edu/Plato/timaeus.html [accessed 10 Nov. 2013], 34A-35A.

beginning of never ceasing and rational life enduring throughout all time', or, in other words, became immortal.[49] Within the soul, God formed the corporeal universe, then days and nights and months, and finally the Sun, the Moon and five planets which moved in their orbits around the Earth, according to Plato, who went on to say that above them was the universe with the number of the stars which was equal with the numbers of souls.[50] Plato also wrote that after death the soul ascends to the stars via the planetary spheres.[51] Aristotle mostly agreed with this model but put much more accent on the movement of the Sun, the Moon and the planets. Discussing the words of Alcmaeon, Aristotle wrote that the soul 'is immortal because it resembles "the immortals," and that this immortality belongs to it in virtue of its ceaseless movement; for all the "things divine," moon, sun, the planets, and the whole heavens, are in perpetual movement' while the prime mover is God.[52] As mentioned above, the images of the Sun and the Moon are in most of the published work on these frescoes presented with rays, which could indicate their movement and could also resonate with the Aristotelian model of the universe.

From my observation of the images, Gabelić's explanation that all the astrological images from the fresco in Lesnovo (see Fig. 4) represent the cosmological character of Christ, can be considered partly correct. The words from Psalm 148 are indeed written in and around this fresco, yet the astrological signs, planets and winged figures are not mentioned in this Psalm. It leads to the conclusion that the cosmology of Plato and Aristotle could offer an extended explanation as to why the images of planets and signs were painted in this fresco.

'Space shuttles' of Plato and Serbian folklore

In the course of my research at various Serbian medieval churches I observed many different representations of the Sun: some had numerous rays, others were personified with angels, kings, or special figures painted inside. Representations of the Moon, however, were less frequent, and every fresco with the personified Moon also included the Sun. Furthermore, all personified images of the Sun and the Moon, which are

---

[49] Plato, *Timaeus*, 36D–36E.
[50] Plato, *Timaeus*, 41E–42A.
[51] Platon, *Država*, trans. Albin Vilnar and Branko Pavlović (Beograd: Beogradski izdavačko-grafički zavod, 1976), 616A–617B.
[52] Aristotle, *On the Soul*, trans. J. A. Smith, book 1, part 2 (2006) at http://classics.mit.edu/Aristotle/soul.mb.txt [accessed 20 March 2017], 405B.

Dragana Van de moortel-Ilić 317

portrayed in Fig. 5, are located in the central, most important, part of the church. In Fig. 5, on the left side, personifications of the Sun are illustrated – the top image is from *The Crucifixion* fresco, the image in the middle is from the west dome of the central nave in Visoki Dečani, and the bottom one is from the Lesnovo monastery. On the right side images of the personified Moon are depicted, in the same order.

Fig. 5. Left: images of the personified Sun, right: images of the personified Moon; top: from *The Crucifixion* fresco and in the middle from the West dome of the central nave in Visoki Dečani; bottom: from the Lesnovo monastery. Images: top: Blago; middle: FB Visoki Dečani; bottom: The Astrologer's Community.

318 An Examination of the Images of the Sun and the Moon in the Visoki Dečani Monastery in Kosovo

To the modern eye, all these images look quite aerodynamic and have one or more rays. They seem to be in motion, and, indeed, as some authors of popular texts have observed, they bear a resemblance to 'space shuttles'. In paintings of the personified Sun and Moon in Serbian churches of the same time period, similar tear–shaped forms are to be seen, as in the South vault of the exonarthex at Bogorodica Ljeviška. The similarity of these images raises the question as to whether the painters of these images might have been inspired by an unusual cosmic phenomenon, such as the appearance of Halley's comet, as mentioned above, whose form might be attributed to the personified Sun and Moon.

Furthermore, it is noticeable that the central part of the church of Visoki Dečani contains the most images of the personified heavenly bodies. Those of *The Crucifixion* fresco are the clearest and most aerodynamic in shape. It is evident that special attention was given to Plato and other Greek philosophers in numerous Serbian medieval churches. This evokes the notion that the central part of the church, which is full of cosmological symbols referring to the ascent and descent of the soul, might be the consequence of the synergy of the medieval and the Platonic view of the immortality of the soul.

I argue that the images of the Sun and the Moon with human figures inside from *The Crucifixion* fresco might have looked familiar to Serbs in the fourteenth century because they reminded them of their pagan gods, as noted above, or they could recognize the 'servants of the Sun and the Moon' in those paintings, with whom they could communicate in the 'mute language' – especially the King, who could observe them directly from his own throne.

Conclusions
The aim of this paper was to put celestial-religious images of the Visoki Dečani monastery into a cultural and contemporary cosmological context by comparing them with other celestial images in Serbian churches constructed and painted in the same period. Most of the celestial images used as references were found in the Bogorodica Ljeviška church and in the Lesnovo monastery during my visit in February 2015.

The methodology I used was a site visit of the Visoki Dečani and Lesnovo monasteries, Bogorodica Ljeviška and other churches in the neighbourhood in order to collect the primary source materials though a personal observation of frescoes with cosmological elements. In order to take photographs of the interior a professional photographer was engaged and the photos were also used as the primary source material. An assessment of the secondary data sources was realised by comparing the observations from the field research with texts and images from the literature. As part of the literature review, academic writings from the last century were used, as well as writings of Plato and Aristotle and texts from the Bible. Subsequently, I placed the findings in the context of the medieval cosmology.

In the central part of the church of Visoki Dečani I found numerous frescoes with a personified Sun and Moon that contain elements similar to those in Lesnovo and Bogorodica Ljeviška, such as framed images with a figure inside, some of them with rays which indicate motion. The conclusion is that details from the personified Sun and Moon in *The Crucifixion* fresco, such as the aerodynamic forms and the appearance of figures, were not specific to Visoki Dečani only.

Therefore, one can argue that the paintings of the aerodynamic frescoes in the central part of the church of Visoki Dečani might be the result of the synergy of the medieval Christian and Hellenistic views of the creation of the universe by God and the immortality of the soul. While the Bible presents the universe as the creation of God, in order to praise him, Plato and Aristotle write about the ensouled cosmos, which includes ensouled planets, the Sun and the Moon, stating that the way to the God is via the planetary spheres. Furthermore, it can be asserted that the figures in the aerodynamic frames in *The Crucifixion* fresco represent souls of the Sun and the Moon, which resonate both – Christian and Hellenistic – views of the cosmos.

NOTES ON CONTRIBUTORS

Editors

Liz Henty left her accountancy career to take the Cultural Astronomy and Astrology MA at Trinity Saint David University of Wales, where she achieved a distinction for her dissertation entitled 'An Examination of Possible Solar, Lunar and Stellar Alignments at the Recumbent Stone Circles of North-East Scotland. After taking some short archaeology courses at Aberdeen University, she is now a PhD Research Student at Trinity Saint David University of Wales, researching the divide between the disciplines of archaeology and archaeoastronomy. She has presented papers at SEAC and the Theoretical Archaeology Group conferences and is a contributor to *Skyscapes in Archaeology* edited by F Silva and N Campion (Oxbow, 2015). She is co-founder and co-editor of the *Journal of Skyscape Archaeology*.

Bernadette Brady has a PhD in Anthropology (2012) and MA in Cultural Astronomy and Astrology (2005). She is currently a tutor in the Sophia Centre for the Study of Cosmology in Culture at the University of Wales Trinity Saint David, UK. Her research interests are in the cultural significant of astrology both historically as well as in contemporary life, the cultural influence of stars and the religious and cultural significant of star phases. Her journal papers include the cultural astronomy contained in examples of Egyptian astronomy (Oxbow, 2012), the orientation of the Solstitial Churches of North Wales (JSA, 2017) and the solar discourse in Cistercian Welsh abbeys (Citeaux, 2016). Apart from journal papers she has also authored *Cosmos, Chaosmos and Astrology* (Sophia Centre Press, 2014). She currently lives in Bristol, UK.

Darrelyn Gunzburg teaches in the Sophia Centre for the Study of Cosmology in Culture, Faculty of Humanities and Performing Arts, at the University of Wales Trinity Saint David. She received her PhD from the University of Bristol (2014) with a thesis entitled 'Giotto's Salone: An Astrological Investigation into the Fresco Paintings of the First Floor Salone of the Palazzo Della Ragione, Padua, Italy'. Her research interests lie in the art historical and visual astronomical exploration of frescos in medieval Italy, the orientation of abbey churches in Wales, the UK, and Europe, and pilgrimage. She is the editor of *The Imagined Sky: Cultural Perspectives* (Equinox, 2016) and has written extensively for *The Art Book* (Wiley-Blackwell) and *Cassone: The International Online Magazine of Art and Art Books.*

Frank Prendergast is an Emeritus Research Fellow in the Dublin Institute of Technology and a post-graduate of Trinity College Dublin. He gained a PhD in 2011 from University College Dublin, specialising in the spatial analysis of Neolithic passage tombs. Recent contributions to interpretative archaeology include publications associated with the discovery of a major Iron Age temple site in Lismullin, Ireland. He has contributed three chapters to the *Handbook of Archaeoastronomy and Ethnoastronomy* published by Springer Reference as well as continuing to write academic papers, invited book chapters and giving public lectures. A more recent venture is in the area of promoting the conservation of the dark sky at landscapes of archaeological importance. In that role, he is a scientific adviser to several heritage organisations. For 2017, he is a faculty member of the International Space University and contributing a Cultural Astronomy element to their Humanities Curriculum.

Fabio Silva is a Marie Skłodowska-Curie Fellow at the Catalan Institute for Human Palaeoecology and Social Evolution (Tarragona, Spain) and a tutor at the Sophia Centre (University of Wales Trinity Saint David), where he is responsible for the 'Skyscapes, Cosmology and Archaeology' module. He is co-founder of the *Journal of Skyscape Archaeology* and received the Fifth Carlos Jaschek Award from SEAC in 2016. He also co-edited several books, including *Skyscapes: The Role and Importance of the Sky in Archaeology* (Oxbow, 2015) and *The Materiality of the Sky* (Sophia Centre Press, 2016).

Contributors

Nasser B. Ayash studied at the Electrical Engineering Department, University of Patras between 2004 and 2010. He is currently enrolled at the University Of Ioannina, School Of Philosophy Department for History And Archaeology. He works as a Language Facilitator and IT Consultant in the development of an integrated software solution in Ministries in Qatar and Greece. His interests include amateur astronomy and linguistics. He has presented archaeoastronomy papers at the Amatuer Astronomy club 'Orion' at the University of Patras, 'Astral Aspects of Minoan and Mycenaean Religion' at the European Society for Astronomy in Culture (SEAC) Conference in Rome 2015, 'Arabic Astronomy in the Middle Ages' at the 9th Greek Amateur Astronomy Conference at Sparti in 2015 and 'Stars in the Minoan Religion' at the Graduate Archaeology Oxford 'GAO' Conference in 2016.

Juan Antonio Belmonte is an astronomer at the Instituto de Astrofísica de Canarias (Tenerife, Spain) where he has lectured history of astronomy and archaeoastronomy and investigates exoplanets, stellar physics and cultural astronomy. He has published or edited a dozen books and authored nearly 200 publications on those subjects. He was Director of the Science and Cosmos Museum of Tenerife (1995–2000), President of SEAC (2005–2011) and President of the Spanish Time Allocation Committee (CAT) of the Canarian observatories (2003–2012). He received the SEAC Carlos Jaschek award in 2012 for his contributions to the discipline. He is now advisory editor of the *Journal for the History of Astronomy*. He has performed extensive research on the astronomical traditions of ancient civilizations, notably ancient Mediterranean cultures and Egypt. Born in Murcia (Spain) in 1962, he studied physics and got his master-thesis in 1986 at Barcelona University and obtained his PhD on Astrophysics at La Laguna University in 1989.

Liana De Girolami Cheney is President of the Association for Textual Scholarship in Art History, presently a Visiting Scholar in Art History at the University of Bari, Italy, and at the University of Coruña, Spain, and emerita Professor of Art History from UMASS Lowell. She received her PhD in Italian and Baroque Art from Boston University, MA. Dr. Cheney has written several books on Italian Renaissance, Mannerism, Pre-Raphaelite Art and Women's Art; notably Botticelli's Neoplatonic Images; Giorgio Vasari's Teachers: Sacred and Profane Art; Giorgio Vasari's Artistic and Emblematic Manifestations; Self-Portraits by Women Painters; Pre-Raphaelitism and Medievalism in the Arts; and Edward Burne-Jones' Mythological Themes. She received an Award for Excellence in Scholarly Research and Publication from SECAC on her publication on 'Leonardo da Vinci's Annunciation: The Holy Spirit', in *Artibus et Historiae* (2011): pp. 1–16.

Felipe Criado-Boado is professor of archaeological research on the Spanish National Research Council (CSIC), director of the CSIC Institute of Heritage Sciences and a specialist on Landscape Archaeology and critical studies of heritage. He is the current president of EAA (European Association of Archaeologists).

J. Anna Estaroth graduated from the University of Wales Trinity Saint David with an MA in Cultural Astronomy and Astrology in 2017. Her dissertation topic was the Clava Cairns in Inverness-shire where she examined the role of the Major lunar limit in locating the ritual centre at Balnuaran of Clava, by exploring the topographical significance of neighbouring river systems. Considerations of darkness and light were a theme that developed from her research.

Benito Vilas Estévez is an archaeologist with an MA in Cultural Astronomy and Astrology from the University of Wales, Trinity Saint David and an MSc in Archaeology from the University of Santiago. He has carried out studies on megalithic astronomy and other issues related to cultural astronomy.

David Fisher has a career in information technology and his lifelong interests in astronomy and archaeology turned into a quest for in-depth analysis, while holding the post of Chief Information Officer for a deepwater archaeological shipwreck recovery company. He devised a method of stratifying artefacts as they were recovered from 1800 feet below sea level. During this time he undertook a PhD in astro-archaeology at the University of Wales, utilising computer technology modelling to assess the astronomical significance of megalithic sites. Papers published on this topic include: 'Employing 3-Dimensional Computer Simulation to Examine the Celestial Dating of Scottish Megalithic Sites' in *Ancient Cosmologies and Modern Prophets*, I. Šprajc, P. Pehani (eds), (Slovene Archaeological Society, 2013) and 'Restructuring the World of Megalithic Sites and Animating Astronomical Phenomena through 3D Computerization' in *Astronomy and Power: How Worlds Are Structured*, M. and B. Rappenglück, N. Campion, F Silva (eds), (BAR, 2016).

Roslyn M. Frank is Professor Emeritus at the University of Iowa. She has published extensively on the history of 'Western' constellations as well as Eurasian skylore, focusing particularly on projections of the belief that humans descended from bears onto the landscape and skyscape. Recent publications include 'Sky Bear Research: Implications for "Cultural Astronomy"' in *Mediterranean Archaeology and Archaeometry* 16, no. 4 (2016): pp. 343–50, and 'Bear Ceremonialism in Relation to Three Ritual Healers' in Enrico Comba and Daniele Ormezzano, eds., *Uomini e Orsi: Morfologia del Selvaggio* (Torino: Accedemia University Press, 2015), pp. 41–122. Her research areas are archaeo- and ethnoastronomy and ethnomathematics as well as cultural linguistics, cognitive linguistics, ethnography and anthropological linguistics with a special emphasis on the Basque language and culture. For additional information, see, http://uiowa.academia.edu/RoslynMFrank.

A. César González García has a PhD in Astrophysics (2003; Groningen, The Netherlands) and he is currently doing research in Cultural Astronomy at the Institute of Heritage Sciences in Santiago de Compostela (Spain). He has done extensive fieldwork on megalithic monuments in The Netherlands, Germany, Bulgaria, the Near East and the Iberian Peninsula. He has also done fieldwork on classical cultures in Anatolia, the Levant and Western Europe. His main research lines are centred on the perception of the sky in classical cultures and megaliths, the possible astronomical and landscape relations of Iron Age sanctuaries and the study of the orientation of Roman cities.

Steven R. Gullberg earned a PhD in astronomy from James Cook University under the supervision of J. McKim Malville. He is an Assistant Professor of Interdisciplinary Studies for the University of Oklahoma where he serves as its College of Professional and Continuing Studies Lead Faculty for Natural Science, as well as being the Lead Faculty for the Master of Arts Integrated Studies Program and the Bachelor of Arts Liberal Studies Program. Steven teaches astronomy and other natural science courses, and is also a member of the International Astronomical Union (IAU). His doctoral research interests focused on the astronomy of the Incas and his Masters research at the University of Oklahoma involved Babylonian astronomy and the Babylonian Astronomical Diaries. His work in archaeoastronomy continues and he has authored a number of associated research papers. He is presently working to further advance the fields of archaeoastronomy and Native American astronomy in the United States.

Tony Hull is Adjunct Professor of Physics and Astronomy at the University of New Mexico. Graduate and undergraduate studies were at the University of Pennsylvania, including an inspirational archaeology course there taught by Fro Rainey. In addition to astrophysics and developing space missions both at NASA and in industry, he has been an active researcher over several decades in Archaeology and Cultural Astronomy. This work has been largely in coastal California and the Cozo Range working with UCLA, and also in Chaco Canyon, NM, principally a comprehensive study of early Navajo manifestations at the eastern end of the Canyon. He chaired the 2011 Maxwell Museum Conference 'Astronomy and Ceremony in the Prehistoric Southwest'. In addition to fieldwork, mapping, recording and evaluating, he has developed theoretical models for cultural astronomy including practical methods for determining cardinality, the date of the equinox and also Monte Carlo criteria for research design.

Stanisław Iwaniszewski is Professor of Archaeology at the Division of Postgraduate Studies at the Escuela Nacional de Antropologia e Historia in Mexico City and archaeologist in the State Archaeological Museum in Warsaw. He specialises in the areas of archaeology and identity, landscape archaeology and archaeoastronomy. He was President of SEAC (1999–2005) and ISAAC (2007–2014). He has presented and published many papers on archaeoastronomy, both on the discipline itself and extensive case studies of monuments in South America and Eastern Europe. In 2015, he authored a monothematic cycle of publications on the theoretical and methodological aspects of research in archaeoastronomy. He is currently researching 'The Lunar Theory among the Maya' and 'The Starry Sky – Animated Sky'.

Tore Lomsdalen has an MA in Cultural Astronomy and Astrology from the University of Wales Trinity Saint David. For his dissertation he conducted fieldwork research on the possible astronomical intentionality behind the prehistoric Mnajdra Temple complex on Malta which was published in 2014; *Sky and Purpose in Prehistoric Malta; Sun, Moon and the Stars of the Temples of Mnajdra* (Sophia Centre Press). His research into the Maltese temples, archaeoastronomy and cosmology continues, and he has held various talks and given presentations on this topic in the last few years. He is currently a PhD student at the University of Malta, affiliated with the Department of Classics and Archaeology, researching cosmology in prehistoric Malta.

J. McKim Malville received his PhD in solar radio physics and for some 25 years published in solar astrophysics, emphasising observations of sun spots, flares, and the corona. During that time he published three books that indicated a growing interest in cultural astronomy: *A Feather for Daedalus*: *Explorations in Science* and Mythology, The Fermenting Universe, and Time and Eternal Change. He has been a professor in the Department of Astrophysical and Planetary Science at the University of Colorado since 1973. After several trips to India, he decided to turn his attention to cultural astronomy and engaged in fieldwork in India, Egypt, the American Southwest, and Peru. He is particularly interested in applying alternate ontologies in the quest to understand what were people thinking when they invested so much time and wealth in their respective skyscapes. In addition to the University of Colorado, he has taught or engaged in research at the High Altitude Observatory, Lockheed Aircraft Corporation, James Cook University, and the universities of Sao Paulo, Oslo, and Wales Trinity Saint David.

Frances Joan Mathien received her PhD from the University of New Mexico and worked for the National Park Service from 1978 through 2005. After retiring from her position as an archaeologist who worked mainly with data from Chaco Canyon, New Mexico, she has continued as a volunteer to research the field schools that were held in the canyon from 1929 through 1942 and again in 1947. Currently she is an Adjunct Assistant Professor with the Department of Anthropology and Research Associate with the Maxwell Museum of Anthropology at the University of New Mexico. Her main research interests include the development of complex societies, exchange networks, sourcing of minerals (e.g., turquoise), and the history of archaeology.

Claude Maumené is an agronomist by training and studies as an independent scholar researching how protohistoric objects and monuments may have incorporated astronomical and calendrical knowledge in different cultural contexts in Western Europe. He uses the myths developed by different religions and cultures, to propose an interpretation of the iconography, and to support various theories. He is a member and administrator of the French Mythology Society and began the Cultural Astronomy and Astrology MA programme at the University of Wales Trinity Saint David in 2016.

Cherilynn Morrow received her PhD in solar astrophysics from the University of Colorado in 1988. She conducted graduate research at the National Center for Atmospheric Research and post-doctoral research at Cambridge University. While serving as a visiting scientist at NASA Headquarters in the early 1990s she began her award-winning work to integrate education and public outreach programmes into scientific research environments in space and Earth science. Her work expanded to integrating the arts and indigenous culture during subsequent leadership roles in science education at the Space Science Institute, the SETI Institute, Georgia State University (GSU), and the Aspen Global Change Institute (AGCI). Dr Morrow has supported interpretive and research programmess at Chaco Culture National Historical Park. This includes directing the Chaco Educator Institute in Astronomy (June 2008) and apprenticing with Andy Munro's archaeoastronomy research team, providing archival research and field support.

Andrew M. Munro received his PhD from James Cook University. He is an Adjunct Professor with the University of Oklahoma's College of Professional and Continuing Studies, and a graduate Project Supervisor for Swinburne University in Australia. He has conducted fieldwork-based archaeoastronomy research at Chaco Culture National Historic Park and outlying Chacoan Great Houses since 2007. His findings provide support for the idea that groups of people with varied cultural traditions collaborated at Chaco Canyon. They also reinforce the magnitude of cultural change at Chaco after 1100 CE.

John L. Ninnemann is Professor Emeritus, Department of Biology and former Dean of Natural and Behavioral Sciences and Professor of Biology, Fort Lewis College. His shows of astronomical photography include Canyon Spirits, Edge of the Cedars State Park, Blanding, UT 2004; Ancient Astronomy in the Southwest, Center of Southwest Studies, Fort Lewis College, Durango, CO February-October, 2012; Ancient Astronomy in the Southwest, Anasazi Heritage Center, Dolores, CO November 2012-March 2013;and One-Room Schoolhouses of Colorado, Mancos Visitors Center, Mancos, CO January-February, 2013.

Marianna Päivikki Ridderstad obtained her PhD in astronomy in 2015. Her doctoral thesis was the first one in the field of archaeoastronomy in Finland. Previously, she had completed a MSc degree in theoretical physics in 2002 and a LicPhil in astronomy in 2011. In archaeoastronomy, her research interests have varied from Minoan astronomical symbolism to the orientations of Neolithic monuments in Finland. Together with Dr Jari Okkonen from the University of Oulu, she discovered the astronomical orientations of the Neolithic Giants' Churches of Ostrobothnia in 2009. In addition to archaeoastronomy, her main subject of interest is astrobiology, which she has taught in the University of Helsinki since 2006.

Florian Schaukowitsch is a Master's student in Visual Computing at the Vienna University of Technology. Currently finalising his thesis work under the supervision of Prof. Michael Wimmer, he has been a semi-regular contributor to Stellarium since 2014, mainly working on graphics-related issues and features. For the work on Stellarium, he has received two stipends from the ESA in the course of their 'Summer of Code in Space' programme.

Lionel Sims was Head of the Department of Anthropology, International Development, International Politics and Refugee Studies at the University of East London until his semi-retirement. He is Vice President of the European Society for Astronomy in Culture (SEAC) and is a lecturer for the Workers' Educational Association (WEA). He is a regular presenter at SEAC and other conferences and author of many inter-disciplinary papers and book chapters combining archaeoastronomy, archaeology, anthropology and mythology. A film of his research, Stonehenge Rediscovered, was commissioned by National Geographic and has been broadcast worldwide. He is a member of the Stonehenge Round Table hosted by English Heritage and the Avebury Sacred Sites Forum hosted by the National Trust in his capacity as leader of the 'King's Drums' summer solstice performance group. He is a Stonehenge and Avebury guide for the tour company 'Ammoun Voyages', collaborates with the Basque cultural organization Jauzarrea and lectures widely.

Dragana Van de moortel-Ilić completed a MA in Cultural Astronomy and Astrology at the University of Wales Trinity Saint David, where she conducted research on cosmological elements in celestial-religious paintings in medieval Serbian churches. Her investigation focused on the paintings of the Sun, the Moon and planets in the Visoki Dečani monastery and the Bogorodica Ljeviška church in Kosovo as well as the Lesnovo monastery in FYR Macedonia. Dragana previously graduated from the Faculty of Natural Sciences and Mathematics at the University of Novi Sad, Serbia, where she held the position of research assistant in chemistry. From 1991 onwards she has worked as a freelance consultant. She lives and works in Ghent, Belgium, and is an active mentor at the Institute for Psychological Astrology and Psychoanalysis, IPAL, in Slovenia and at the Faculty of Astrological Studies, UK.

Michael Wimmer is Associate Professor and head of the Rendering and Modeling Group at the Institute of Computer Graphics and Algorithms at Vienna University of Technology (TU Wien).

Georg Zotti holds a PhD in Computer Sciences from Vienna University of Technology University and B.Sc. in Astronomy from Vienna University. He combines research interests in the history of astronomy, astronomical instruments (esp. the Astrolabe), Virtual Realities, Virtual Archaeology and Archaeo-astronomy, forming what can be termed Virtual Archaeoastronomy. While researching the possible astronomical orientation of Neolithic Circular Ditch Systems in Lower Austria, he joined the Stellarium project in 2010 and enhances its applicability mostly towards simulation of historical and prehistorical astronomical events and skyscapes. He is currently one of the SEAC Vice Presidents.

BACK ISSUES OF CULTURE AND COSMOS
http://www.cultureandcosmos.org/backIssues.html

Contents, Vol. 1 no 1 (Spring/Summer 1997)
Robin Heath *An Astronomical Basis for Solar Hero Myths*; **Norris Hetherington** *Ancient Greek Cosmology and Culture: a Historiographical Review*; **Alan Weber** *The Development of Celestial Journey Literature, 1400–1650*; **Ken Negus** *Kepler's Tertius Interveniens*; **John Durant** and **Martin Bauer** *British Public Perceptions of Astrology: an Approach from the Sociology of Knowledge.*

Contents Vol. 1 no 2 (Autumn/Winter 1997)
Otto Neugebauer *On the History of Wretched Subjects*; **Nick Kollerstrom** *The Star Zodiac of Antiquity*; **Robert Zoller** *The Hermetica as Ancient Science*; **Edgar Laird** *Christine de Pizan and Controversy Concerning Star Study in the Court of Charles V*; **Jürgen G.H. Hoppman** *The Lichtenberger Prophecy and Melanchthon's Horoscope for Luther*; **Elizabeth Heine** *W.B.Yeats: Poet and Astrologer.*

Contents Vol. 2 no 1 (Spring/Summer 1998)
J. McKim Malville and **R.N. Swaminathan** *People, Planets and the Sun: Surya Puja in Tamil Nadu, South India*; **Carlos Trenary** *Yaxchilan Lintel 25 as a Cometary Record*; **Graziella Federici Vescovini** *Biagio Pelacani's Astrological History for the Year 1405*; **Frank McGillion** *The Influence of Wilhelm Fliess' Cosmobiology on Sigmund Freud*; **Nicholas Campion** *Sigmund Freud's Investigation of Astrology.*

Contents Vol. 2 no 2 (Autumn/Winter 1998)
Giuseppe Bezza *Astrological Considerations on the Length of Life in Hellenistic, Persian and Arabic Astrology*; **Angela Voss** *The Music of the Spheres: Marsilio Ficino and Renaissance harmonia*; **Robert Zoller** *Marc Edmund Jones and New Age Astrology in America.*

Contents Vol. 3 no 1 (Spring/Summer 1999)
Michael R. Molnar *Firmicus Maternus and the Star of Bethlehem*; **Roger Beck** *The Astronomical Design of Karakush, a Royal Burial Site in Ancient Commagene: an Hypothesis*; **Chantal Allison** *The Ifriqiya Uprising Horoscope from* On Reception *by Masha'alla, Court Astrologer in the Early 'Abassid Caliphate.*

Contents Vol. 3 no 2 (Autumn/Winter 1999)
Robin Waterfield *The Evidence of Astrology in Classical Greece*; **Remo Catani** *The Polemics on Astrology 1489-1524*; **Claudia Rousseau** *An Astrological Prognostication to Duke Cosimo de Medici of Florence.*

Contents Vol. 4 no 1 (Spring/Summer 2000)
Patrick Curry *Historical Approaches to Astrology*; **Edgar Laird** *Heaven and the Sphaera Mundi in the Middle Ages*; **George D. Chryssides** *Is God a Space Alien? The Cosmology of the Raëlian Church.*

Contents Vol. 4 no 2 (Autumn/Winter 2000)
David J. Ross *The Bird, The Cross, and the Emperor: Investigations into the Antiquity of The Cross in Cygnus*; **Angela Voss** *The Astrology of Marsilio Ficino: Divination or*

Science?; **Patrick Curry** *Astrology on Trial, and its Historians: Reflections on the Historiography of 'Superstition'.*

Contents Vol. 5 no 1 (Spring/Summer 2001)
Demetra George: *Manuel I Komnenos and Michael Glykas: A Twelfth-Century Defence and Refutation of Astrology,* Part I; **Richard L. Poss:** *Stars and Spirituality in the Cosmology of Dante's Commedia.*

Contents Vol. 5 no 2 (Autumn/Winter 2001)
Arkadiusz Sołtysiak *The Bull of Heaven in Mesopotamian Sources*; **Demetra George** *Manuel I Komnenos and Michael Glykas: A Twelfth-Century Defence and Refutation of Astrology,* Part 2; **Garry Phillipson** and **Peter Case** *The Hidden Lineage of Modern Management Science: Astrology, Alchemy and the Myers-Briggs Type Indicator.*

Contents Vol. 6 no 1 (Spring/Summer 2002)
Ari Belenkyi *A Unique Feature of the Jewish Calendar—Deĥiyot*; **Demetra George** *Manuel I Komnenos and Michael Glykas: A Twelfth-Century Defence and Refutation of Astrology,* Part 3; **Germana Ernst** *The Sky in a Room: Campanella's Apologeticus in defence of the pamphlet* De siderali fato vitando; **Tommaso Campanella** *Apologia for the opuscule on* De siderali fato vitando.

Contents Vol. 6 no 2 (Autumn/Winter 2002)
Jesse Krai *Rheticus' Poem 'Concerning the Beer of Breslau and the Twelve Signs of the Zodiac'*; **Anna Marie Roos** *Israel Hiebner's Astrological Amulets and the English Sigil War*; **Nicholas Campion** *Surrealist Cosmology: André Breton and Astrology.*

Contents Vol. 7 no 1 (Spring/Summer 2003) GALILEO'S ASTROLOGY
Nick Kollerstrom *Foreword: Galileo as Believer*; **Nicholas Campion** *Introduction: Galileo's Life and Work*; **Antonio Favaro** *Galileo, Astrologer*; **Germana Ernst** *Astrology and Prophecy in Campanella and Galileo*; **Nick Kollerstrom** *Galileo as an Astrologer*; **Antonino Poppi** *On Trial for Astral Fatalism: Galileo Faces the Inquisition*; **Guiseppe Righini** *Galileo's Horoscope for Cosimo II de Medici*; **Mario Biagioli** *An Astrologico-Dynastic Encounter; Galileo's Correspondence; Galileo's Letter to Dini, May 1611; On the Character of Sagredo: Galileo's judgements upon his nativity; Galileo's Horoscopes for his Daughters; Rome, 1630*; **Bernadette Brady** *Four Galilean Horoscopes: An Analysis of Galileo's Astrological Techniques; A Sonnet by Galileo.*

Contents Vol. 7 no 2 (Autumn/Winter 2003)
Günther Oestmann *Tycho Brahe's Geniture*; **Bernard Eccles** *Astrological physiognomy from Ptolemy to the present day*; **James Brockbank** *Planetary signification from the second century until the present day*; **Julia Cleave** *Ficino's Approach to Astrology as Reflected in Book VII of his Letters.*

Contents Vol. 8 no 1 and 2 (Spring/Summer, Autumn/Winter 2004)
Valerie Shrimplin *Organising INSAP*; **Rolf Sinclair** *Foreword: INSAP IV in Oxford: A Summary*; **Nicholas Campion** *Introduction: The Inspiration of Astronomical Phenomena*; **Hubert A. Allen, Jr.** *Hawkins' Way: Remembering Astronomer Gerald S. Hawkins*; **Hubert A. Allen, Jr. and Terry Edward Ballone** *Star Imagery in Petroglyph National Monument*; **Mark Butterworth** *Astronomy and the Magic Lantern*; **Ann Laurence**

Caudano *Sun, Moon, and Stars on Kievan Rus Jewellery (10th–13th Centuries)*; **Nicholas Campion** *The Sun is God*; **Anne Chapman-Rietschi** *Cosmic Gardens*; **Deborah Garwood** *Paris Solstice*; **N. J. Girardot** *Celestial Worlds In the Work of Self-Taught Visionary Artists With Special Reference to Howard Finster's Vision of 1982*; **John G. Hatch** *Desire, Heavenly Bodies, and a Surrealist's Fascination with the Celestial Theatre*; **Holly Henry** *Bertrand Russell in Blue Spectacles: His Fascination with Astronomy*; **Ronald Hicks** *Astronomy and the Sacred Landscape in Irish Myth*; **Chris Impey** *Why Are We So Lonely?*; **Bernd Klähn** *The Aberration of Starlight and/in Postmodernist Fiction*; **Nick Kollerstrom** *How Galileo dedicated the moons of Jupiter to Cosimo II de Medici*; **Arnold Lebeuf** *Dating the five Suns of Aztec cosmology*; **Andrea D. Lobel** *Trailing the Paper Moon: Astronomical Interpretations of Exodus 12:1–2*; **Stephen C. McCluskey** *Wordsworth's 'Rydal Chapel' and the Astronomical Orientation of Churches*; **David Madacsi** *Sky: Atmospheres and Aesthetic Distance in Planetary and Lunar Environments*; **Daniel R. Matlaga** *A Journey of Celestial Lights: The Sky as Allegory in Melville's* Moby Dick; **Paul Murdin** *Representing the Moon*; **R. P. Olowin** *Robinson Jeffers: Poetic Responses to a Cosmological Revolution*; **David W. Pankenier** *A Brief History of Beiji (Northern Culmen)*; **Richard Poss** *Poetic Responses to the Size of the Universe: Astronomical Imagery and Cosmological Constraints*; **Barbara Rappenglück** *The material of the solid sky and its traces in cultures*; **Brad Ricca** *The Night of Falling Stars: Reading the 1833 Leonid Meteor Storm*; **Patricia Ricci** *Lux ex Tenebris: Etienne-Louis Boullée's Cenotaph for Sir Isaac Newton*; **Sarah Richards** *Die Planetentheorie: its uses and meanings for the Saxon mining communities and the culture of the Dresden Court 1553–1719*; **William Saslaw and Paul Murdin** *The Double Apollos of Istrus*; **Petra G. Schmidl** *Dusk and Dawn in Medieval Islam; On the Importance of Twilight Phenomena with Some Examples of Their Representations in Texts and on Instruments*; **Valerie Shrimplin** *Borromini and the New Astronomy: the elliptical dome*; **Joshua Stein** *Cicero's Use of Astronomy as Proof of the Existence of the Gods*; **Antje Steinhoefel** *Art and Astronomy in the Service of Religion: Observations on the Work of John Russell (1745–1806)*; **Burkard Steinrücken** *An interpretation of the 'Sky Disc of Nebra' as an icon for a bronze age planetarium mechanism with parallels to the moving world-soul in Plato's* Timaeus; **Gary Wells** *Daumier and The Popular Image of Astronomy*.

Contents Vol. 9 no 1 (Spring/Summer 2005)
Gennadij Kurtik and Alexander Militarev *Once more on the origin of Semitic and Greek star names: an astronomic-etymological approach updated*; **Prudence Jones** *A Goddess Arrives: Nineteenth Century Sources of the New Age Triple Moon Goddess*; **Louise Curth** *Astrological Medicine and the Popular Press in Early Modern England*.

Contents Vol. 9 no 2 (Autumn/Winter 2005)
Marinus Anthony van der Sluijs *A Possible Babylonian Precursor to the Theory of ecpyrōsis*; **Liz Greene** *Did Orphic Beliefs Influence the Development of Hellenistic Astrology?*; **Ariel Cohen** *Astronomical Luni-Solar Cycles and the Chronology of the Masoretic Bible*; **Tayra Lanuza-Navarro** *An Astrological Disc from the Sixteenth Century*; **J.C. Holbrook** *Celestial Navigators and Navigation Stories*.

Contents Vol. 10 no 1 and 2 (Spring/Summer, Autumn/Winter 2006)
Lucia Dolce *Introduction: The worship of celestial bodies in Japan: politics, rituals and icons*; **Lucia Dolce** *The State of the Field: A basic bibliography on astrological cultic practices in Japan*; **Hayashi Makoto** *The Tokugawa Shoguns and Yin-yang knowledge*

(onmyōdō); **John Breen** *Inside Tokugawa religion: stars, planets and the calendar-as-method*; **Mark Teeuwen** *The imperial shrines of Ise: An ancient star cult?*; **Lilla Russell-Smith** *Stars and Planets in Chinese and Central Asian Buddhist Art from the Ninth to the Fifteenth Centuries*; **Matsumoto Ikuyo** *Two Mediaeval Manuscripts on the Worship of the Stars from the Fujii Eikan Collection*; **Tsuda Tetsuei** *The Images of Stars and Their Significance in Japanese Esoteric Buddhist Art*; **Meri Arichi** *Seven Stars of Heaven and Seven Shrines on Earth: The Big Dipper and the Hie Shrine in the Medieval* Period; **Gaynor Sekimori** *Star Rituals and Nikko Shugendô*; **Meri Arichi** *The front cover image: Myōken Bosatsu.*

Contents Vol. 11 no 1 and 2 (Spring/Summer, Autumn/Winter 2007)
Micah Ross *A Survey of Demotic Astrological Texts*; **Francis Schmidt** *Horoscope, Predestination and Merit in Ancient Judaism*; **Stephan Heilen** *Ancient Scholars on the Horoscope of Rome*; **Joanna Komorowska** *Philosophy among Astrologers*; **Wolfgang Hübner** *The Tropical Points of the Zodiacal Year and the* Paranatellonta *in Manilius' Astronomica*; **Aurelio Pérez Jiménez** *Hephaestio and the Consecration of Statues*; **Robert Hand** *Signs as Houses (Places) in Ancient Astrology*; **Dorian Gieseler Greenbaum** *Calculating the Lots of Fortune and Daemon in Hellenistic Astrology*; **Susanne Denningmann** *The Ambiguous Terms* ἑῴα *and* ἑσπερία, ἀνατολή, *and* ἑῴα *and* ἑσπερία δύσις **Joseph Crane** *Ptolemy's Digression: Astrology's Aspects and Musical Intervals*; **Giuseppe Bezza** *The Development of an Astrological Term – from Greek* hairesis *to Arabic* ḥayyiz; **Deborah Houlding** *The Transmission of Ptolemy's Terms: An Historical Overview, Comparison and Interpretation.*

Contents Vol. 12 no 1 (Spring/Summer 2008)
Liz Greene *Is Astrology a Divinatory System?*; **James Maffie** *Watching the Heavens with a 'Rooted Heart': The Mystical Basis of Aztec Astronomy*; **J.C. Holbrook** *Astronomy and World Heritage.*

Contents Vol. 12 no 2 (Autumn/Winter 2008)
Mark Williams *Astrological Poetry in late medieval Wales: the case of Dafydd Nanmor's 'To God and the planet Saturn'*; **Scott Hendrix** *Choosing to be Human: Albert the Great on Self Awareness and Celestial Influence*; **Graham Douglas** *Luis Vilhena and the World of Astrology.*

Contents Vol. 13 no 1 (Spring/Summer 2009)
Josefina Rodríguez-Arribas *Astronomical and Astrological Terms in Ibn Ezra's Biblical Commentaries: A New Approach*; **Andrew Vladimirou** *Michael Psellos and Byzantine Astrology in the Eleventh Century*; **Marinus Anthony van der Sluijs** *The Dragon of the Eclipses—A Note*; **Patrick Curry** *Response to Liz Greene's 'Is Astrology a Divinatory System?'*

Contents Vol. 13 no 2 (Autumn/Winter 2009)
Liz Greene *Mystical Experiences Among Astrologers*; **Peter Pesic** *How the Sun Stood Still: Old English Interpretations of Joshua and the Leap Year*; **Doina Ionescu** *Virginia Woolf and Astronomy*; **Carlos Ziller Camenietzki** and **Luis Miguel Carolino** *Astrologers at War: Manuel Galhano Lourosa and the Political Restoration of Portugal, 1640–1668*; **Nick Campion** *Astrology's Role in New Age Culture: A Research Note*

Contents Vol. 14 no 1 and 2 (Spring/Summer, Autumn/Winter 2010)
Dorian Gieseler Greenbaum *Introduction*; **Friederike Boockmann** *Johann Kepler's Horoscope Collection*; **J. Cornelia Linde (trans.)** *Helisaeus Röslin's Delineation of Kepler's Birthchart, 1592*; **J. Cornelia Linde and Dorian Greenbaum (trans.)** *David Fabricius and Kepler on Kepler's Personal Astrology, 1602*; **Dorian Greenbaum (trans.)** *Kepler's Delineation of his Family's Astrology*; **J. Cornelia Linde and Dorian Greenbaum (trans.)** *Kepler and Michael Mästlin on their Son's Nativities, 1598*; **J. Cornelia Linde and Dorian Greenbaum (trans.)** *Kepler's Methods of Astrological Interpretation for Rudolf II, 1602*; **J. Cornelia Linde and Dorian Greenbaum (trans.)** *Kepler's Astrological Interpretation of Rudolf II by Traditional Methods, 1602*; **J. Cornelia Linde and Dorian Greenbaum (trans.)** *Kepler's Letter to an Official on Rudolf II and Astrology, 1611*; **J. Cornelia Linde and Dorian Greenbaum (trans.)** *Excerpts from Kepler's Correspondence and Interpretation of Wallenstein's Nativity, 1624–1625*; **J. Cornelia Linde and Dorian Greenbaum (trans.)** *The Nativities of Mohammed and Martin Luther, 1604*; **J. Cornelia Linde and Dorian Greenbaum (trans.)** *The Nativity of Augustus*; **John Meeks** *Introduction: Kepler and the Art of Weather Prognostication*; **John Meeks (trans.)** *Kepler's Weather Calendar of 1618*; **John Meeks (trans.)** *Excerpts from Kepler's Weather Calendar of 1619*; **Patrick J. Boner (trans.)** *Astrology on Trial: Kepler, Pico and the Preservation of the Aspects De stella nova: Chapters 7–9*; **J. Cornelia Linde and Dorian Greenbaum (trans.)** *On Directions*; **J. Cornelia Linde and Dorian Greenbaum (trans.)** *David Fabricius and Kepler on Astrological Theory and Doctrine, 1602*; **J. Cornelia Linde and Dorian Greenbaum (trans.)** *David Fabricius and Kepler on Fabricius's Directions, 1603–1604*; **J. Cornelia Linde and Dorian Greenbaum (trans.)** *On Aspects, 1602*; **Appendix** *A Selection of Kepler's Handwritten Charts*

Contents Vol. 15 no 1 (Spring/Summer 2011)
Miguel Querejeta *On the Eclipse of Thales, Cycles and Probabilities*; **Nicholas Campion** *The Shock of the New: The Age of Aquarius*; **Alejandro Gangui** *The Barolo Palace: Medieval Astronomy in the Streets of Buenos Aires*; **Nicholas Campion and John Frawley** *Research Note: A Horoscope by André Breton*

Contents Vol. 15 no 2 (Autumn/Winter 2011)
Liz Greene *Heavenly Hosts: Angelic Intermediaries as Soul-Gates*; **Pamela Armstrong** *Ritual Ornamentation—From the Secular to the Religious*; **Paul Cheshire** *William Gilbert: Macrocosmal Astrologer in an Age of Revolution*; **Sylwia Konarska-Zimnicka** *Astrologia Licita? Astrologia Illicita? The Perception of Astrology at Kraków University in the Fifteenth Century*; **John Frawley** *Research Note: William Blake and Antares*

Contents Volume 16 no 1 and 2 (Spring/Summer, Autumn/Winter 2012)
Nicholas Campion *Editorial: The Inspiration of Astronomical Phenomena*; **Chris Impey** *The Inspiration of Astronomical Phenomena*; **Ulisses Barres de Almeida** *What are these sparks of infinite clarity? And what am I? So I pry*; **Michael Hoskin** *William Herschel's Wonderful Decade, 1781–1790*; **Francis Ring** *The Bath Philosophical Society and its influence on William Herschel's career*; **Roberta J.M. Olson** and **Jay M. Pasachoff** *The Comets of Caroline Herschel, Sleuth of the Skies at Slough*; **V.F. Polcaro** and **A. Martocchia** *Guidelines for a social history of Astronomy*; **Euan MacKie** *A new look at the astronomy and geometry of Stonehenge*; **Leonid Marsadolov** *Archaeoastronomical Aspects of the Archaeological Monuments of Siberia*; **Christian Etheridge** *A systematic re-evaluation of the sources of Old Norse astronomy*; **Aidan Foster** *Hierophanies in the*

Vinland Sagas: Images of a New World; **Inga Elmqvist Söderlund** *Inspiration from antique heroic deeds: Hercules as an astronomer*; **Patricia Aakhus** *Astral Magic and Adelard of Bath's Liber Prestigiorum; or Why Werewolves Change at the Full Moon*; **David Pankenier** *Astrology for an Empire: The 'Treatise on the Celestial Offices' (ca. 100 BCE)*; **Steven Renshaw** *The Inspiration of Subaru as a Symbol of Values and Traditions in Japan*; **Daniel Armstrong** *Citing The Saucers: Astronomy, UFOs and a persistence of vision*; **Alberto Cappi** *The concept of gravity before Newton*; **Paul Murdin** *Artilleryman to head of state— how astronomy inspired Francois Arago*; **Paolo Molaro** and **Alberto Cappi** *Edgar Allan Poe's cosmology in Eureka*; **Voula Saridakis** *For 'the present and future happiness of my dear Pupils': The Astronomical and Educational Legacy of Margaret Bryan*; **Michael Rowan–Robinson** *The invisible universe*; **Arnold Wolfendale** *The Inter-Relation of the Visual Arts and Science in General and Astronomy in Particular*; **Lynda Harris** *Changing Images of the Milky Way during the Greco-Roman and Medieval Periods*; **Lucia Ayala** *The Universe in images: Iconography of the Plurality of Worlds*; **Tayra M. Carmen Lanuza-Navarro** *Astrological culture before its public: the representation of astrology in Golden Age Spanish Theatre*; **Emily Urban** *Depicting the Heavens: The Use of Astrology in the Frescoes of Rome*; **Michael Mendillo** *The Artistic Portrayal of the Medicean Moons in Early Astronomical Charts, Books and Paintings*; **Rolf Sinclair** *Howard Russell Butler: Painter Extraordinary of Solar Eclipses*; **Beatriz Garcia, Estela Reynoso, Silvina Pérez Alvarez** and **Rubén Gabellone** *Inspiration of Astronomy in the movies: a history of a close encounter*; **Gary Wells** *The Moon in the Landscape: Interpreting a Theme of 19th Century Art*; **Clive Davenhall** *The Space Art of Scriven Bolton*; **Matthew Whitehouse** *Astronomical Organ Music*; **Aaron Plasek** *Between Scientists, Writers and Artists: Theorising and Critiquing Knowledge-Production at the Interstices between Disciplines*; **Merja Markkula** *The Way I See the Stars: fibre art inspired by astrobiology*; **Govinda Sah** *Beyond the Notion*; **Gisela Weimann** *Above all the stars*; **Courtney Wrenn** *Nebulae (emission / absorption)*; **Toby MacLennan** *Presentation of Playing the Stars*; **Felicity Spear**, *Extending vision: sky-situated knowledge and the artist's eye*; **Vanessa Stanley** *Surveillance-Surveillance-Surveillance*; **Jim Cogswell** *Molecular Delirium*.

Contents Vol. 17 no 1 (Spring/Summer 2013)
Clifford J. Cunningham and **Günther Oestmann** *Classical Deities in Astronomy: The Employment of Verse to Commemorate the Discovery of the Planets Uranus, Ceres, Pallas, Juno, and Vesta*; **Dorian Knight** *A Reinvestigation Into Astronomical Motifs in Eddic Poetry*; **Karen Smyth** *'I specially note their Astronomie, philosophie, and other parts of profound or cunning art': The Use of Cosmos Registers by Chaucer and Others*; **Kirk Little** *Spellbound: The Astrological Imagination of Washington Irving*; **Guiliano Masola** and **Nicola Reggiani** *Σελήνη Τοξότη: Business and Astrology in the Papyri*; **Reinhard Mussik** *Research Note:* Weltall, Erde, Mensch *and Marxist Cosmology in East Germany*

Contents Vol. 17 no 2 (Autumn/Winter 2013)
Daniel Brown *The Experience of Watching: Place Defined by the Trinity of Land-, Sea-, and Skyscape*; **Pamela Armstrong** *Skyscapes of the Mesolithic/Neolithic Transition in Western England*; **Olwyn Pritchard** *North as a Sacred Direction? Traces of a Prehistoric North-South Route Across Pembrokeshire*; **Tore Lomsdalen** *The Islandscape of the Megalithic Temple Structures of Prehistoric Malta*; **Fernando Pimenta, Nuno Ribeiro, Anabela Joaquinito, António Félix Rodrigues, Antonieta Costa** and **Fabio Silva** *Land, Sea and Skyscape: Two Case Studies of Man-made Structures in the Azores Islands.*

Contents Vol. 18 no 1 (Spring/Summer 2014)
César Esteban *Struggling for Interdisciplinarity: Reflections of an Astrophysicist Working in Cultural Astronomy*; **Ronald Hutton** *Prehistoric British Astronomy: Whatever Happened to the Earth and Sun?*; **Nick Kollerstrom** *Galileo and the Astrological Prophecy of Manuel Rosales*; **Clive Davenhall** *Dr Katterfelto and the Prehistory of Astronomical Ballooning*; **Nicholas Campion** *Celestial Art: An Interview with Geoff MacEwan*.

Contents Vol. 18 no 2 (Autumn/Winter 2014)
Roger Beck *The Ancient Mithraeum as a Model Universe. Part 2*; **Helena Avelar** and **Charles Burnett** *The Interpretation of a Horoscope Cast by Abraham the Jew in Béziers for a child born on 29 November 1135: An Essay in Understanding a Medieval Astrologer*; **Lindsay Starkey** *Creation, Providence, and the Limits of Human Knowledge of the World: Mellin de Saint-Gelais and John Calvin on Astrology*; **Scott Hendrix** *The Contextual Rationality of Galileo's Astrology*; **Richard Angelo Bergen** *Paradise Lost and the Descent of Urania: from Astrology to Allegory*; **R. Hakan Kırkoğlu** *Ilm-i nudjum and 18th century Ottoman Court Politics*; **Graham Douglas** *Trystes Cosmologiques: When Lévi-Strauss met the Astrologers*.

Contents Vol. 19 no 1 and 2 (Autumn/Winter, Spring/Summer 2015)
Mike Harding *The Meanings of Magic*; **José Manuel Redondo** *The Celestial Imagination: Proclus the Philosopher on Theurgy*; **Liz Greene** *The God in the Stone: Gemstone Talismans in Western Magical Traditions*; **Claire Chandler** *Investigating the Magical Practice found in PGM (Greek Magical Papyri) XIII*; **M.E. Warlick** *Alchemy and the Transgendering of Mercury*; **Karen Parham** *Teleological and Aesthetic Perfection in the Aurora Consurgens*; **Alison Greig** *Angelomorphism and Magical Transformation in the Christian and Jewish Traditions*; **Christine Broadbent** *Celestial Magic as the 'Love Path': The Spiritual Cosmology of Ibn 'Arabi*; **Hereward Tilton** *Bells and Spells: Rosicrucianism and the Invocation of Planetary Spirits in Early Modern Germany*; **Joscelyn Godwin** *Astral Ascent in the Occult Revival*; **Sue Lewis** *The Transformational Techniques of Huber Astrology*; **Jane Burton** *Ancient Necromantic Rituals in Contemporary Celestial Magic*; **Lilan Laishley** *South Indian Ritual Dispels Negative Karma in the Birth Chart*.

Contents Vol. 20 no 1 and 2 (Autumn/Winter, Spring/Summer 2016)
Juan Antonio Belmonte *Cosmic Landscapes in Ancient Egypt: A Diachronic Perspective*; **Kim Malville** *Passages between Worlds: Heaven, Earth, and the Underworld in the Andean Cosmos*; **Harold H. Green** *The Zenith Sun as Organizing Principle of the Constructed Sacred Space and Calendrics of Central Mexico*; **Stanisław Iwaniszewski** *Communicating with the Ancestors in the Spiritual Landscape at Yaxchilán, Chiapas, Mexico*; **Shon Hopkin** *The Joining of Heaven and Earth in Mormon Sacred Texts and Temples*; **Joanna Popielska-Grzybowska** *Some Remarks on the Sky in the Ancient Egyptian Pyramid Texts*; **Scott Hendrix** *From the Margins to the Image of 'The Most Christian Science': Astrology and Theology from Albert the Great to Marsilio Ficino*; **Gerardina Antelmi** *Poetry Creation as Space of Union between Natural and Supernatural: A Reading of* The House of Fame; **Edina Eszenyi** *Thunderbolt: Shaping the Image of Lucifer in the Cinquecento Veneto*; **Alexander Cummins** *The Worldly Faces of the Heavens: Nature and Seventeenth-Century English Astrological Images*.

www.ingramcontent.com/pod-product-compliance
Lightning Source LLC
Chambersburg PA
CBHW042113100526
44587CB00025B/4031